Tribology Handbook

Volume I

Tribology Handbook
Volume I

Edited by **Irving Russo**

CLANRYE
INTERNATIONAL

New Jersey

Published by Clanrye International,
55 Van Reypen Street,
Jersey City, NJ 07306, USA
www.clanryeinternational.com

Tribology Handbook: Volume I
Edited by Irving Russo

© 2015 Clanrye International

International Standard Book Number: 978-1-63240-501-2 (Hardback)

Printed in the United States of America.

Contents

Preface

The word tribology derives from a Greek root, which essentially means "I rub" in classic Greek. Tribology is a subject of science and engineering which studies about the interaction of surfaces in relative motions through principles such as friction, lubricants, and wear & tear. British physicist David Tabor has been credited for coining the term Tribology. This subject is an interdisciplinary field of mechanical engineering and material sciences. When one material slides over another, the entire product is affected by complex tribological interactions.

The tribological interactions of solid material surfaces, when exposed to other material or environmental substances, might result in loss of materials from the surface, which is called "wear". Major types of wear include abrasion, adhesion and cohesion, corrosion, erosions, etc. Tribology also plays a very crucial role in manufacturing. New areas in the field of tribology such as nanotribology, biotribology and green tribology have emerged since the 1990s.

To avoid or reduce the effects of wear, techniques such as surface modification which falls under the surface engineering processes, such as lubrication and coating exist. Some very important theories related to reduction of wear are contact theory, fluid lubrication, terotechnology, horst Czichos's approach, etc.

The different chapters in this book cover different aspects of Tribology. I would like to express my sincere appreciation to the authors of this book for their excellent contributions and their efforts involved in the publication process. I thank all our authors for allocating much of their scarce time to this project. Not only do we appreciate their participation, but also their adherence as a group to the time parameters set for this publication.. I do believe that the contents in this book will be helpful to many researchers in this field around the world.

Editor

Parametric Investigations at the Head-Disk Interface of Thermal Fly-Height Control Sliders in Contact

Sripathi V. Canchi,[1] David B. Bogy,[1] Run-Han Wang,[2] and Aravind N. Murthy[2]

[1] *Computer Mechanics Laboratory, Mechanical Engineering, University of California, Berkeley, CA 94720, USA*
[2] *HGST, a Western Digital Company, San Jose, CA 95135, USA*

Correspondence should be addressed to Sripathi V. Canchi, sripathi.canchi@cal.berkeley.edu

Academic Editor: Bo Liu

Accurate touchdown power detection is a prerequisite for read-write head-to-disk spacing calibration and control in current hard disk drives, which use the thermal fly-height control slider technology. The slider air bearing surface and head gimbal assembly design have a significant influence on the touchdown behavior, and this paper reports experimental findings to help understand the touchdown process. The dominant modes/frequencies of excitation at touchdown can be significantly different leading to very different touchdown signatures. The pressure under the slider at touchdown and hence the thermal fly-height control efficiency as well as the propensity for lubricant pickup show correlation with touchdown behavior which may be used as metrics for designing sliders with good touchdown behavior. Experiments are devised to measure friction at the head-disk interface of a thermal fly-height control slider actuated into contact. Parametric investigations on the effect of disk roughness, disk lubricant parameters, and air bearing surface design on the friction at the head-disk interface and slider burnishing/wear are conducted and reported.

1. Introduction

In order to realize higher magnetic storage densities in hard disk drives, it is necessary to reduce and control the read-write head-to-media spacing, or, equivalently, the physical spacing/clearance separating the head from the disk. Current hard disk drive (HDD) products operate with subnanometer clearances using the thermal fly-height control (TFC) technology, where the TFC heater locally deforms the region around the read-write head of the slider bringing it closer to the disk. The head-to-disk clearance can therefore be adjusted by changing the power supplied to the TFC heater.

A touchdown test is used to calibrate the TFC heater power to the clearance. The heater power required to make the head contact the disk lubricant surface, that is, the touchdown power (TDP), is first determined, and the heater power is then reduced to retract the head away from the disk in order to achieve a target clearance. Accurate TDP detection is therefore a key enabling step to using the TFC technology. Inaccurate TDP detection can severely compromise the drive

performance: if the actual clearance is too high, the recording performance suffers, and if the actual clearance is too low, it increases the probability for head-disk contact, which leads to unwanted head wear, thus compromising drive reliability.

Research studies on the slider-disk contact interactions and the resulting slider vibrations have been of interest to the HDD community for a long time. Research investigations on the traditional (non-TFC) slider dynamics were motivated by the need to design low flying sliders while mitigating the contact-induced slider vibrations and slider-lubricant interactions, and extensive literature exists on these topics [1–8]. After the introduction of TFC sliders, several researchers have tried to understand and explain the contact and touchdown behavior of TFC sliders owing to its importance in HDD spacing calibration. It has been shown through experiments and simulation that in addition to the slider air bearing surface (ABS) design the disk lubricant and suspension design play an important role in the slider touchdown process [9]. Numerical and analytical simulations accounting for the nonlinear forces at the head-disk interface (HDI) have also successfully explained the strong vibration dynamics of the slider close to the TDP and the subsequent suppression of vibrations for powers higher than the TDP [10–12], and they qualitatively support the experimental results reported for certain ABS designs [12, 13]. A full understanding of the touchdown behavior of TFC sliders is still lacking, and it continues to be an active topic of research.

In this work, acoustic emission (AE) sensors and laser doppler vibrometers (LDVs) are used in spin stand experiments focused on touchdown detection and slider dynamics studies. The similarities and differences in the touchdown signature for different slider ABS designs are highlighted. Parametric investigations are conducted to explain the dependence of friction and head wear on disk roughness and lubrication condition. In conjunction with recent work on the interactions of TFC sliders with the disk lubricant during touchdown/contact [14, 15], these results help develop a better understanding of the touchdown process of TFC sliders and the complex interactions at the HDI.

2. Experiments

Experiments are conducted on a spin stand equipped with an AE sensor to detect contact and an LDV to detect vertical flexure motions of the head gimbal assembly. Three different ABS designs mounted on the same suspension are used in this study. The TDP (i.e., power required to achieve zero clearance or contact with the disk lubricant) is determined experimentally by supplying the TFC heater with a square pulse lasting 70 ms with increasing power. The AE signal standard deviation is monitored during each power pulse, and the power at which the AE signal standard deviation crosses a specified threshold (set to be 20% above baseline) is recorded as the TDP.

3. Results and Discussion

3.1. Touchdown Behavior/Characteristics. The touchdown plots for three different ABS designs are shown in Figure 1 where ABS-3 shows a favorable "sharp touchdown" while ABS-1 shows an unfavorable "gradual touchdown," and has a slow rise in AE signal for increasing TFC power. ABS-2 has touchdown performance which falls between ABS-1 and ABS-3. A sharp touchdown behavior is preferred as it gives a well-defined estimate of the exact power at which contact with the disk lubricant is achieved.

The sharp touchdown for ABS-3 is characterized by strong individual spikes in the time history of the AE signal while the gradual touchdown for ABS-1 shows a uniformly increased AE signal during the TFC pulse as shown in Figure 2. (The AE signal on these plots have been shifted by 1 V to show them clearly.)

In addition, tests in overpush (i.e., TFC power above the TDP) reveal that ABS-3 with "sharp touchdown" shows an "overshoot behavior" with very strong AE signal at powers slightly above the TDP and a subsequent suppression of AE signal when the power increases into overpush as seen in Figure 3. In contrast, ABS-1 with the "gradual touchdown" shows no "overshoot" behavior but a gradual increase in

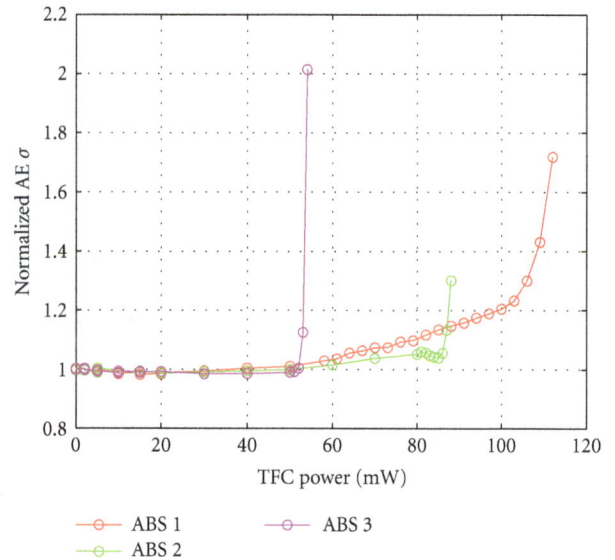

FIGURE 1: Touchdown plots for the three different ABS designs.

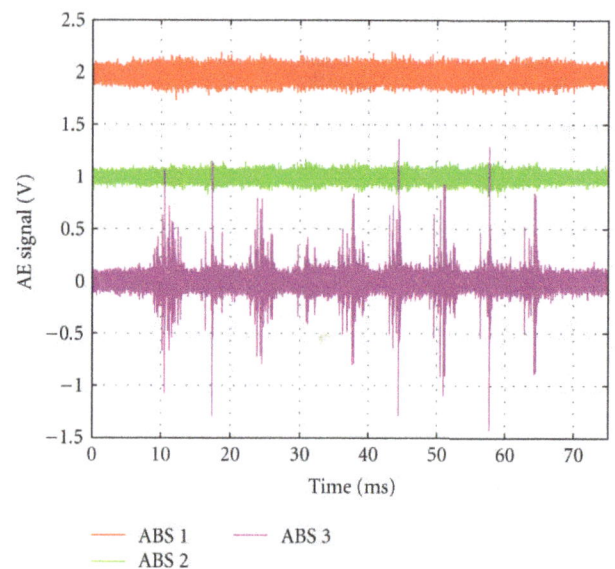

FIGURE 2: Touchdown signature for the three different ABS designs (plots offset by 1 V for clarity).

AE detected contact with overpush. ABS-2 has behavior in between those of ABS-3 and ABS-1.

Simulations for these three ABS designs show that the increasing sharpness of touchdown correlates with decreasing pressure under the TFC protrusion at touchdown, increased TFC efficiency, and lower TDP (Table 1). It is also observed that lubricant pickup is higher for ABS-3 compared to ABS-2 or ABS-1.

While an explanation of the physical mechanism behind the different ABS touchdown signatures requires a further comprehensive study, important correlations may be drawn from these results. The strong individual spikes for ABS-3 are evidence of a snap-in/snap-out behavior, which is an indication of the spontaneous instability of the slider

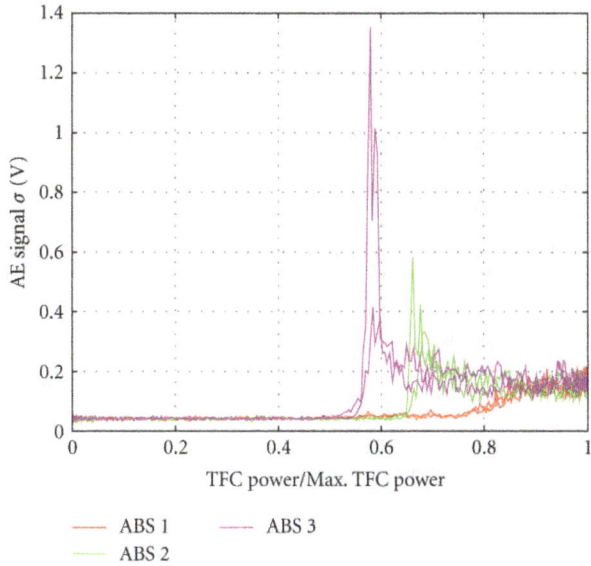

FIGURE 3: Contact behavior in overpush for the three different ABS designs.

TABLE 1: Simulated results for the three different ABS designs.

ABS design	Simulation TFC efficiency nm/mW	Touchdown power mW	Pressure at touchdown atm.
ABS-1	0.108	96	60
ABS-2	0.119	91	38
ABS-3	0.145	69	27

during touchdown. It is highlighted that the circumferential locations of the spikes do not remain fixed upon repetition of the touchdown test and hence are not caused by disk defects as such. The snap-in/snap-out process causes a higher level of head-disk lubricant interaction, which correlates well with the higher amount of lubricant pickup for ABS-3. It may also be reasoned that the lower pressure at the TFC bulge for ABS-3 makes it more susceptible for lubricant transfer from disk to the slider in comparison to ABS designs with higher pressure at the TFC bulge, such as ABS-1 and ABS-2.

3.2. Analogous Experimental Results. A separate set of experiments with ABS-2 reveals that the touchdown can be sharp or gradual depending on the disk RPM (or the linear velocity) as shown in Figure 4. Specifically, a lower disk RPM increases the touchdown sharpness and a higher disk RPM degrades the touchdown sharpness. It is also observed that the touchdown performance degrades for a burnished slider (as shown by the dashed lines in Figure 4), where the slider is burnished in a controlled fashion by increasing the TFC power above the TDP on a separate disk track. The results for the 7200 RPM case on Figure 4 highlight the possibility of false TDP detection: for the same 20% AE threshold, the unburnished case shows a gradual AE rise until about 95 mW, but the sharp AE rise at 102 mW is the actual TDP.

However, the burnished case reads a false TDP at 93 mW owing to the gradual rise in AE that occurs before the sharp AE rise marking touchdown. The time history of the AE signal is similar to that observed with the different ABS designs; namely, strong individual AE spikes appear for the 3600 RPM case with sharp touchdown, and a uniformly increased AE signal appears for the 7200 RPM case with gradual touchdown (Figure 5). These analogous results for ABS-2 provide a way to probe the same HDI under different disk RPM to understand the changes that occur in the AE and LDV signals for "sharp" and "gradual" touchdown signatures as well as in overpush.

3.3. LDV Spectrum and AE Signal Content. Experiments are conducted with ABS-2 to simultaneously capture the AE signal and the LDV signal in order to identify the frequencies that correspond to the flexure and slider vertical motions and to see how they appear in the AE signal. The tests are conducted with the power increased above the TDP (i.e., into overpush). For this ABS-2 design, the simulated air bearing frequencies at 5400 RPM and 1 nm minimum spacing are 142 kHz (roll), 167 kHz (pitch-1), and 324 kHz (pitch-2).

For each of the overpush tests conducted at different RPM, Figure 6 shows the profile of the applied TFC voltage and the time history of the LDV signal on one plot, and the corresponding joint time frequency spectrum of the LDV signal on a separate plot. As seen on the figures, the TFC voltage increases from zero to the target overpush value (between 50 ms to 100 ms), remains at this value (between 100 ms to 200 ms), and decreases back to zero (between 200 ms to 250 ms). During the overpush regime, the slider vibrations are excited due to contact, and the dominant excitation frequencies are identified on the plots showing the joint time frequency spectrum. It is observed that for 3600 RPM the lower frequencies (notably 139, 148, and 165 kHz which are close to simulated roll and pitch-1 air bearing frequencies) are dominant, while for the 7200 RPM the higher air bearing frequency 321 kHz (corresponding to pitch-2 air bearing frequency) is dominant. This result indicates that the mechanism/nature of touchdown and contact at the HDI is significantly altered by the disk RPM. Specifically in this case, the excited vibration modes at contact are shifted from the lower frequency pitch-1 and roll modes at lower disk RPM, to the higher frequency pitch-2 modes at higher disk RPM. The pitch-1 nodal line passes through the trailing end of the slider, and its excitation in general is associated with stronger motions of the entire slider body. Such excitation may, in fact, be responsible for the stronger "sharp touchdown" signature. In contrast, the pitch-2 nodal line lies closer to the leading end of the slider, and its excitation only causes the vertical bouncing motion of the trailing end, resulting only in localized contact at the TFC bulge location, and, therefore, leading to the weaker "gradual touchdown" signature. It is surmised that, in an analogous fashion, the touchdown process excites different vibrations modes for the three different ABS designs presented in Section 3.1, hence causing the very different sensitivities at touchdown.

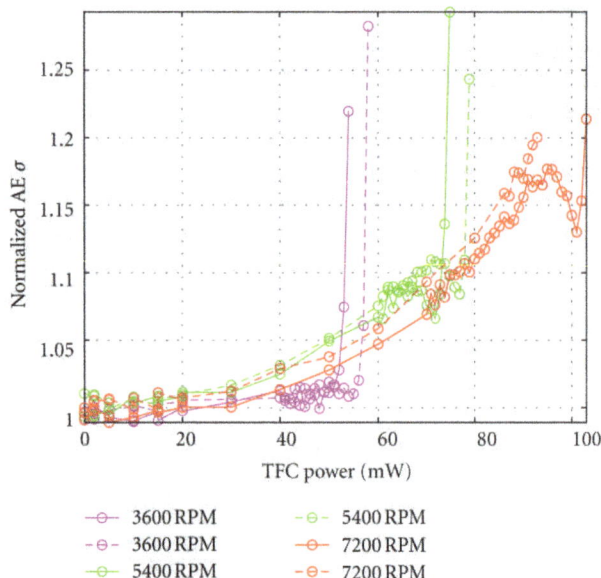

FIGURE 4: Touchdown plots for ABS-2 at different disk RPM (dashed lines for a burnished slider).

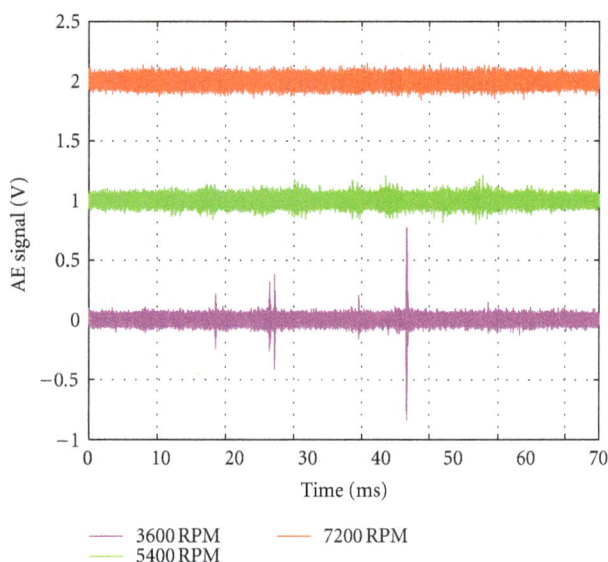

FIGURE 5: Touchdown signature for ABS-2 at different disk RPM (plots offset by 1 V for clarity).

The components of the AE signal at the different frequencies observed in the LDV signal are plotted in Figure 7 to observe how they change as a function of the TFC power. (The cumulative effect of adding all these components would result in the plot shown in Figure 4.) At 3600 RPM the touchdown is marked by the sharp rise in the 148 kHz component and there are no components that show gradual rise. At 5400 RPM, the 148 kHz component shows a gradual rise, but touchdown is marked by a sharp rise in the 321 kHz component. At 7200 RPM, there are no components with a sharp rise, and the 321 kHz component shows a gradual rise.

These results indicate that at 3600 RPM the contact is dominated by the slider's pitch-1 and roll motions (together with any suspension-related motions that give rise to frequency peaks in the 65–100 kHz region.) At 7200 RPM, contact is mainly dominated by the vibration of the slider at the pitch-2 frequency. At 5400 RPM, the interaction is a combination of the above two modes: as the TFC powers increase, a gradual rise in the 148 kHz component occurs first, but a strong vibration in the pitch-2 mode (321 kHz) eventually marks touchdown.

These results are in agreement with recent studies that show that at close spacing and at the onset of lubricant-contact, the in-plane shear forces and friction can destabilize the slider for certain ABS designs resulting in vibrations dominantly occurring at suspension and lower air bearing frequencies (60–200 kHz in our case), while stronger contact with the disk causes slider vibrations with higher frequency content (above 200 kHz) [9].

3.4. *Friction Measurements in Contact.* Friction forces at the HDI become important during contact conditions and may in fact play a dominant role in HDI performance and slider dynamics. Friction-induced slider wear, as well as disk lubricant redistribution and disk overcoat damage, needs to be examined carefully to explore future designs that can accommodate a certain level of head-disk contact.

Experiments are devised to measure the friction forces in the downtrack direction during contact and overpush conditions by instrumenting a strain gage on the fixture which holds the head gimbal assembly on the spin stand. The voltage signal from the strain gage may be converted into force measurement by determining the calibration constant for the strain gage. Such a calibration is performed using the usual technique: by noting the strain gage voltage signal corresponding to different standard forces, which are applied by suspending standard weights from the fixture, and subsequently fitting a linear curve.

Once the TDP is determined on the test track, the TFC is powered with a voltage profile having 100 ms dwell time at the maximum power. It is noted that strain gages have a low bandwidth, and several experiments reveal that a dwell time of at least 100 ms is necessary to allow the strain gage to respond to the friction force and give good, repeatable measurements. All experiments are conducted with ABS-2 on a reference "standard disk" unless specified otherwise.

3.4.1. *Friction, AE, and Slider Bouncing in Contact.* Figures 8(a) and 8(b) show the TFC voltage profile and the resulting slider bouncing (displacement and velocity), AE detected contact, and the friction force, for 10 mW and 20 mW overpush, respectively. It is evident that slider bouncing and AE signal remain high throughout the overpush region for 10 mW overpush case, and they get suppressed (after an initial overshoot region) for the 20 mW overpush case. The friction force measured by the strain gage, however, continues to increase with the amount of overpush indicating a higher level of interference and contact for larger overpush powers even though the AE detected contact and slider

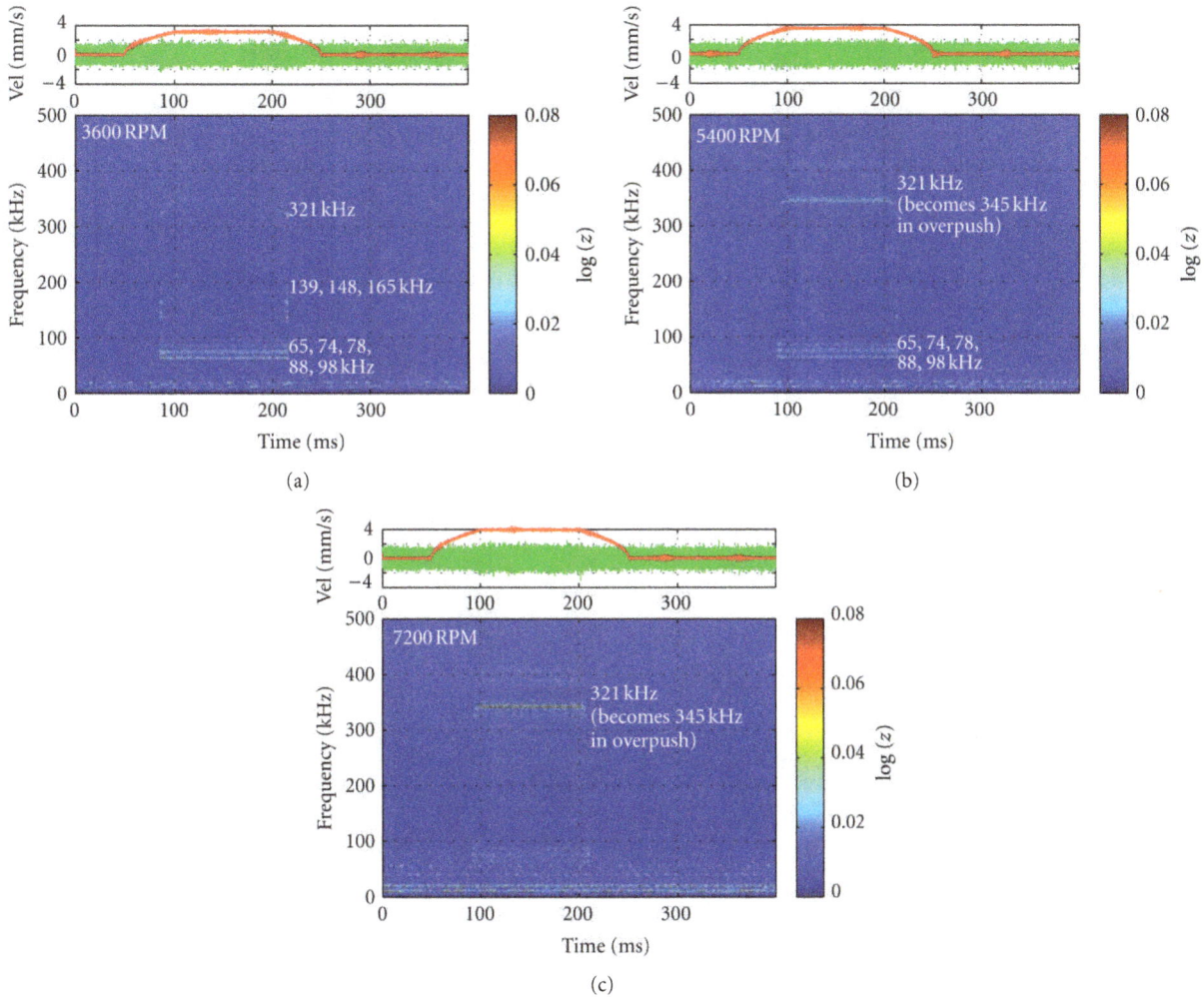

FIGURE 6: Applied TFC voltage profile, vertical velocity time history, and the joint time frequency spectrum of the vertical velocity for ABS-2 at different disk RPM.

dynamics get suppressed. (It is noted that the amplitude of the AE signal in the suppressed state is noticeably higher than the baseline AE signal with no TFC power, implying a certain amount of contact).

3.4.2. Effect of Disk Roughness. Disk roughness plays an important role in HDI performance. The combined slider and disk roughness affect the nominal physical spacing at the HDI, the magnitude of interaction forces (intermolecular/adhesive, etc.), and the actual area of contact, thereby influencing the magnitude of contact and friction forces. A parametric study is conducted with three disk types: disks A, B, and C with decreasing roughness, in that order, and with surface roughness parameters tabulated in Table 2, where R_q is the root mean square roughness, R_p is the maximum peak height, and R_v is the maximum valley depth. These disks are coated with ZTMD lubricant of nominal 12 Å thickness.

First, several tests are conducted using the same slider to determine the TDP on a standard disk and on each of the disks A, B, and C. Table 2 presents the change in

the TDP (i.e., δTDP) on each of the disks A, B, and C compared to the TDP on a standard disk. The roughness of the standard disk is similar to that of disk A. This difference in TDP is converted into a clearance gain value (i.e. a gain in clearance from that on a standard disk) using a conversion factor of 0.119 nm/mW, which is the TFC efficiency estimated for ABS-2 from simulations. Figure 9 shows the same information in graphical form and highlights the linear relationship between the disk roughness (R_p or R_q) and clearance gain. Since the thermal protrusion comes into contact with the peaks of the roughness, the relationship between the clearance gain and R_p (Figure 9(b)) is of importance, and it is seen that for every 1 nm decrease in R_p there is a 0.8 nm actual gain in clearance at the HDI for the range of surface roughness values considered in these experiments.

Next, the dependence of friction on the disk roughness is investigated by conducting a "friction test" on each of the three disk types. A new (unburnished) slider is flown on a fresh test track, the TDP is determined, and the TFC heater is then supplied with the power profile with a 100 ms

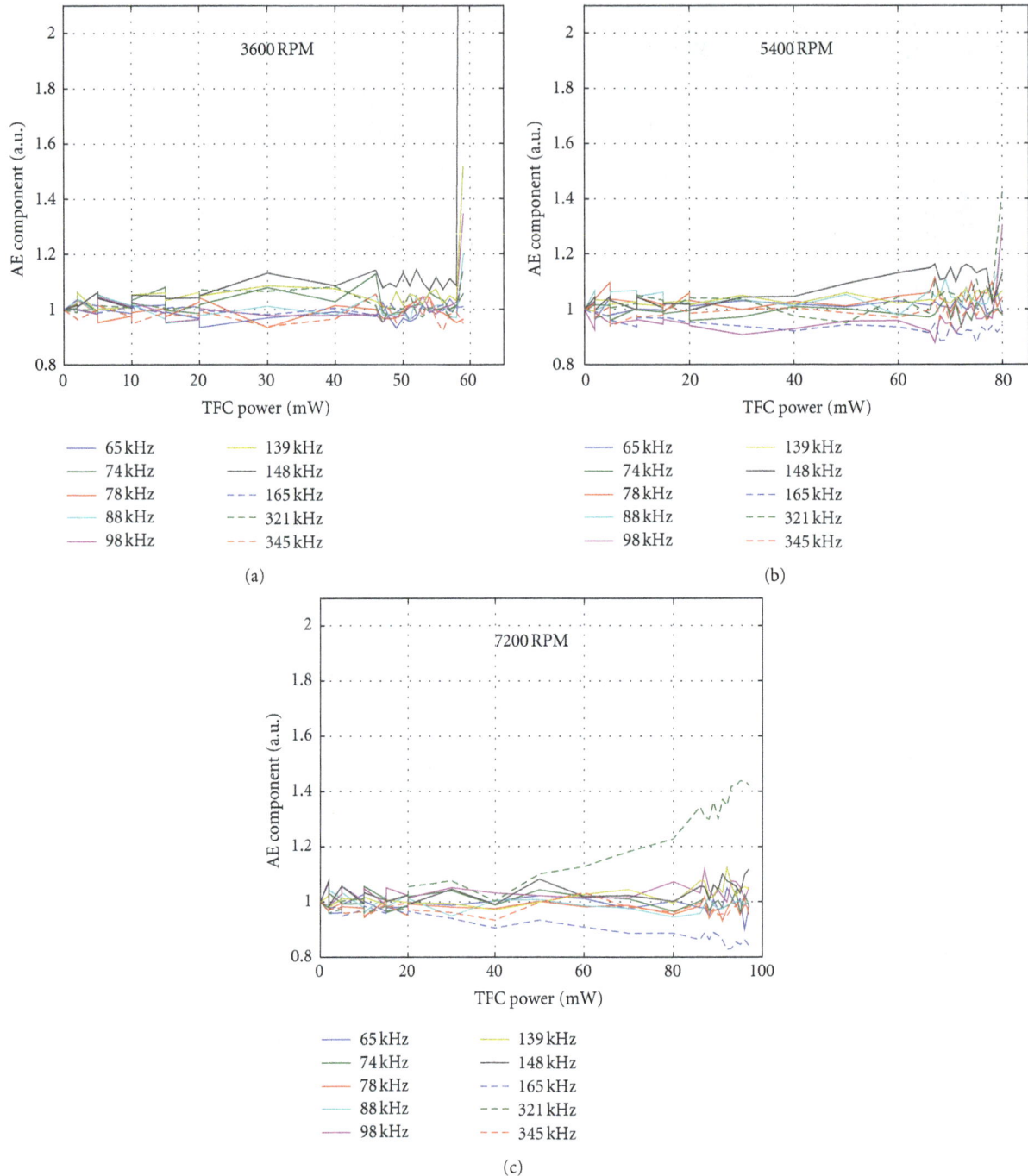

Figure 7: AE signal components in the touchdown plot for ABS-2 at different disk RPM.

dwell time (as shown in Figure 8). The peak TFC power is increased from TDP to a maximum of TDP + 50 mW in 5 mW increments, and it is then similarly decreased back to TDP. The average friction measured by the strain gage at each power step is tabulated. All tests are conducted on the same disk track. The measured friction values are plotted as those for the "unburnished" case. The same slider, which is now deemed "burnished" because of the overpush testing, is flown on an adjacent track and the friction test is repeated to obtain friction values for the "burnished" case. Figure 10(a)

shows a representative plot for the strain gage measured friction values as a function of overpush power supplied to the TFC heater.

A quadratic curve passing through the origin is fit to the friction measurements for the "unburnished" and "burnished" cases, and the slope of this curve at the 10 mW overpush point is used to obtain the "friction (μN) per milliwatt of overpush power" value. Figure 10(b) plots these friction values measured on the three disk types (A, B, and C) based on experiments conducted with three new sliders

TABLE 2: Disk roughness parameters and its effect on TDP/clearance gain.

Disk	R_p nm	R_q nm	R_v nm	δTDP mW	Clearance gain nm
A	2.02	0.49	1.87	0.08	0.01
B	1.89	0.36	1.47	0.97	0.12
C	1.00	0.24	1.11	6.71	0.84

TABLE 3: Effect of TFC efficiency (ABS/heater design) on friction.

ABS design	Simulation TFC efficiency nm/mW	Friction unburnished μN/mW	Friction burnished μN/mW
ABS-A (ABS-1)	0.108	9	17
ABS-B	0.111	24	31
ABS-C (ABS-2)	0.119	37	54
ABS-D (ABS-3)	0.145	56	67

on each disk type. While there is no particular trend relating the measured friction and surface roughness, the friction is higher for the burnished slider compared to the unburnished slider in all tests. Slider burnishing wears and smoothens the slider surface, increasing the actual contact area between the thermal protrusion and the disk, thereby resulting in the slightly higher friction force.

In order to directly compare the friction values between the three disk types, another set of experiments is conducted by flying the same "burnished" slider on the three disk types in succession. The slider is burnished in a controlled fashion separately before use in this test. The friction against the overpush power is plotted in Figure 10(c) using data from two "burnished" heads. It is concluded based on these results that within the range of disk roughness considered in this work there is no significant effect of the disk surface roughness on the measured friction.

3.4.3. Effect of Lubricant Parameters. Friction tests are conducted to determine the effect of lubricant type/bonding on the friction in contact. Disks with three different lubricant type/bonding ratios are used: Lube A (61% bonded ratio, 10.5 Å), Lube A (69% bonded, 10.5 Å), and Lube B (82% bonded, 12 Å). Figure 11(a) shows the friction measured for the three media for the "unburnished" and "burnished" slider cases (based on three experiments each). The friction values are comparable for the unburnished sliders on all three disks types. While the friction values for the burnished and unburnished sliders are comparable on the disks with Lube A 61% and 69% bonded ratio, the friction for the burnished slider on the disk with Lube B 82% bonded ratio is relatively higher than that for the unburnished slider. This result is consistent with results for the change in TDP occurring because of a friction test, where the TDP change after and before a friction test is a measure of slider burnishing. As shown in Figure 11(b), the highest burnishing (indicated by highest δTDP) occurs to a new "unburnished" slider on the Lube B 82% bonded disk. As a result of greater slider burnishing on this disk, the friction is higher when a subsequent test is conducted with this burnished slider.

A direct comparison of friction values is reported in Figure 11(c) based on tests conducted in succession on the three different disks using two "burnished" sliders, and it shows marginally higher friction values on the disk with Lube B 82% bonded ratio.

Friction tests are conducted to understand the effect of the mobile part of the lubricant on friction and slider burnishing. The disk with Lube A 10.5 Å 61% bonded fraction is delubed by immersing it in a solution of Vertrel XF solution

to remove the mobile lubricant. The delubed disk has a lubricant thickness of 6 Å (bonded lubricant). Figure 12(a) shows the measured friction on the lubed and delubed disks for the unburnished and burnished slider cases. The friction values for the unburnished as well as burnished slider on the lubed disk are similar and comparable to the friction value measured for the unburnished slider on the delubed disk. However, the friction is substantially higher for the burnished slider on the delubed disk. Figure 12(b) shows that slider burnishing (indicated by δTDP after a friction test) is higher for tests conducted on the delubed disk implying that an unburnished slider is substantially burnished on this disk type, and the friction is higher for the subsequent test conducted with such a burnished slider. These results highlight the important role of the mobile part of the lubricant in reducing friction and slider burnishing, thereby increasing the reliability of an HDI with contact.

3.4.4. Effect of TFC Efficiency. The thermal protrusion size and shape make a significant difference in the slider's touchdown and contact behavior. The friction during contact for different slider ABS/heater designs is plotted in Figure 13 for the unburnished and burnished cases, and the same data is tabulated in Table 3 together with each design's TFC efficiency estimated from simulations. It is observed that the friction forces increase as the TFC efficiency increases and may be explained by the following argument. For the same amount of overpush power, a higher TFC efficiency slider will have greater level of interference (because of a larger protrusion). As a result, the effective contact area is larger for the higher TFC efficiency case, reflecting in a higher measured friction.

3.4.5. Effect of Disk RPM. The similarities between the touchdown plot and contact signature of ABS-2 at different disk RPMs to those of ABS designs with different TFC efficiencies are highlighted in Section 3.2. Particularly, it is shown that at a higher RPM ABS-2 behaves like a design with low TFC efficiency (showing a gradual touchdown plot), and at a lower RPM ABS-2 behaves like a design with high TFC efficiency (showing a sharp touchdown plot).

The friction results from tests with ABS-2 at different RPM are consistent with the above analogy and with the results presented in Section 3.4.4. Figure 14 shows that the friction increases as the disk RPM decreases; that is, when ABS-2 is made to behave like a slider with high TFC

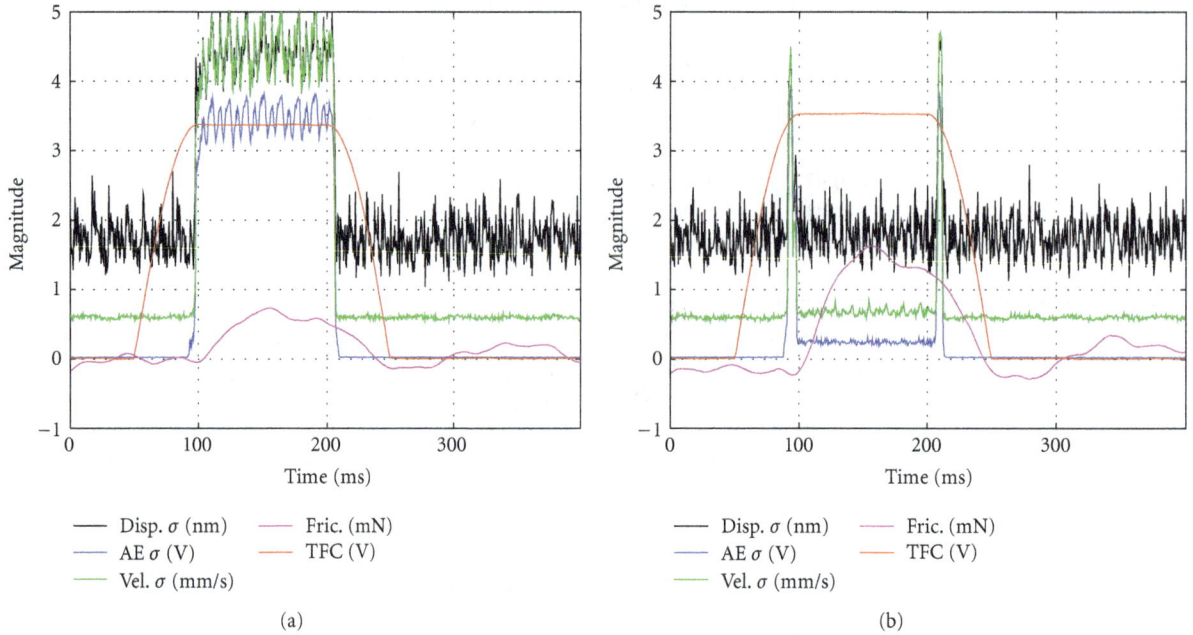

FIGURE 8: Time history of TFC power, vertical displacement, vertical velocity, AE signal, and friction. (a) 10 mW overpush, (b) 20 mW overpush.

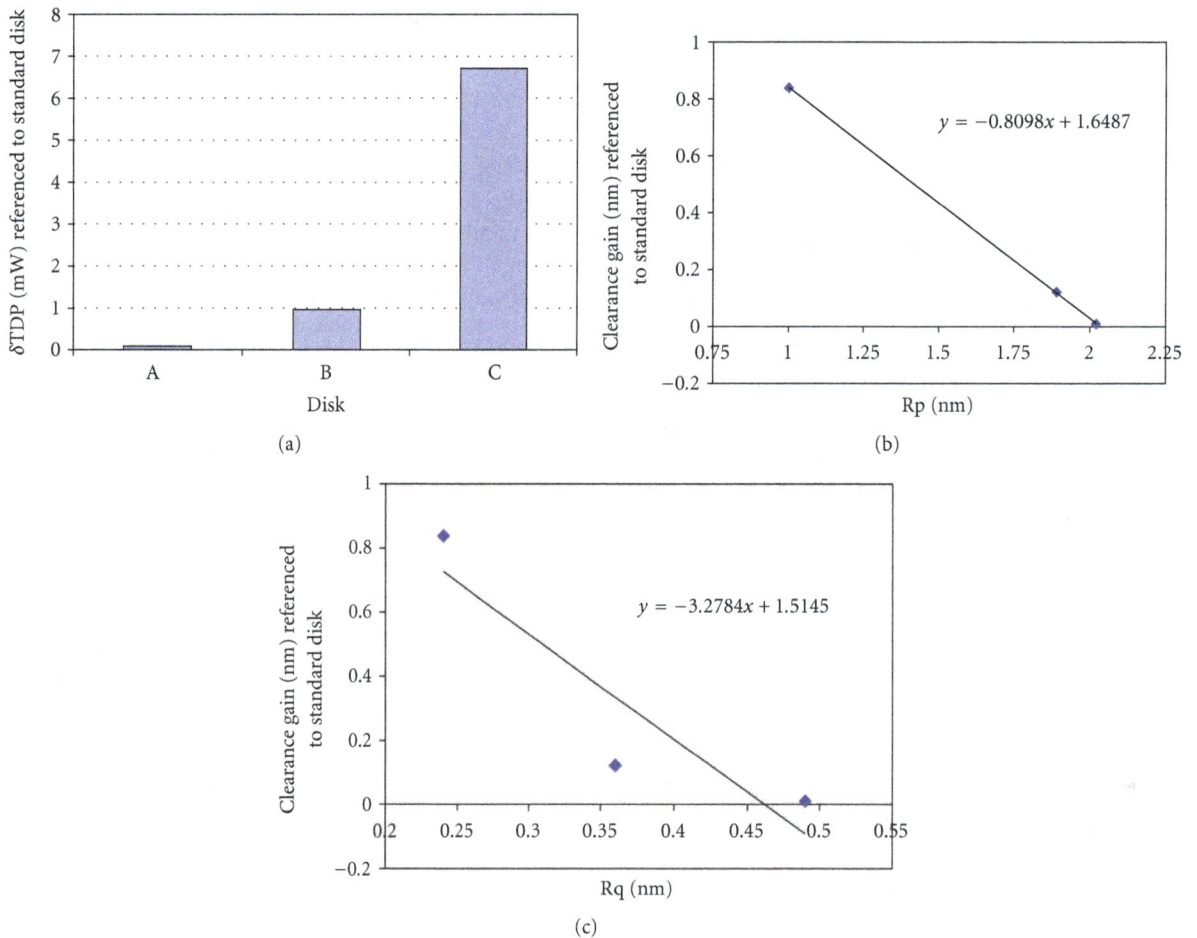

FIGURE 9: Effect of disk roughness on clearance.

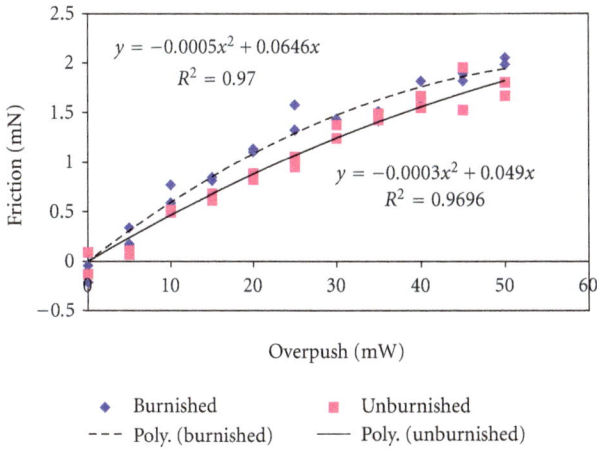

(a)

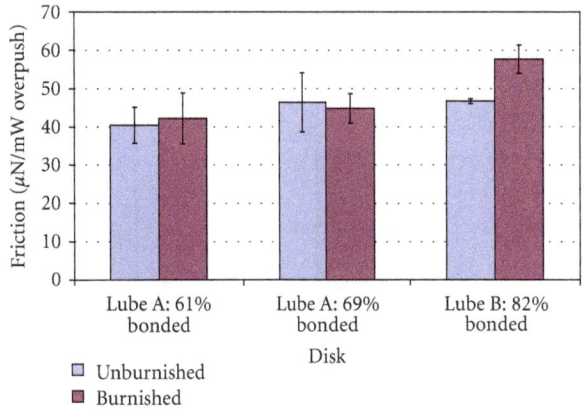

(a)

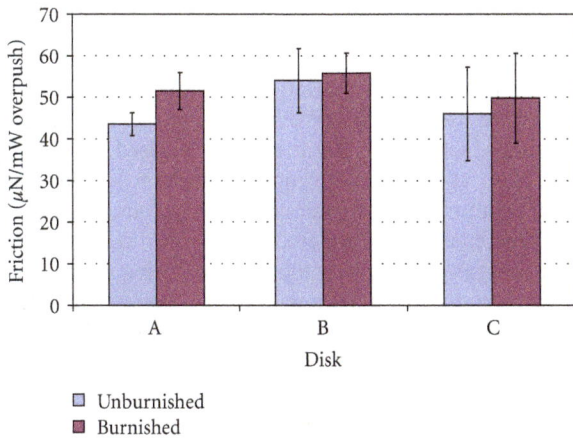

(b)

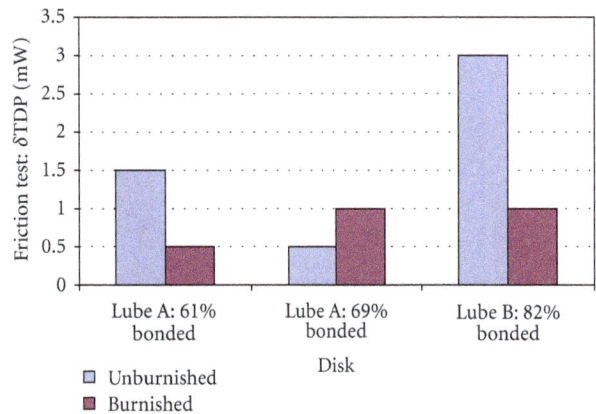

(b)

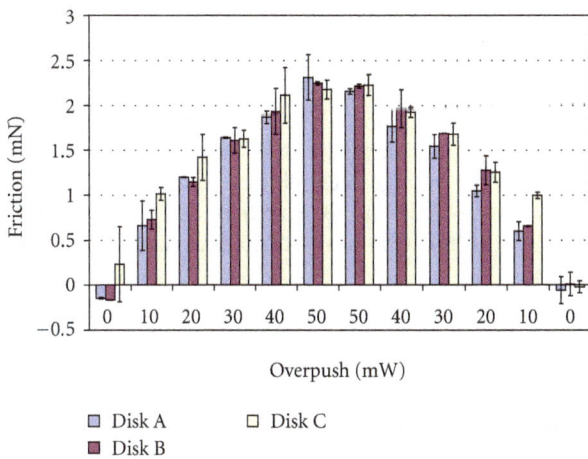

(c)

FIGURE 10: Effect of disk roughness on friction.

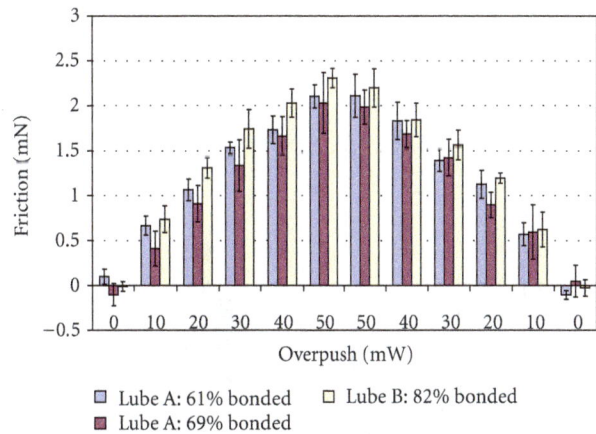

(c)

FIGURE 11: Effect of lubricant parameters on friction and slider burnishing.

efficiency by decreasing RPM, it exhibits the characteristic sharp touchdown plot and higher friction.

4. Conclusion

The touchdown behavior of TFC sliders is investigated through experiments. Certain sliders exhibit a sharp rise of

AE signal at touchdown when the power is increased in milliwatt steps while others show a gradual rise making it difficult to exactly define the TDP to milliwatt resolution. An analogous behavior occurs when the disk RPM is changed for a particular slider ABS. It is found that the dominant modes/frequencies of excitation at touchdown are significantly different in these cases leading to the very different touchdown signatures. Particularly, the sharp touchdown

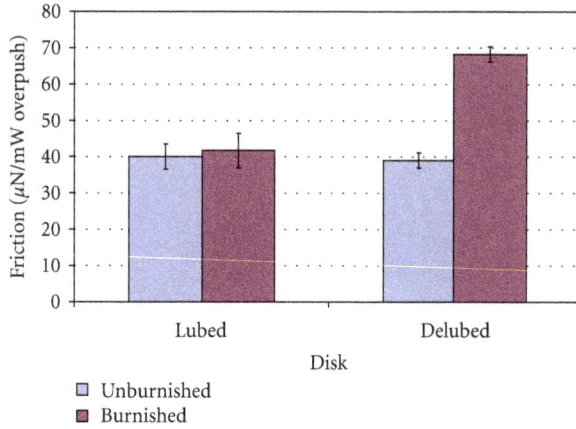

(a)

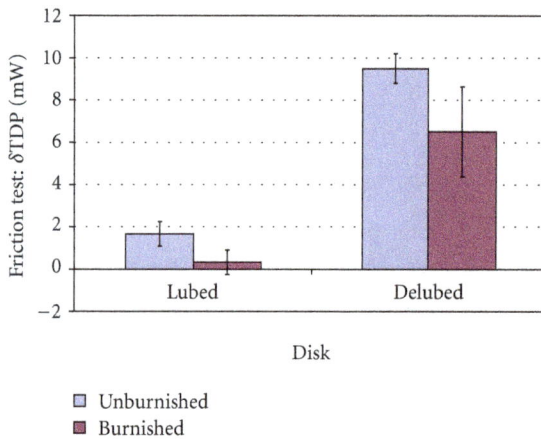

(b)

FIGURE 12: Effect of mobile lubricant on friction and slider burnishing.

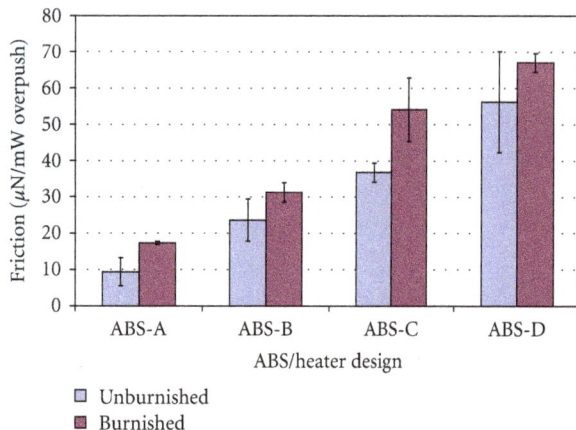

FIGURE 13: Effect of TFC efficiency (ABS/heater design) on friction.

case is characterized by strong individual contact events as observed in the AE signal, and the dominant excitation occurs at frequencies that correspond to the slider's first pitch and roll modes in addition to suspension related-frequencies. In contrast, the gradual touchdown case is characterized by a uniform rise in AE signal over the duration of the TFC

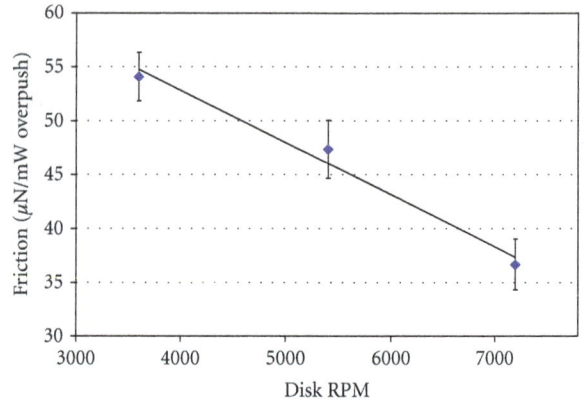

FIGURE 14: Effect of disk RPM on friction for ABS-2.

pulse, and the dominant excitation occurs at the slider's second pitch mode. The pressure under the TFC protrusion at touchdown, the TFC efficiency, and the propensity for lubricant pickup show correlation with touchdown behavior and may be used as metrics for designing sliders with good touchdown features. Experiments are devised to measure the friction at the HDI during TFC-induced contact, and several parametric investigations are carried out. In the range of parameter values considered, the disk surface roughness does not significantly affect the friction during contact. The mobile part of the lubricant plays an important role in reducing friction as well as slider burnishing. A burnished slider shows a higher friction value than an unburnished slider because of an increase in effective/actual contact area, and, for the same reason, sliders with higher TFC efficiency show higher friction compared to sliders with lower TFC efficiency.

Acknowledgments

This work was supported by the Computer Mechanics Laboratory at the University of California at Berkeley and Hitachi Global Storage Technologies. The authors thank Q. Dai, L. Dorius, XC. Guo, B. Marchon, and R. Waltman for their support and helpful discussions during the course of this work.

References

[1] B. Knigge and F. E. Talke, "Dynamics of transient events at the head/disk interface," *Tribology International*, vol. 34, no. 7, pp. 453–460, 2001.

[2] C. M. Mate, P. C. Arnett, P. Baumgart et al., "Dynamics of contacting head-disk interfaces," *IEEE Transactions on Magnetics*, vol. 40, no. 4, pp. 3156–3158, 2004.

[3] J. Xu, H. Kohira, H. Tanaka, and S. Saegusa, "Partial-contact head-disk interface approach for high-density recording," *IEEE Transactions on Magnetics*, vol. 41, no. 10, pp. 3031–3033, 2005.

[4] J. D. Kiely and Y. T. Hsia, "Slider dynamic motion during writer-induced head-disk contact," *Microsystem Technologies*, vol. 14, no. 3, pp. 403–409, 2008.

[5] Q. Dai, B. E. Knigge, R. J. Waltman, and B. Marchon, "Time evolution of lubricant-slider dynamic interactions,"

IEEE Transactions on Magnetics, vol. 39, no. 5, pp. 2459–2461, 2003.

[6] Q. Dai, F. Hendriks, and B. Marchon, "Washboard effect at head-disk interface," *IEEE Transactions on Magnetics*, vol. 40, no. 4, pp. 3159–3161, 2004.

[7] B. Marchon, X. C. Guo, A. Moser, A. Spool, R. Kroeker, and F. Crimi, "Lubricant dynamics on a slider: 'the waterfall effect'," *Journal of Applied Physics*, vol. 105, no. 7, Article ID 074313, 2009.

[8] C. M. Mate, B. Marchon, A. N. Murthy, and S. H. Kim, "Lubricant-induced spacing increases at slider-disk interfaces in disk drives," *Tribology Letters*, vol. 37, no. 3, pp. 581–590, 2010.

[9] Q. H. Zeng, C. H. Yang, S. Ka, and E. Cha, "An experimental and simulation study of touchdown dynamics ," *IEEE Transactions on Magnetics*, vol. 47, pp. 3433–3436, 2011.

[10] J. Zheng and D. B. Bogy, "Investigation of flying-height stability of thermal fly-height control sliders in lubricant or solid contact with roughness," *Tribology Letters*, vol. 38, no. 3, pp. 283–289, 2010.

[11] J. Zheng and D. Bogy, "Dynamic instability of thermal-flying-height-control sliders at touchdown," *Microsystem Technologies*, vol. 18, no. 9-10, pp. 1319–1322, 2012.

[12] S. Vangipuram Canchi and D. B. Bogy, "Thermal fly-height control slider instability and dynamics at touchdown: explanations using nonlinear systems theory," *Journal of Tribology*, vol. 133, no. 2, Article ID 021902, 2011.

[13] S. Vangipuram Canchi and D. B. Bogy, "Slider dynamics in the lubricant-contact regime ," *IEEE Transactions on Magnetics*, vol. 46, pp. 764–769, 2010.

[14] S. Vangipuram Canchi and D. B. Bogy, "Slider-lubricant interactions and lubricant distribution for contact and near contact recording conditions," *IEEE Transactions on Magnetics*, vol. 47, no. 7, pp. 1842–1848, 2011.

[15] S. Vangipuram Canchi and D. B. Bogy, "Experiments on slider lubricant interactions and lubricant transfer using TFC sliders," *Microsystem Technologies*, vol. 18, pp. 1517–1523, 2012.

Mechanical and Tribological Properties of PVD-Coated Cemented Carbide as Evaluated by a New Multipass Scratch-Testing Method

M. Fallqvist,[1] R. M'Saoubi,[2] J. M. Andersson,[2] and M. Olsson[1]

[1] *Department of Material Science, Dalarna University, 781 88 Borlänge, Sweden*
[2] *R&D Materials and Processes, Seco Tools AB, 737 82 Fagersta, Sweden*

Correspondence should be addressed to M. Fallqvist, mfa@du.se

Academic Editor: Shyam Bahadur

A new test method based on multipass scratch testing has been developed for evaluating the mechanical and tribological properties of thin, hard coatings. The proposed test method uses a pin-on-disc tribometer and during testing a Rockwell C diamond stylus is used as the "pin" and loaded against the rotating coated sample. The influence of normal load on the number of cycles to coating damage is investigated and the resulting coating damage mechanisms are evaluated by posttest scanning electron microscopy. The present study presents the test method by evaluating the performance of $Ti_{0.86}Si_{0.14}N$, $Ti_{0.34}Al_{0.66}N$, and $(Al_{0.7}Cr_{0.3})_2O_3$ coatings deposited by cathodic arc evaporation on cemented carbide inserts. The results show that the test method is quick, simple, and reproducible and can preferably be used to obtain relevant data concerning the fatigue, wear, chipping, and spalling characteristics of different coating-substrate composites. The test method can be used as a virtually nondestructive test and, for example, be used to evaluate the fatigue and wear resistance as well as the cohesive and adhesive interfacial strength of coated cemented carbide inserts prior to cutting tests.

1. Introduction

Although CVD (chemical vapour deposition) and PVD (physical vapour deposition) coatings have been used in order to increase the lifetime and performance of cutting tools for more than 30 years there is still an intensive research interest to improve the performance of existing coating materials and to develop new coating materials and coating processes. Even though many test methods have been developed to simulate the contact conditions prevailing during metal cutting conditions, cutting tests are still the most reliable methods in order to characterize the properties and performance of coatings-substrate systems aimed for metal cutting operations [1, 2]. However, cutting tests including in-depth post-test characterization of the worn cutting tools can be both costly and time-consuming. Consequently, laboratory tests, although performed at room temperature, can preferably be used in order to evaluate specific properties such as the cohesive and adhesive strength, intrinsic abrasion resistance, and so forth of specific coating systems. As a result, a wide range of laboratory techniques are available to assess the mechanical and tribological characteristics of CVD and PVD coatings for metal cutting tools [3–13].

Today, the scratch test is probably the most common and used test to increase the understanding of the mechanical response of a coating-substrate system exposed to an external load [12]. In the conventional scratch test a Rockwell C diamond stylus (120° cone with 200 μm radius) is drawn across the coated substrate surface under an increasing normal load until some kind of well-defined coating failure occurs, frequently associated with localized chipping or spalling [3]. The corresponding normal load is termed the critical load and is often used as a measure of the coating mechanical strength or "practical adhesion." The critical load can be detected by on-line monitoring of the friction force and acoustic emission signals and/or by post-test examination using optical or scanning electron microscopy [12–18].

One drawback with the conventional scratch test is the need for such high normal loads that the deformation behaviour of the substrate material in many cases controls the mechanical response of the coating-substrate system,

Mechanical and Tribological Properties of PVD-Coated Cemented Carbide as Evaluated by a New Multipass Scratch-Testing Method

13

TABLE 1: Deposition process parameters and thickness for the coatings investigated.

	$Ti_{0.86}Si_{0.14}N$	$Ti_{0.34}Al_{0.66}N$	$(Al_{0.7}Cr_{0.3})_2O_3$
Thickness (μm)	2.5	2.5	3.5
Cathode	$Ti_{0.80}Si_{0.20}$	$Ti_{0.33}Al_{0.67}$	$Al_{0.70}Cr_{0.30}$
Atmosphere	N_2	N_2	O_2
Pressure (Pa)	4	4	1
Bias (V)	-30	-50	-100
Temp. (°C)	450	450	550

TABLE 3: Critical normal loads of the coatings investigated for Rockwell C diamond styli with radius 50 μm and 200 μm, respectively.

	$Ti_{0.86}Si_{0.14}N$	$Ti_{0.34}Al_{0.66}N$	$(Al_{0.7}Cr_{0.3})_2O_3$
$F_{N,C1}$, $R = 50\,\mu$m [N]	19 ± 1	23 ± 1	17 ± 1
$F_{N,C2}$, $R = 50\,\mu$m [N]	22 ± 1	25 ± 1	20 ± 1
$F_{N,c1}$, $R = 200\,\mu$m [N]	60 ± 1	68 ± 1	51 ± 1
$F_{N,C2}$, $R = 200\,\mu$m [N]	66 ± 2	74 ± 2	55 ± 3

TABLE 2: Characteristic properties of the investigated coatings in the as-deposited condition.

	$Ti_{0.86}Si_{0.14}N$	$Ti_{0.34}Al_{0.66}N$	$(Al_{0.7}Cr_{0.3})_2O_3$
R_a (nm)	87 ± 7	77 ± 5	174 ± 10
R_z (μm)	5.1 ± 0.5	5.1 ± 0.9	7.2 ± 1
Droplets (no/100 μm^2)	6.3 ± 0.9	4.9 ± 0.7	14.3 ± 2
Area ratio droplets (%)	4.6 ± 0.8	2.7 ± 0.6	18.8 ± 4
Craters (no/100 μm^2)	0.4 ± 0.05	0.4 ± 0.06	0.1 ± 0.02
Area ratio craters (%)	0.7 ± 0.2	0.8 ± 0.2	0.3 ± 0.1
Hardness, composite (GPa)	22 ± 1.5	20 ± 1	19 ± 2
E-modulus composite (GPa)	420 ± 30	500 ± 40	380 ± 36
Hardness, coating (GPa)	38 ± 4	30 ± 3	24 ± 2
E-modulus coating (GPa)	500 ± 70	570 ± 80	440 ± 60
H_{coat}/E_{coat}	0.08	0.05	0.06
Thickness (μm)	2.5	2.5	3.5

*The H/E ratio is 0.04 (17.1/426) for the cemented carbide.

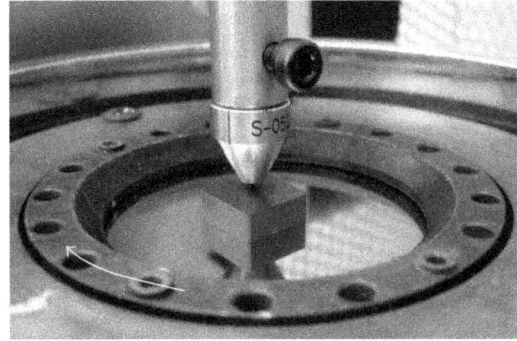

FIGURE 1: Instrumental test set-up used in the present study. The arrow indicates the rotating direction.

The objectives of the present study are to:

(a) propose a new test concept for multi-pass scratch testing using a pin-on-disc tribometer;

(b) evaluate the new test concept by studying the multi-pass fatigue and wear properties of three different PVD coatings used for metal cutting applications.

The new test approach is based on a three-step procedure aimed at evaluating the mechanical and tribological properties of CVD and PVD coatings under repeated scratching contact conditions. In the first step, conventional single pass scratch testing using a scratch tester is carried out to determine the critical normal load(s) for coating failure. In the second step, the data obtained from the conventional scratch test are used to perform multi-pass scratch testing at several sub-critical normal loads using a pin-on-disc tribometer. Finally, post-test examination of the tested samples is performed using scanning electron microscopy and energy dispersive X-ray spectroscopy in order to determine the major coating failure mechanisms.

2. Experimental

2.1. Materials. Commercial cemented carbide inserts with a composition of 94 wt% WC and 6 wt% Co and hardness 1700 $HV_{0.020}$ were used as substrate material in the present study. Prior to coating deposition the substrate surfaces were polished to a surface finish of $R_a \approx 100$ nm. Three different PVD coatings, $Ti_{0.86}Si_{0.14}N$, $Ti_{0.34}Al_{0.66}N$ and

including the adhesion of the coating to the underlying substrate. This is to a large extent due to the fact that the coating deposition processes including substrate cleaning and pre-treatment of today result in excellent adhesion.

An alternative test, able to evaluate the mechanical properties including the fatigue resistance of CVD and PVD coatings is the repetitive impact test [5, 19–21]. In this test the coated substrate is exposed to an oscillating indenter, usually a ceramic or a cemented carbide ball, impacting the surface at a constant maximum force during a large number of impacts, typically 10^6. The major drawbacks with this test are the large contact area and the fact that compressive stresses dominates while in a typical tribological contact coating failure is mainly due to the presence of tensile stresses [19, 20].

Consequently, a more realistic testing approach would be a test based on multi-pass scratch testing performed at lower loads, that is, loads significantly lower than the critical normal loads obtained in the conventional scratch test. In the multi-pass scratch test, a constant normal load lower than the critical load, is applied to the same scratch track for a large number of repeated cycles [22–27].

TABLE 4: Schematic illustrations summarizing the observed coating damage mechanisms during the circular multi-pass scratch testing.

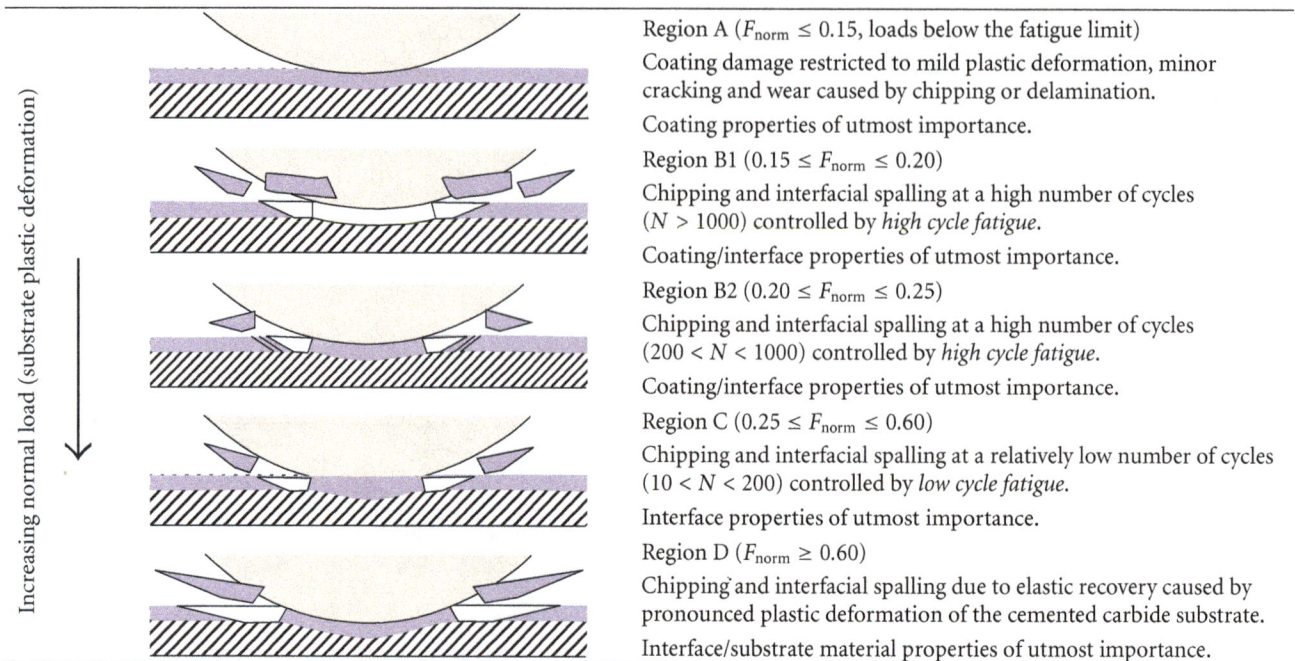

Region A ($F_{norm} \leq 0.15$, loads below the fatigue limit)

Coating damage restricted to mild plastic deformation, minor cracking and wear caused by chipping or delamination.

Coating properties of utmost importance.

Region B1 ($0.15 \leq F_{norm} \leq 0.20$)

Chipping and interfacial spalling at a high number of cycles ($N > 1000$) controlled by *high cycle fatigue.*

Coating/interface properties of utmost importance.

Region B2 ($0.20 \leq F_{norm} \leq 0.25$)

Chipping and interfacial spalling at a high number of cycles ($200 < N < 1000$) controlled by *high cycle fatigue.*

Coating/interface properties of utmost importance.

Region C ($0.25 \leq F_{norm} \leq 0.60$)

Chipping and interfacial spalling at a relatively low number of cycles ($10 < N < 200$) controlled by *low cycle fatigue.*

Interface properties of utmost importance.

Region D ($F_{norm} \geq 0.60$)

Chipping and interfacial spalling due to elastic recovery caused by pronounced plastic deformation of the cemented carbide substrate.

Interface/substrate material properties of utmost importance.

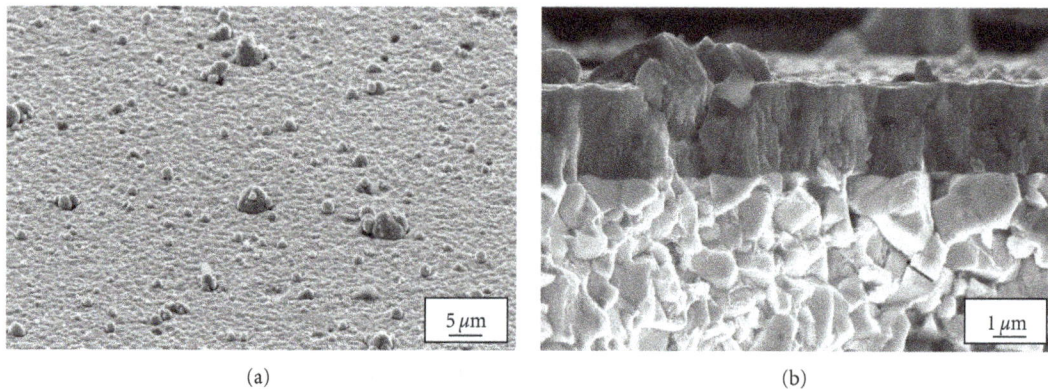

FIGURE 2: Surface morphology (a) and fractured cross-section (b) of $Ti_{0.34}Al_{0.66}N$ coating.

$(Al_{0.7}Cr_{0.3})_2O_3$, representing state of the art PVD coatings for metal cutting applications were included in the study. The coatings were deposited using cathodic arc evaporation with a coating growth rate of approximately 2 μm/h using the deposition process parameters according to Table 1. When depositing the $Ti_{0.86}Si_{0.14}N$ and $(Al_{0.7}Cr_{0.3})_2O_3$ coatings a thin $Ti_{0.34}Al_{0.66}N$, layer (thickness 0.5 μm) was deposited in order to promote the adhesion to the underlying substrate. It should be noted that the chemical composition of the coatings has been determined by combining results from energy dispersive X-ray spectroscopy (EDX), Rutherford backscattering spectrometry (RBS) and elastic recoil detection analysis (ERDA).

2.2. *Coating Characterization.* The surface morphology, microstructure, and element composition of the as-deposited coatings were investigated using a 3D optical surface

profilometer (WYKO NT-9100) and a ZEISS Ultra 55 field emission Gun Scanning Electron Microscope (FEG-SEM) equipped with an Oxford INCA Energy Dispersive X-Ray spectroscopy (EDX) system. The presence of defects (surface irregulaties), that is, droplets and craters in the as-deposited coating surfaces was evaluated using stereological methods (disector counting rule and point grid method) [28]. Only defects larger than 0.4 μm were considered in five representative areas (30×50 μm) in the centre of each sample.

2.3. *Mechanical Testing.* The Vickers hardness values of the coatings and the coating-substrate composites were measured using a Micro Combi Tester (CSM Instruments), using a load of 200 mN and 4000 mN, respectively. The indents using the lower load were all done in flat areas free from visible defects, resulting in hardness values for the defect free coating material. Loading and unloading were

Mechanical and Tribological Properties of PVD-Coated Cemented Carbide as Evaluated by a New Multipass
Scratch-Testing Method

15

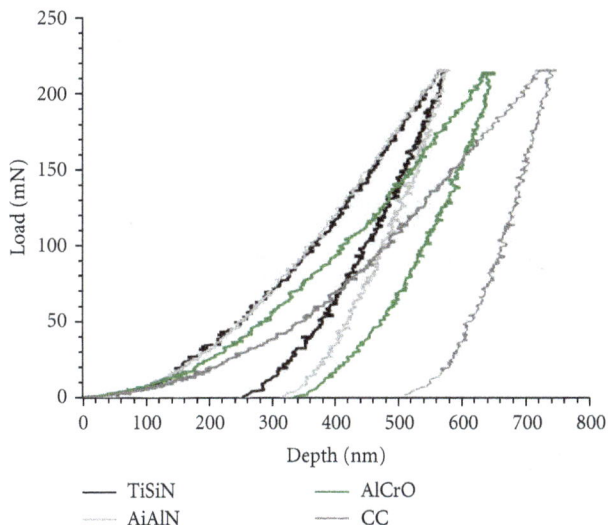

FIGURE 3: Indentation load-displacement curves obtained for the uncoated and the PVD coated cemented carbide substrates investigated.

performed for 30 s, respectively, with a holding period of 15 s at the maximum load. By using the obtained indentation curves (load versus displacement) the hardness and Young's modulus were calculated according to Oliver-Pharr [29].

The Micro Combi Tester was also used in order to evaluate the resistance to scratch induced failure of the coatings. Single pass scratch testing under increasing normal load (0–30 N) with a scratch length of 10 mm and a scratching velocity of 10 mm/min was performed using a 50 μm radius Rockwell C diamond stylus. Besides, conventional single pass scratch testing, using the CSM Instruments Revetest, was performed using a 200 μm radius Rockwell C diamond stylus. A normal load range of 0–100 N, a scratch length of 10 mm and a scratching velocity of 10 mm/min were used in the experiments. During scratch testing the friction force and acoustic emission (A.E.) signals were continuously recorded.

In the scratch tests, two critical normal loads ($F_{N,C1}$ and $F_{N,C2}$, resp.) were used as a measure of the coating cohesive and interfacial adhesive strength and were taken as the normal load corresponding to the first local coating failure ($F_{N,C1}$) and continuous substrate exposure ($F_{N,C2}$) as determined by post-test characterization using SEM.

2.4. Tribological Testing. Circular multi-pass scratch testing was performed in order to evaluate the fatigue and wear behaviour of the coatings under repeated scratching contact conditions. The experiments were performed using a pin-on-disc test equipment (CSM High Temperature Tribometer), see Figure 1, but instead of a pin or a ball commonly used in standard sliding wear tests a Rockwell C diamond stylus used in conventional scratch testing (using a scratch tester) was used. The pin-on-disc test set-up makes it possible to significantly reduce the extensive time associated with multi pass scratch testing for a large number of cycles (>10) using conventional scratch testing equipment.

The circular multi-pass tests were carried out using a 50 μm radius Rockwell C diamond stylus, a speed of 90 mm/min and with a stepwise (loading step 2 N) increasing normal load from 3 N up to the critical normal load, $F_{N,C2}$, as obtained from the conventional single pass scratch tests. The maximum number of cycles was set to 2000. The radius of the circular scratch track was 1.5, 2.0, or 2.5 mm. It should be noted that as long as the diamond stylus used in the experiment is not damaged no influence from a variation in circular scratch track radius on coating failure tendency could be detected within the radius range and test parameters used.

All tests were performed in ambient air (20–22°C, 25-26% RH) and repeated two times. During testing the friction coefficient was continuously recorded and, in combination with post-test characterisation of the wear track using optical microscopy, used in order to determine the number of cycles to coating failure. After testing all wear tracks were carefully characterized by SEM and EDX in order to evaluate the influence of normal load and number of scratching cycles on the prevailing coating failure mechanisms.

3. Results

3.1. Coating Morphology and Microstructure. Figures 2(a) and 2(b) show the surface morphology and microstructure of the as-deposited $Ti_{0.34}Al_{0.66}N$ coating investigated. As can be seen, the as-deposited coating surface shows a significant roughness due to the generation of μm- and sub-μm-sized droplets during the deposition process. Of these, the larger droplets (>3–5 μm in diameter) typically show an interdroplet distance in the range of ~15–25 μm, that is, they constitute a nonnegligible area fraction of the coating. Also, μm and sub-μm-sized craters contribute to the observed roughness. EDX-analysis of the droplets shows that these display a higher metal content, especially in the core, as compared with the surrounding coating matrix. The above observations are common for all coatings investigated. However, when comparing the three coatings, see Table 2, it is observed that the $(Al_{0.7}Cr_{0.3})_2O_3$ coating shows a significantly rougher surface as compared with the $Ti_{0.86}Si_{0.14}N$ and $Ti_{0.34}Al_{0.66}N$ coatings. The reason for this is mainly due to a large number of large, up to 5-6 μm in diameter, droplets as compared to the other two coatings. The surface roughness and defect density are about the same for the $Ti_{0.86}Si_{0.14}N$ and $Ti_{0.34}Al_{0.66}N$ coatings, with slightly higher values for the $Ti_{0.86}Si_{0.14}N$ coating.

3.2. Mechanical Properties

3.2.1. Hardness. Figure 3 shows representative load-displacement curves for the coatings. From these curves, the corresponding Young's modulus, E, and hardness values can be calculated, see Table 2. As can be seen, the $Ti_{0.86}Si_{0.14}N$ coating displays the highest hardness while the $(Al_{0.7}Cr_{0.3})_2O_3$ coating displays the lowest hardness. The $(Al_{0.7}Cr_{0.3})_2O_3$ coating also displays the lowest E-modulus while the $Ti_{0.34}Al_{0.66}N$ coating displays the highest E-modulus.

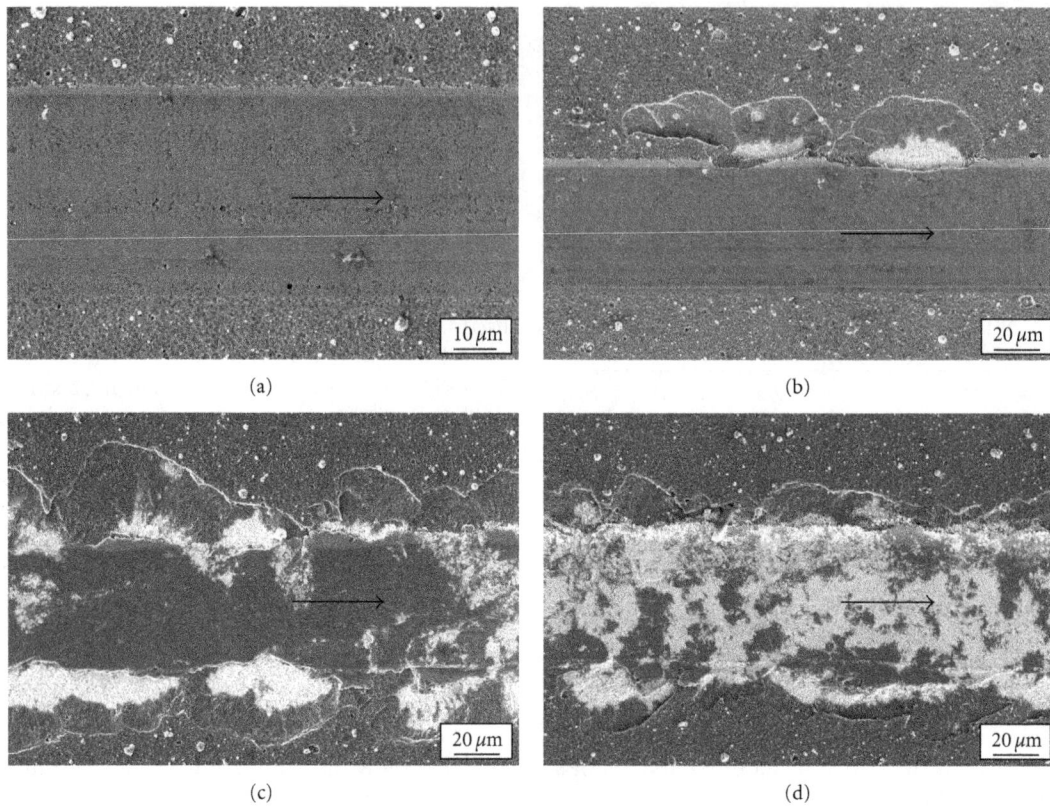

FIGURE 4: SEM micrographs illustrating the major coating failure mechanisms in the single-pass scratch test by using a diamond stylus with radius $R = 50\,\mu m$. (a) Plastic deformation and minor cracking (no coating material detached), (b) chipping (cohesive failure) in combination with local interfacial spalling (adhesive failure), (c) continuous interfacial spalling (adhesive failure), and (d) continuous interfacial spalling in combination with extensive coating fragmentation at normal loads above $F_{N,C2}$ for $Ti_{0.34}Al_{0.66}N$ coating. The arrows indicate the scratching direction.

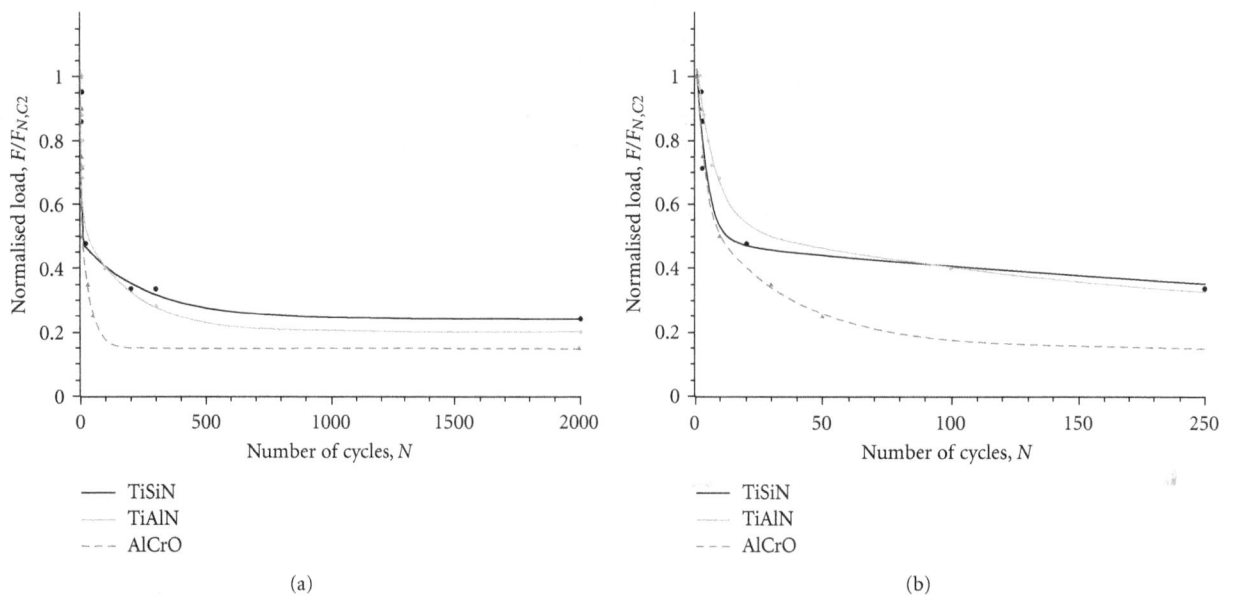

FIGURE 5: Normalised load-number of cycles curves for the coatings investigated showing the normalised normal load for continuous interfacial spalling.

Mechanical and Tribological Properties of PVD-Coated Cemented Carbide as Evaluated by a New Multipass
Scratch-Testing Method

17

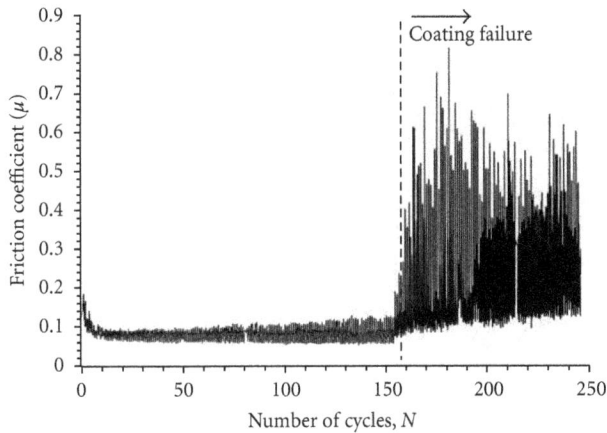

FIGURE 6: Friction coefficient-number of cycles curve for the $Ti_{0.34}$ $Al_{0.66}N$ coating illustrating the increase in friction when coating failure occurs.

3.2.2. Single Pass Scratch Test Behaviour.

Table 3 shows the critical normal loads, $F_{N,C1}$ and $F_{N,C2}$, as obtained in the conventional single pass scratch test of the coatings investigated. As expected, the larger $200\,\mu m$ radius diamond stylus results in significantly higher critical normal loads. As can be seen, the $Ti_{0.34}Al_{0.66}N$ coating displays the highest critical normal loads and the $(Al_{0.7}Cr_{0.3})_2O_3$ coating the lowest critical normal loads. Also, the difference between the two critical normal load values, that is, $F_{N,C1}$ and $F_{N,C2}$, is for all coatings relatively small.

Examination of the scratch tested samples revealed that a number of coating failure and detachment mechanisms could be identified. Minor cracking and chipping of small, sub-μm, coating fragments in connection to droplets and craters were sporadically observed already at very low normal loads ($\approx$1-2 N) dependent on the size and type of defect. However, these failures were all very limited in size and restricted to the local defects. With increasing normal load the coating-substrate composites displayed the following major scratch induced damage mechanisms; plastic deformation and minor cracking (no coating material detached) within the scratch, chipping (cohesive failure) in combination with local interfacial spalling(adhesive failure) outside the scratch, and continuous interfacial spalling(adhesive failure) outside the scratch, see Figure 4. The characteristics of the spalled off areas reveal that the observed adhesive failure is associated with the elastic recovery which occurs upon unloading behind the diamond stylus as it scratches the surface [30]. Consequently, plastic and elastic deformation of the cemented carbide substrate plays an important role for coating damage and as a result the tendency to chipping and spalling will decrease with decreasing normal load, that is, decreasing plastic deformation of the cemented carbide substrate. For normal loads above $F_{N,C2}$, extensive plastic deformation within the scratch will occur resulting in coating fragments being pressed or squeezed into the cemented carbide substrate surface.

3.2.3. Multipass Scratch Test Behaviour.

Figure 5 shows the results from the multi-pass scratch test where the influence of

number of scratching cycles (N) and load (normalised with respect to $F_{N,C2}$ as obtained in the single pass scratch test) on the onset of continuous interfacial spalling along the circular wear track is presented. As can be seen, although minor differences exist, all coatings display a similar appearance with an increasing number of cycles to coating failure with decreasing normal load until a critical load, corresponding to a "fatigue limit", is reached below which no major coating failure such as chipping and spalling is obtained up to 2000 cycles. Further, it should be mentioned that the values of the critical loads are obtained by combining detailed characterisation of the scratch tracks with detection of changes in the acquired friction curves. An increase in friction, see Figure 6, indicates that coating failure occurs.

Examination of the circular wear tracks after different combinations of normal load and number of cycles shows that the major failure mechanisms observed in the conventional single pass scratch tests also are present in the multi-pass scratch tests, see Figures 7 and 8. Further, Figure 8(c) shows a crack obtained in the substrate/coating interface at the rim of the scratch. However, the multi-pass test also result in two new types of coating failure controlling the coating life time at a large number of cycles.

For normal loads below the "fatigue limit" gradual wear of the coating occurs. Two different wear mechanisms can be distinguished both acting on a sub-μm scale and typically after a large number of cycles. First, chipping of fine coating fragments is observed in connection to fatigue cracks obtained after a critical number of cycles, see Figure 9(a). Secondly, delamination of thin plate-like fragments, originating from a thin tribo film formed by compacting and sintering of sub-μm coating fragments at the diamond tip-coating interface, see Figure 9(b), may also take place.

Consequently, for a combination of low loads, not resulting in pronounced deformation of the cemented carbide substrate, and large number of cycles the intrinsic tribological properties of the coating controls the life of the coating. It should be noted that the defects (droplets and craters) present in the coating do not necessarily result in any extensive coating wear at low normal loads. Also, the observed wear rate is relatively low; that is, the maximum reduction in coating thickness after 2000 cycles does not exceed 15% of the coating thickness for the coatings investigated.

For normal loads slightly above the fatigue limit large scale interfacial spalling resulting in complete exposure of the cemented carbide substrate within the scratch is observed after a large number of repeated contacts ($N > 1000$), see Figure 10. Consequently, fatigue controlled crack propagation at the coating-substrate interface has a significant influence on coating failure.

Figure 11 shows a coating failure map where the influence of normalised normal load and number of scratching cycles on the major damage mechanisms for the $Ti_{0.34}Al_{0.66}N$ coating is illustrated. In the map, the transition from a region (grey) corresponding to plastic deformation, minor cracking, and wear to a region corresponding to continuous interfacial spalling (dashed) is represented by a relatively

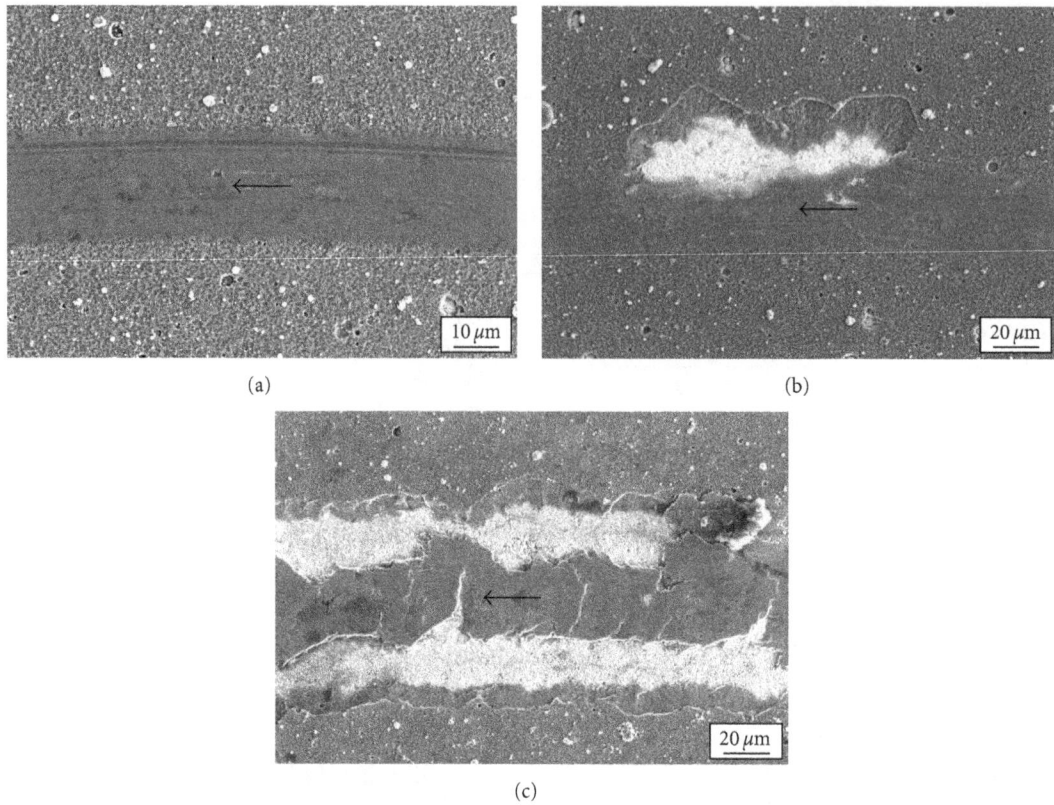

FIGURE 7: SEM micrographs illustrating the major coating failure mechanisms in the multi-pass scratch test. (a) Plastic deformation and minor cracking (no coating material detached), (b) chipping (cohesive failure) in combination with local interfacial spalling (adhesive failure), and (c) continuous interfacial spalling (adhesive failure) for $Ti_{0.34}Al_{0.66}N$ coating. The arrows indicate the scratching direction.

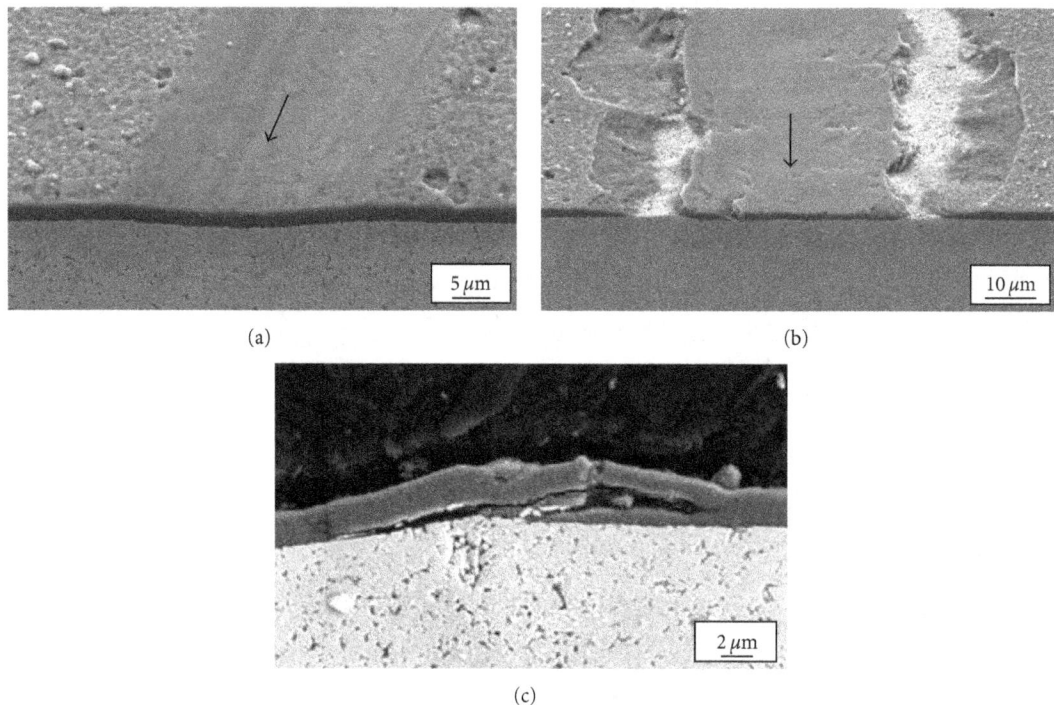

FIGURE 8: Cross-sections of the circular wear tracks illustrating (a) plastic deformation and minor cracking (no coating material detached), (b) continuous interfacial spalling (adhesive failure), and (c) interfacial cracking at the rim of a scratch track for $Ti_{0.34}Al_{0.66}N$ coating. The arrows indicate the scratching direction.

Mechanical and Tribological Properties of PVD-Coated Cemented Carbide as Evaluated by a New Multipass
Scratch-Testing Method

19

FIGURE 9: Wear mechanisms observed in the multi-pass scratch test. (a) Chipping of fine coating fragments in connection to transverse cracking and (b) delamination of thin plate-like fragments (fractured cross section) originating from a thin tribo film for $Ti_{0.34}Al_{0.66}N$ coating. The arrows indicate the scratching direction.

FIGURE 10: Complete exposure of the cemented carbide substrate due to continuous interfacial spalling (adhesive failure) within the circular wear track after a large number of repeated contacts ($N > 1000$) for $Ti_{0.34}Al_{0.66}N$ coating. The arrows indicate the scratching direction.

narrow band corresponding to chipping and local interfacial spalling. The narrow band signifies a quick transition from plastic deformation, minor cracking and wear via local interfacial spalling to continuous interfacial spalling with increasing normal load.

Region A ($F_{norm} \leq 0.15$). For very low normalised normal loads the coating displays just a mild plastic deformation as revealed by the flattening of the as-deposited surface morphology in the scratch. With increasing normal load the plastic deformation extends through the coating into the substrate and with increasing number of cycles the coating starts to display fatigue induced cracking and mild wear caused by chipping or delamination of fine coating fragments.

Region B ($0.15 \leq F_{norm} \leq 0.25$). For normalised normal loads below 0.20 complete interfacial spalling will occur at a high number of cycles ($N > 1000$). For normalised normal loads slightly above 0.20 coating damage by chipping and local interfacial spalling starts to occur at a critical number of cycles ($350 < N < 1000$). Consequently, coating damage is affected by crack propagation within the coating and along the coating-substrate interface and *high cycle fatigue* controls the damage. With increasing normalised normal load, the critical number of cycles to chipping and spalling will rapidly decrease ($dN/dF \approx -1 \cdot 10^3$) until a normalised normal load around 0.25 is reached where a transition to *low cycle fatigue* is obtained.

Region C. ($0.25 \leq F_{norm} \leq 0.60$): For normalised normal loads above 0.30 the critical number of cycles to chipping and spalling is relatively low, typically less than 250–350. In this region the normal load is high enough causing extensive deformation of the cemented carbide substrate and the critical number of cycles to chipping and spalling is to a large extent controlled by the stress state and defect density at the coating-substrate interface promoting crack propagation during repeated cyclic loading.

Region D. Coating damage controlled chipping and interfacial spalling mainly due to elastic recovery caused by pronounced plastic deformation of the cemented carbide substrate during the scratching process. This failure occurs for normalised normal loads $F_{norm} \geq 0.60$ and is restricted to a low number of cycles, typically less than 10. The decrease in number of cycles to local and interfacial spalling with increasing normal load is very low ($dN/dF \approx -1$) in this region.

Further, retesting of the $Ti_{0.34}Al_{0.66}N$ coating after long term testing shows equal results as compared to former test, see Figure 12. The differences are most evident at higher

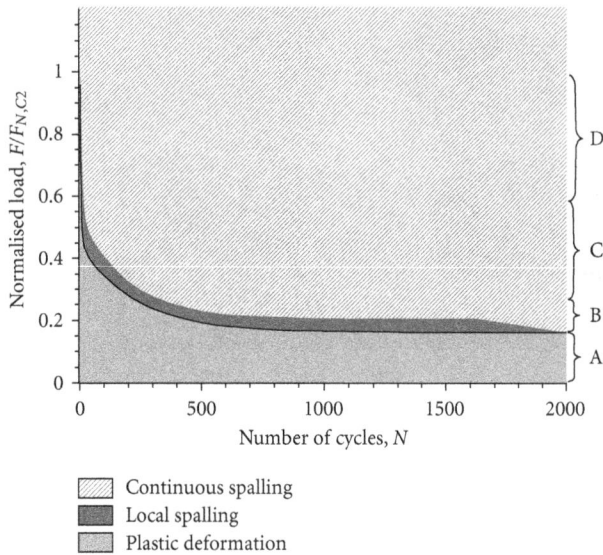

FIGURE 11: Coating failure map of the $Ti_{0.34}Al_{0.66}N$ coating as obtained from the circular multi-pass scratch test. The map illustrates the influence of normalized normal load and number of cycles on dominant failure modes, that is, plastic deformation, cracking and wear, local spalling, and continuous spalling. In the map, four different regions, A, B, C, and D, are indicated.

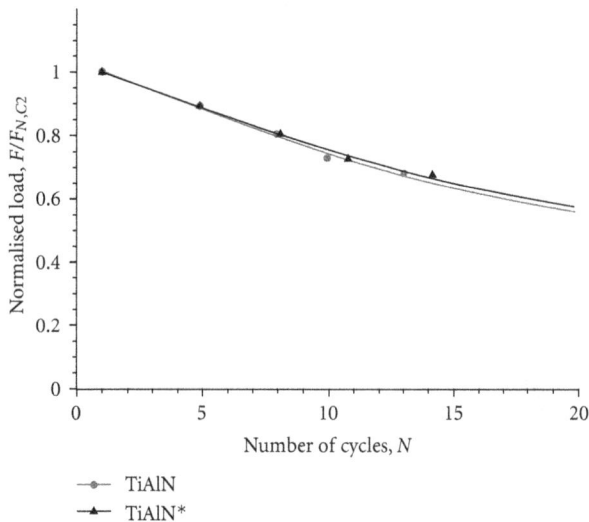

FIGURE 12: Normalised load-number of cycles curves for the $Ti_{0.34}Al_{0.66}N$ coating showing the normalised normal load for continuous interfacial spalling. The * indicates retesting after long term testing of the other coatings investigated.

loads but are subtle which indicates that the technique is reproducible.

4. Discussion

4.1. Proposed Test Method. In the present study, circular multi-pass scratch testing using a pin-on-disc equipment is proposed as a test method to evaluate the fatigue, wear, chipping, and spalling behaviour of thin hard coatings under

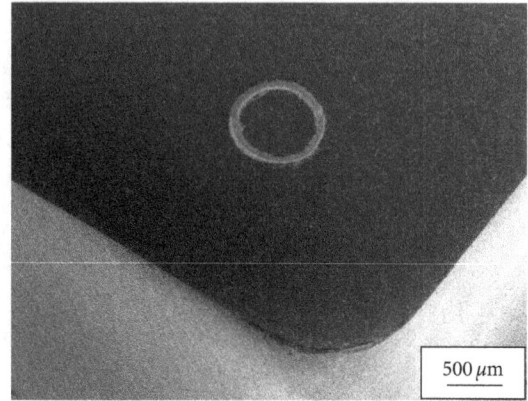

FIGURE 13: Circular wear track near the cutting edge of a coated cemented carbide insert illustrating the possibility to evaluate small test regions using the present test method.

repeated/cyclic sliding contact conditions. The results show that the test method is quick, simple and reproducible and can preferably be used to obtain relevant data concerning the mechanical and tribological properties of different coating-substrate composites. Other advantages of the proposed test method are.

(i) The test method utilizes a commercially available tribometer, an equipment widely used for tribological testing and present at many material laboratories.

(ii) The test method uses small, simple flat samples making it possible to easily evaluate for example, the effect of different types of pre- and post-coating deposition surface treatment processes, and so forth. Also, the small sample size facilitates post-test microscopy and surface analysis.

Besides, the use of a small diamond stylus as in the present study offers the possibility to characterise the coating at a very small, close to single asperity, scale. Also, the small size of the test region (the radius of the circular scratch track can be as small as 0.5 mm) makes it possible to perform tests within small restricted areas. Consequently, the test method can be used as a virtually non-destructive test if the active region of the investigated sample is avoided. For example, it can be used to evaluate the cohesive and adhesive interfacial strength, fatigue, and wear resistance of coated cemented carbide inserts prior to cutting tests in order to correlate the obtained data to the tool life, see Figure 13.

Consequently, circular multi-pass scratch testing is a fast and powerful complement to conventional single pass scratch testing being one of the most common tests for the mechanical characterization of thin hard coatings. In combination with post-test scanning electron microscopy the proposed test method will give information about not only the critical normal loads and coating damage mechanisms at single pass scratch testing and a low number of loading cycles but also the fatigue and wear properties at a high number of loading cycles, $N > 1000$. Compared with the common used impact tests used for evaluating the

Mechanical and Tribological Properties of PVD-Coated Cemented Carbide as Evaluated by a New Multipass
Scratch-Testing Method

21

fatigue properties of thin hard coatings the proposed test is believed to more closely simulate the contact conditions and prevailing stress states in sliding contact tribosystems although the number of cycles are significantly lower than in the impact tests [5, 19–21].

It may be questioned whether the test approach, which comprises a large number of loading cycles, will result in extensive wear of the diamond stylus. Steinmann et al. found that unless the diamond tip shows extensive chipping due to "half-crown" shaped cracking the influence of diamond tip wear on the critical normal load is not significant [31]. However, as stated by the authors regular microscopic observations of the diamond tips are of utmost importance. Also, the use of high quality natural single-crystal diamond tips with the [100] direction oriented along the z-axis can substantially increase their life time. Furthermore, in order to optimize the life time the diamond stylus should be rotated 180° before catastrophic damage is reached. Besides the above mechanical wear, the diamond tip may, depending on the contact conditions and the surrounding environment, suffer from tribo-oxidation and graphitization [32], which will influence on the wear resistance as well as the friction at the scratching (sliding) interface. Consequently, high scratching (sliding) speeds should preferably be avoided. In the present study, two high quality diamond styli with guaranteed spherical geometry were used and the wear/damage of these was restricted to mild polishing and initial "half-crown" shaped cracking after completing the test series.

4.2. Coating Damage Mechanisms. When comparing the results from the conventional single pass scratch test with the circular multi-pass scratch test it can be concluded that both tests display similar major coating failure mechanisms as long as the normal loads in the multi-pass test is relatively high, typically $F_{norm} \geq 0.60$, and is restricted to a low number of cycles, typically less than 10. Within this region coating failure is due to: plastic deformation and minor cracking (no coating material detached), chipping (cohesive failure) in combination with local interfacial spalling(adhesive failure), and continuous interfacial spalling(adhesive failure) outside the scratch. Also, the characteristics of the spalled off areas reveals that the adhesive failures in both tests are associated with the elastic recovery which occurs behind the diamond stylus as it scratches the surface [30]. However, while the conventional single pass scratch test mainly gives information of the mechanical response of the coating-substrate composite, the circular multi-pass scratch test also gives information of the tendency to surface fatigue and wear when exposed to a large number of contact cycles. Besides, it also makes it possible to determine a "fatigue limit" for the coating-substrate composite. In Table 4 and Figure 14 the observed mechanisms are illustrated by simplified schematic drawings and discussed in some detail.

The ranking of the coatings investigated is, keeping in mind that all coatings were deposited with identical interfacial properties, believed to be controlled by the differences in mechanical properties and defect density. Thus, the ranking with respect to spalling resistance obtained at low normal loads, that is, $Ti_{0.86}Si_{0.14}N > Ti_{0.34}Al_{0.66}N > (Al_{0.7}Cr_{0.3})_2O_3$, can be explained by the fact that the $Ti_{0.86}Si_{0.14}N$ coating displays the highest hardness and $(Al_{0.7}Cr_{0.3})_2O_3$ the lowest hardness, a high hardness reducing the tendency to plastic deformation of the substrate material and thus the stresses at the coating-substrate interface. Increasing normal load will increase the stresses at the interface and consequently the strength of the interface region will have an increased impact on the tendency to spalling. Thus the ranking displayed at normalized normal load above 0.5, that is, $Ti_{0.34}Al_{0.66}N > Ti_{0.86}Si_{0.14}N > (Al_{0.7}Cr_{0.3})_2O_3$, can be explained by the fact that the $Ti_{0.34}Al_{0.66}N$ displays the lowest defect density and $(Al_{0.7}Cr_{0.3})_2O_3$ displays the largest defect density of the coatings investigated, see Figure 5 and Table 2. At even higher normalized normal loads the tendency to plastic deformation of the cemented carbide substrate will increase and consequently have a strong impact on the tendency to spalling. As a result the different coatings will display similar spalling characteristics, see Figure 5.

5. Conclusions

In the present study a new test method based on multi-pass scratch testing is presented and used in order to evaluate the mechanical and tribological properties of arc evaporated coatings of $Ti_{0.86}Si_{0.14}N$, $Ti_{0.34}Al_{0.66}N$ and $(Al_{0.7}Cr_{0.3})_2O_3$ deposited on cemented carbide inserts. Based on the obtained results the following conclusions can be drawn.

(i) The multi-pass test method utilizes a commercially available pin-on-disc tribometer, an instrument widely used for evaluating the friction and sliding wear characteristics of materials, present in many tribology laboratories making the test easily accessible.

(ii) The multi-pass scratch test method is quick, simple, and reproducible and can preferably be used to obtain relevant information concerning the fatigue, wear, chipping, and spalling characteristics of different coating-substrate composites.

(iii) The simple and small sample size makes the test cheap and easy to combine with post-test microscopy and surface analysis. Also, the small size of the test region (the radius of the circular scratch track can be as small as 0.5 mm) makes it possible to perform tests within small restricted areas; that is, the test can be regarded as a virtually non-destructive test.

(iv) As long as the wear/damage of the diamond tip is restricted to mild polishing and initial "half-crown" shaped cracking the effect on the scratching process and the resulting critical normal loads is negligible.

(v) Of the coatings investigated, the $Ti_{0.34}Al_{0.66}N$ coating displays the best performance under low cycle fatigue conditions while the harder $Ti_{0.86}Si_{0.14}N$ shows the best performance under high cycle fatigue conditions.

Contact condition in Region A.

i) Chipping of fine coating fragments due to surface fatigue induced cracking.

ii) Delamination of tribo film of compacted and sintered fine coating fragments.

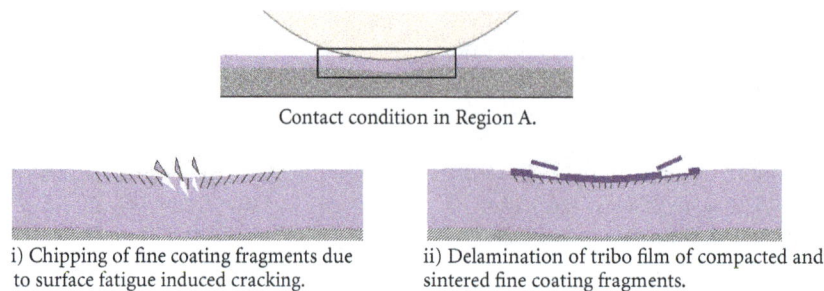

FIGURE 14: Wear mechanisms observed in Region A corresponding to normalised normal loads < 0.2.

(vi) The intrinsic coating properties, including the defect density, have a significant impact on the fatigue properties of the coatings. Consequently, the presence of defects in the coating and the coating-substrate interface decreases the low cycle fatigue properties while the combination of coating defects and a low coating hardness decreases the high cycle fatigue.

Acknowledgment

The financial support for this work from the National Graduate School in Materials Science is gratefully acknowledged.

References

[1] P. A. Dearnley and E. M. Trent, "Wear mechanisms of coated carbide tools," *Metals Technology*, vol. 9, no. 2, pp. 60–75, 1982.

[2] P. A. Dearnley, "Rake and flank wear mechanisms of coated and uncoated cemented carbides," *Journal of Engineering Materials and Technology*, vol. 107, no. 1, pp. 68–82, 1985.

[3] Europeen Standard, Advanced technical ceramics—methods of test for ceramic coatings—Part: Determination of adhesion and other mechanical failure modes by scratch test, prEN1071-3:2000:E, CEN Management Centre, Brussels, Belgium, 42, 2000.

[4] J. L. Vossen, "Measurements of film-substrate bond strengths by laser spallation," in *Adhesion Measurements of Thin Films, Thick Films and Bulk Coatings*, K. L. Mittal, Ed., vol. 640 of *ASTM Special Technical Publication*, pp. 122–133, American Society for Testing Materials, Philadelphia, Pa, USA, 1978.

[5] O. Knotek, B. Bosserhoff, A. Schrey, T. Leyendecker, O. Lemmer, and S. Esser, "A new technique for testing the impact load of thin films: the coating impact test," *Surface and Coatings Technology*, vol. 54-55, pp. 102–107, 1992.

[6] M. Fallqvist, M. Olsson, and S. Ruppi, "Abrasive wear of multilayer κ-Al₂O₃-Ti(C,N) CVD coatings on cemented carbide," *Wear*, vol. 263, no. 1–6, pp. 74–80, 2007.

[7] H. Czichos, S. Becker, and J. Lexow, "Multilaboratory tribotesting: results from the Versailles advanced materials and standards programme on wear test methods," *Wear*, vol. 114, no. 1, pp. 109–130, 1987.

[8] S. C. Lim and M. F. Ashby, "Wear-Mechanism maps," *Acta Metallurgica*, vol. 35, no. 1, pp. 1–24, 1987.

[9] D. Klafke, "Towards a tribological reference test—fretting test," in *Proceedings of the 14th International Colloquium Tribology*, pp. 1839–1846, Esslingen, Germany, January 2004.

[10] S. Baragetti, G. M. la Vecchia, and A. Terranova, "Variables affecting the fatigue resistance of PVD-coated components," *International Journal of Fatigue*, vol. 27, no. 10–12, pp. 1541–1550, 2005.

[11] F. Ledrappier, Y. Gachon, C. Langlade, and A. B. Vannes, "Surface fatigue behaviour mapping of PVD coatings for mechanical purposes," *Tribotest*, vol. 11, no. 4, pp. 333–343, 2005.

[12] K. Holmberg and A. Matthews, *Coating Tribology: Properties, Mechanisms, Techniques and Applications in Surface Engineering*, vol. 56 of *Tribology and Interface Engineering*, 2009.

[13] Å. Kassman, S. Jacobson, L. Erickson, P. Hedenqvist, and M. Olsson, "A new test method for the intrinsic abrasion resistance of thin coatings," *Surface and Coatings Technology*, vol. 50, no. 1, pp. 75–84, 1991.

[14] A. Matthews, "Methods for assessing coating adhesion," *Le Vide, Les Couches Minces*, pp. 7–15, October 1988.

[15] S. J. Bull, D. S. Rickerby, A. Matthews, A. Leyland, A. R. Pace, and J. Valli, "The use of scratch adhesion testing for the determination of interfacial adhesion: the importance of frictional drag," *Surface and Coatings Technology*, vol. 36, no. 1-2, pp. 503–517, 1988.

[16] P. J. Burnett and D. S. Rickerby, "The relationship between hardness and scratch adhesion," *Thin Solid Films*, vol. 154, no. 1-2, pp. 403–416, 1987.

[17] M. T. Laugier, "An energy approach to the adhesion of coatings using the scratch test," *Thin Solid Films*, vol. 117, no. 4, pp. 243–249, 1984.

[18] S. J. Bull, "Failure modes in scratch adhesion testing," *Surface and Coatings Technology*, vol. 50, no. 1, pp. 25–32, 1991.

[19] K. D. Bouzakis and A. Siganos, "Fracture initiation mechanisms of thin hard coatings during the impact test," *Surface and Coatings Technology*, vol. 185, no. 2-3, pp. 150–159, 2004.

[20] K. D. Bouzakis, A. Asimakopoulos, N. Michailidis et al., "The inclined impact test, an efficient method to characterize coatings' cohesion and adhesion properties," *Thin Solid Films*, vol. 469-470, pp. 254–262, 2004.

[21] M. Stoiber, M. Panzenböck, C. Mitterer, and C. Lugmair, "Fatigue properties of Ti-based hard coatings deposited onto tool steels," *Surface and Coatings Technology*, vol. 142–144, pp. 117–124, 2001.

[22] S. Bennet, A. Matthews, J. Valli, A. J. Perry, S. J. Bull, and W. D. Sproul, "Multi-pass scratch testing at sub-critical loads," *Tribologia*, vol. 13, pp. 16–24, 1994.

[23] C. Liu, Q. Bi, and A. Matthews, "Tribological and electrochemical performance of PVD TiN coatings on the femoral head of Ti–6Al–4V artificial hip joints," *Surface and Coatings Technology*, vol. 163-164, pp. 597–604, 2003.

[24] J. Stallard, S. Poulat, and D. G. Teer, "The study of the adhesion of a TiN coating on steel and titanium alloy substrates using a multi-mode scratch tester," *Tribology International*, vol. 39, no. 2, pp. 159–166, 2006.

Mechanical and Tribological Properties of PVD-Coated Cemented Carbide as Evaluated by a New Multipass
Scratch-Testing Method

23

[25] S. J. Bull and D. S. Rickerby, "Multi-pass scratch testing as a model for abrasive wear," *Thin Solid Films*, vol. 181, no. 1-2, pp. 545–553, 1989.

[26] M. G. Gee, "Low load multiple scratch tests of ceramics and hard metals," *Wear*, vol. 250, no. 1–12, pp. 264–281, 2001.

[27] M. Fallqvist and M. Olsson, "The influence of surface defects on the mechanical and tribological properties of VN-based arc-evaporatedcoatings," submitted to. *Wear*.

[28] L. M. Karlsson and L. M. Cruz-Orive, "The new stereological tools in metallography: estimation of pore size and number in aluminium," *Journal of Microscopy*, vol. 165, no. 3, pp. 391–415, 1992.

[29] W. C. Oliver and G. M. Pharr, "An improved technique for determining hardness and elastic modulus using load and displacement sensing indentation experiments," *Journal of Materials Research*, vol. 7, no. 6, pp. 1564–1583, 1992.

[30] S. J. Bull, "Failure mode maps in the thin film scratch adhesion test," *Tribology International*, vol. 30, no. 7, pp. 491–498, 1997.

[31] P. A. Steinmann, Y. Tardy, and H. E. Hintermann, "Adhesion testing by the scratch test method: the influence of intrinsic and extrinsic parameters on the critical load," *Thin Solid Films*, vol. 154, no. 1-2, pp. 333–349, 1987.

[32] P. John, N. Polwart, C. E. Troupe, and J. I. B. Wilson, "The oxidation of (100) textured diamond," *Diamond and Related Materials*, vol. 11, no. 3–6, pp. 861–866, 2002.

3

A Clue to Understand Environmental Influence on Friction and Wear of Diamond-Like Nanocomposite Thin Film

Sukhendu Jana,[1] **Sayan Das,**[1] **Utpal Gangopadhyay,**[1] **Anup Mondal,**[2] **and Prajit Ghosh**[1]

[1] *Meghnad Saha Institute of Technology, Techno India Group, Kolkata 700150, India*
[2] *Department of Chemistry, Bengal Engineering and Science University, Howrah 711103, India*

Correspondence should be addressed to Prajit Ghosh; dr_p_ghosh@rediffmail.com

Academic Editor: Qian Wang

The wear and friction of diamond-like nanocomposite (DLN) film have been investigated in air with different relative humidity (RH), under deionized (DI) water and saline solution. The structure of the film has been characterized by Fourier transform infrared (FTIR), Raman spectroscopy, and scanning electron microscope (SEM). The result shows two interpenetrating network structure: a–C:H and a–Si:O, and they are interpenetrated by Si–C bonding. The tribological performance has been measured using ball-on-disc tribometer with tungsten carbide ball as counterbody at 10 N normal load. Results show that with increasing relative humidity (RH) from 35% to 80%, the coefficient of friction (COF) increases gradually from 0.005 to 0.074, whereas with increasing RH the wear factor decreases from 9.8×10^{-8} mm^3/Nm and attains a minimum value of 2.7×10^{-8} mm^3/Nm at 50% RH. With further increase of RH the wear factor increases again. Moreover, in DI water and especially in saline solution, both the COF and wear factor have been found to be significantly low. A clue has been interpreted to understand environmental dependency, considering the effect of surface dangling bonds, charge transfer, and chemical interactions.

1. Introduction

Diamond-like nanocomposite (DLN) film comprises of two amorphous interpenetrating network structures: one is "diamond-like" (a-C:H) network, and the other is "glass-like" (a-Si:O) network [1–4]. The presence of a-Si:O network as a reinforcement matrix distinguishes the DLN film from conventional diamond-like carbon (DLC) film [5, 6]. The material possesses a number of unique bulk and surface properties like hardness with flexibility [7], thermal stability [7, 8], corrosion and wear resistance [8, 9], biocompatibility [10, 11], and so forth. The low residual stress and good adherence to any type of substrates make DLN film a potent material for a variety of tribological applications.

It is believed that the tribological properties of the film are not only inherent properties of the film. They also strongly depend on the surrounding environment and counterbody. Many researchers have reported the dependency of environment on tribological behavior of DLC film [12, 13], though limited reports have been published particularly for DLN film. Neerinck et al. have reported that the COF of the DLN

film against steel ball at 50% RH varied from 0.04 to 0.08 and remains less than 0.1 even at 90% RH. They have also observed that film wear factor was extremely low under water [14]. Scharf et al. have also reported that COF increased from 0.02 to 0.2 when the RH changed from 1% to 50% [15]. However, continuous variation of RH and presence of ionic solution effect on tribological properties have not been yet reported.

In the present study, the wear factor and COF of DLN film on glass substrate have been studied using standard ball-on-disc tribometer with a tungsten carbide (WC) ball as counterface material under different ambients like air with different RH, under DI water, and saline solution. A clue for environmental dependency on tribological properties of DLN film has been interpreted, considering the films surface dangling bond passivation with environmental species and hence interaction with counterbody.

2. Experimental Details

2.1. Synthesis of DLN Film. The plasma-assisted chemical vapour deposition (PACVD) system exploited in the present

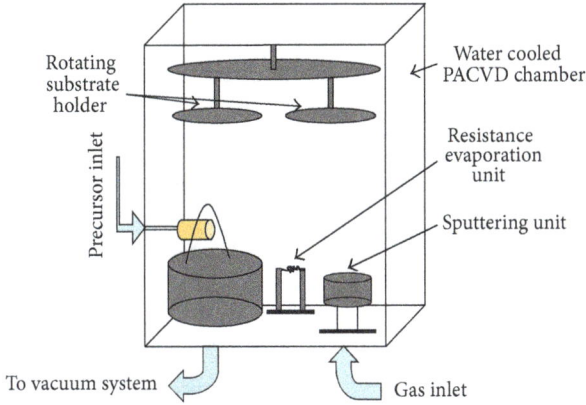

FIGURE 1: Schematic diagram of PACVD system for DLN thin film deposition.

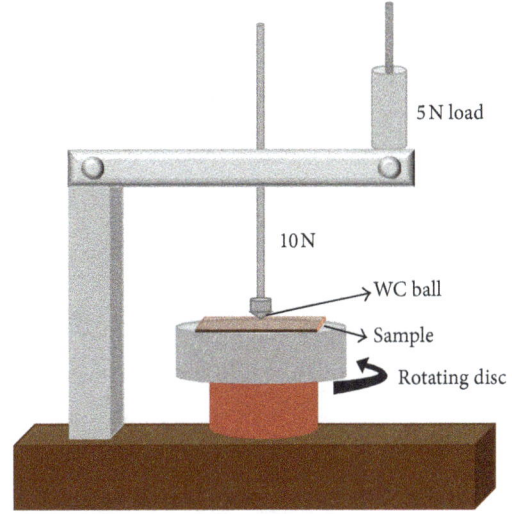

FIGURE 2: Schematic diagram of ball-on-disc tribometer with 10 N normal load and tungsten carbide ball.

study has been illustrated schematically in Figure 1. The substrates were cleaned by acetone and the alcohol with ultrasonic vibration, followed by drying in a nitrogen jet. Cleaned substrates were loaded into the PACVD chamber followed by evacuation up to a pressure of about 10^{-5} mbar. Thereafter, argon gas was introduced into the chamber via a mass flow controller until the chamber pressure reached approximately 10^{-4} mbar. The samples were further cleaned *in situ* by argon plasma for 5 minutes prior to DLN film deposition. During deposition, the filament current was maintained at about 100 A, with a voltage of about 10 V. The thermoionic electrons emitted from the filament were drawn towards the ground at zero potential by applying a voltage of about −142 V to the floating filament with respect to ground. The precursor flow was adjusted via a needle valve and by using gravity control. The precursor ejection head was adjusted below the filament, such that on evaporation the precursor molecules come in the path of thermionic electrons. The precursor vaporized due to low pressure and high temperature near the filament in electron atmosphere which enhanced ionization of vapor molecules by collision. The RF substrate bias power supply was concurrently switched on. The precursor ions formed the stable plasma (the plasma current could be adjusted from 0.2 A to 20 A), and the ions accelerated towards the substrates due to the negative DC substrate bias induced by the RF power. A crossed magnetic field was applied using an electromagnet to induce spiraling motion, in order to increase the path length of ions. For the present study, DLN film was deposited on glass substrates in a specially designed PACVD system, using a liquid precursor (2,3,4-triphenyl nonamethyl pentasiloxane) containing C, H, Si, and O as constituents. A typical growth condition has been given in Table 1. During deposition, sputtering and resistance evaporate units were shut down. The film was deposited on glass for 1 hr.

2.2. Tribological Measurement. The wear and friction characterization was performed using ball-on-disc tribometer (Figure 2). The samples were fixed on a metal disc which rotated with 200 rpm. A WC ball of 4 mm diameter with 10 N

normal load touched the film surface. The distance between WC ball and center of the disc was 4 mm. So the diameter of the wear track was 8 mm, and the linear speed of the ball over the film surface was 0.08 m/s. At the initial stages of wearing process, the COF was usually high but after a few revolutions, the COF decreased and reached a low and stable value. Again it increased abruptly at the instant; the film was completely worn off, resulting from the contact of the WC ball tip with the substrate surface. The wear test was performed at room temperature. Each wear test was carried out until the WC ball reached the base substrate, and each observation was measured three times and confirmed.

The wear factor is defined as the amount of volume (mm^3) wear out per unit normal force (N), per unit sliding distance (m). In ideal case, the cross section of the wear track is a circular curvature due to WC ball as counterbody as has been shown in Figure 3. But in our experiment, the film thickness was very much smaller than the wear track width (Figure 4), and the wear test was performed until completely worn out of the film. So a rectangular cross section of the wear track has been assumed. The total surface area of the wear track is $\pi(r_1^2 - r_2^2)$ where r_1 and r_2 are the external and internal track radius, respectively.

The total worn-out volume is $\pi(r_1^2 - r_2^2)t$ where t is the thickness of the film. The coating wear factor k is expressed as

$$k = \frac{\pi\left(r_1^2 - r_2^2\right)t}{Fs}. \tag{1}$$

Here, F is the normal force (10 N), and s is the total sliding distance over DLN film. The internal and external wear track radius was measured by a high resolution optical microscope (Figure 4). A high temperature thermal adhesive tape was attached before the film deposition on glass substrate to measure the film thickness. The thickness of the film was estimated using a stylus surface profilometer. The resolution of the instrument was 1 nm, and the film thickness is 1093 nm.

TABLE 1: Deposition condition for DLN film.

Bias voltage	Plasma current	Precursor flow rate	Ar flow rate	Magnetic field	RPM of substrate	Working pressure
550 V	0.8 A	0.756 mL/10 mins	50 mL/min	120 Gauss	3	$7.5E-5$ mbar

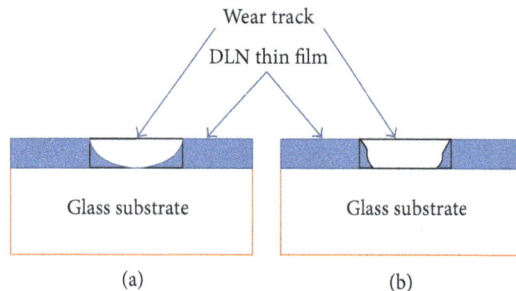

FIGURE 3: Schematic diagram of cross-sectional wear track of DLN film for (a) ideal case and (b) assumed.

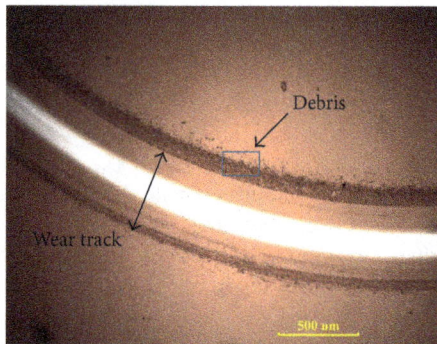

FIGURE 4: Optical microscope image of the wear track of the 1.09 μm thick DLN film at 35% RH ambient shows the track width 483 μm.

The total number of revolutions divided by the measured thickness of the film is the wear rate (cycles/micron) of film.

3. Results and Discussion

3.1. Film Structure Analysis. In order to study the local bonding of C, H, Si, and O sites of the DLN film, the FTIR spectrum of a representative film was recorded from 400 to 4000 cm^{-1} range with 4 cm^{-1} resolution and 100 scan. The corresponding FTIR trace has been shown in Figure 5. The spectra shows a strong Si–O absorption peak which had appeared around the wave number of 1100 cm^{-1} [12]. In the range of 682–885 cm^{-1}, broad absorption band is found which corresponds to Si–C, Si–CH$_3$ fragments [16, 17]. The mode at 760 cm^{-1} is due to Si–CH$_3$ rocking/wagging or Si–C stretching [16, 17]. The absorption band at ~857 cm^{-1} is attributed to (SiH$_2$)$_n$ bending [12, 13].

An absorption band has appeared in the range of 2000–2300 cm^{-1} which corresponds to Si–H and Si–H$_2$ stretching. The C=C stretching vibration has shown its presence with an absorption band from 1372 to 1970 cm^{-1}, but only a broad spectrum from 1370 to 1650 cm^{-1} is clearly seen in the IR

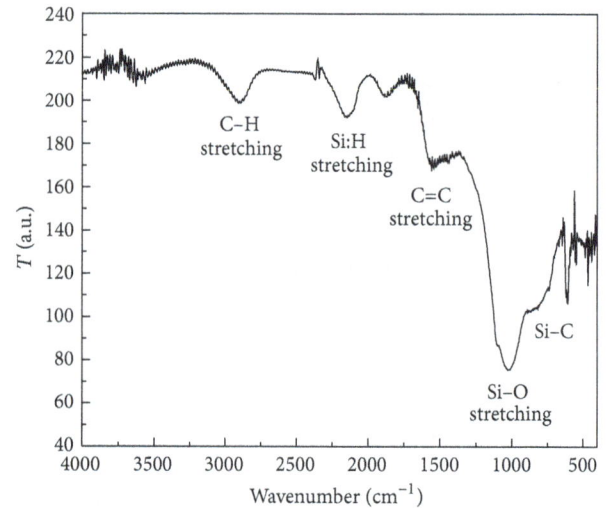

FIGURE 5: FTIR spectrum of the film on glass substrate shows typical diamond-like nanocomposite structure.

spectra, and some C=C bands might have been covered up by the moisture noise, which occurs at around 1885 cm^{-1}. A broad spectrum due to C–H stretching has occurred at 2750–3100 cm^{-1}. [16, 17]. The FTIR trace of the film shows two networks: one is diamond-Like C:H network and another is Si:O network, and they have been interpenetrated with the Si–C bonding depicting the typical nature of DLN film.

In order to understand the structure of DLN film the Raman spectrum was investigated in the wavenumber ranging from 400 to 2500 cm^{-1}. Raman spectrum analysis of DLN film wasdeconvoluted into two Gaussian peaks: the G peak and D peak by curve fitting (Figure 6). The Raman spectrum corresponding to single crystal diamond has a sharp peak located at 1332 cm^{-1} [18], while that for single crystal graphite has a sharp peak at 1580 cm^{-1} [18, 19]. The diamond peak has associated with the fourfold sp^3 hybridized O$_h$ symmetry, while the graphite peak has associated with the threefold sp^2 coordination [18–21]. The G peak, which may occur from C=C sp^2 stretching vibration of olefinic or conjugated carbon chains, is attributed to the relative motion of sp^2 hybridized carb,on and the down shift of the G peak is related to bond angle disorder [22–24]. The D peak is attributed to the disordered breathing motion of sixfold aromatic rings [18, 22] Thus, the peak positions in the Raman spectrum of DLN film and the intensity ratio of D/G peak are the most important parameters to understand the bulk properties of the DLN film for tribological purpose. There are two bands around 1355 cm^{-1} (D peak) and 1524 cm^{-1} (G peak). The position of D peak at 1355 cm^{-1} and down shift of the G line indicate the increase in the number/size of sp^2 C=C bonds with bond angle disorder in the DLN film deposited on glass

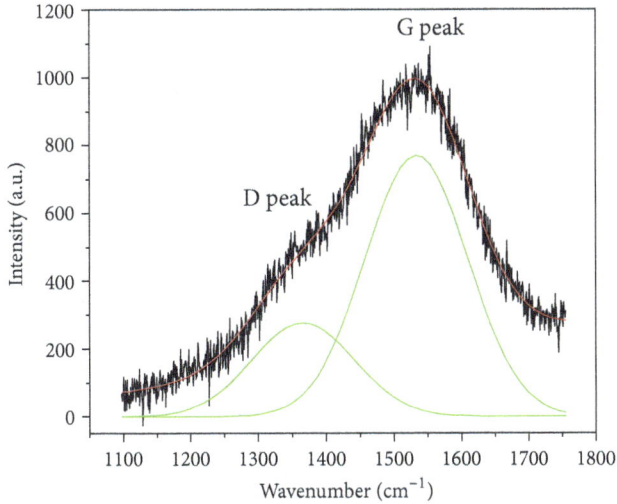

FIGURE 6: Raman spectrum of DLN film on glass substrate deconvoluted at two Gaussian peaks: D peak and G peak.

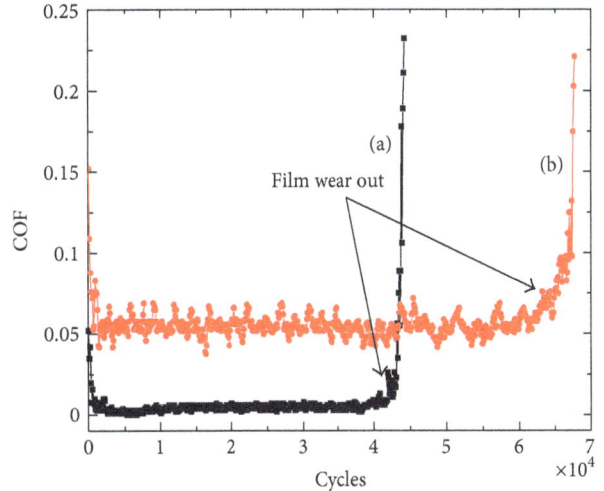

FIGURE 8: Measurement of the coefficient of friction, in a ball-on-disk test with 10 N load, of DLN film in air with (a) 35% RH and (b) 60% RH.

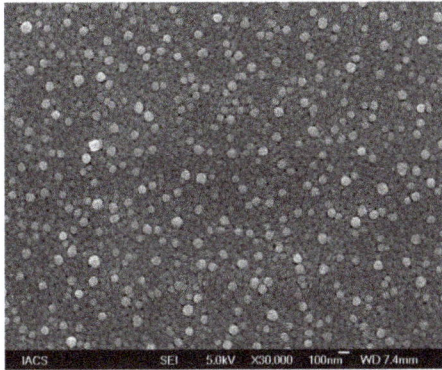

FIGURE 7: FE-SEM image of DLN film surface on glass substrate shows granular like matter distributed randomly.

substrate. Again the intensity ratio I_D/I_G varies inversely with the graphite cluster size [22, 23]. In this respect the ratio, I_D/I_G, may be indirectly related to sp^3/sp^2 intensity, and it is observed to be 0.35 from Gaussian fitting after base line correction.

Figure 7 shows the FE-SEM image of DLN film deposited on glass substrate. Before capturing the image, a very thin layer of gold film was deposited on film surface to nullify the charging effect. The film comprised of amorphous granular like matter, and they were distributed randomly. The average size of these grains was around 75 nm.

3.2. Wear and Friction Characterization. The COF and wear factor of DLN film deposited on glass substrate in different ambient condition like air environment for different relative humidity starting from 35% RH to 80% RH under water and saline solution were measured. The COF for a particular wear test was not fixed at a particular value. It was varying within a certain range. The average COF values were obtained by averaging the friction coefficients measured in tribometer during the wearing process of the film. Figure 8 shows the

measurement of COF of DLN film against WC ball in presence of air with 35% RH and 60% RH. The results indicate that, initially the COF was high, after few cycles it reached a stable value, and finally it increased abruptly after the completely wearing out of the film. The DLN film after deposition exhibited surface undulation in the nanoscale (Figure 7). When the TC ball in the tribometer started to slide over the surface of DLN film, the top of the granular structure was gradually flattened due to wearing effect, and the track surface became smoother resulting in the fall of COF from the initial value. The obvious reason was that the crests come in close proximity to the sliding ball surface. Consequently the dangling bonds present on DLN surface and on the crests came in closest proximity of the ball surface. The ball being metallic, electron transfer occurred, and relatively strong interaction and adhesion set in between the DLN film and the ball surface. This strong interaction helped to wear down the crests during sliding of ball.

When the surface became flatter, the sliding ball surface came in close proximity to much larger number of dangling bonds than only on the crest to show the following effects. As the number of interacting dangling bonds increased, the top layer of DLN surface stuck to the ball resulting in high wear factor of DLN film. Furthermore the peeled off fragments came on the wear track to induce abrasive wear which in turn enhanced the wearing of the film [15, 25].

The data from Figure 9 suggests that although the COF increased gradually with increase in relative humidity (RH), the value of COF was very low (<0.08) even in high humid condition (RH 80%). Another interesting observation was that the COF under DI water was less than that in high humid air (RH 80%) and was further reduced to 0.021 in saline solution which was close to that in 50% RH condition. The data collected from the wearing process have been listed in Table 2, and the corresponding results on the wear factor for different ambient conditions have been shown in Figure 10.

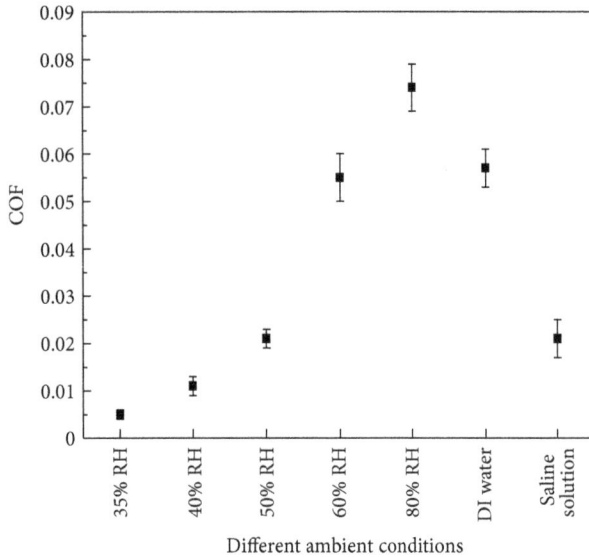

FIGURE 9: COF of DLN film on glass substrate in different ambient that is, 35% RH air; 40% RH air; 50% RH air; 60% RH air; 80% RH air; DI Water; 0.9% NaCl aqueous soln.

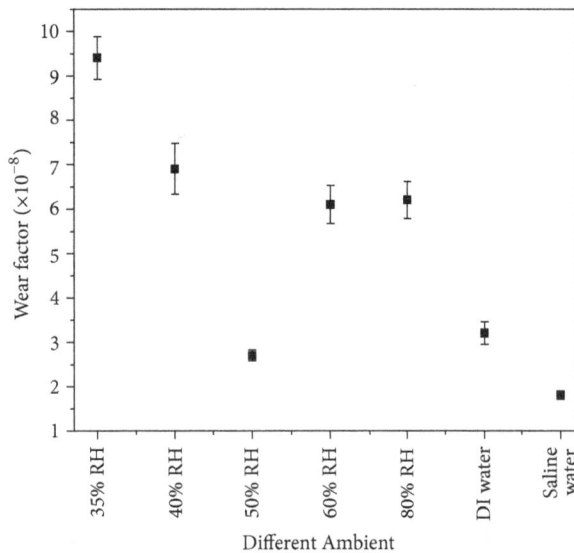

FIGURE 10: Wear factor of DLN film on glass substrate in different ambient that is, 35% RH air; 40% RH air; 50% RH air; 60% RH air; 80% RH air; DI Water; 0.9% NaCl aqueous soln.

In this context it might be note that the wear factor of DLN film decreased with increase in humidity of air from RH 35% to RH 40% and attained a sharply low value at RH 50%. Further observation revealed that the wear factor rose again at humidity higher than RH 50%, attained relatively higher values again at RH 60%, and remained more or less unaffected by humidity till RH 80%. But under DI water and under 0.9% NaCl solution the wear factor maintained an extremely low value (Table 2 and Figure 10). As discussed previously, our observations can be summarized from Figures 9 and 10 and Table 2 as

(a) both the COF and wear factor were low under 0.9% NaCl solution;

(b) the COF was high but wear factor was relatively low under DI water;

(c) the COF was very low and wear factor rather high in low humidity air with RH < 40%;

(d) the COF and wear factor both were high at high humid condition at RH 80%.

Above-mentioned frictional and wear behavior could be explained by the consideration of electrical and chemical effects during the process of measurements.

Further Figure 10 shows that with increasing RH from 35% to 80% the wear factor has first decreased, touched minima, and then increased. The effect of passivation of surface dangling bond by water molecules in ambient conditions is the primary reason behind this phenomenon. The water vapor in the air passivated [26] the dangling bonds on the contact surface of the DLN film through adsorption and/or dissociation mechanisms [26]. This passivation reduced the adhesion interaction with the DLN film and ball surface. This mechanism caused by the presence of water vapor was applicable not only for the DLN coating surface but also for the transferred layer on the contact surface of the ball. When the percentage of water vapor was low (in 35% RH) in the test chamber, a strong adhesion interaction grew between counter faces, since the passivation of dangling bonds on DLN surface by water vapour was less. This caused higher value of wear factor at less humid air. The COF remained low, as the surface of the DLN film became smoother, and the debris size was in the order of 2–4 micrometer as seen by high resolution optical microscope (Figure 11).

It was observed that when the ball-on-disc test was started in ambient air with 50% RH, maximum number of dangling bonds got passivated via adsorption of water molecules, and interaction between counter faces was very low which led to low wear factor of DLN film. Adsorption of water molecule on DLN surface also led to dissociation and oxidation of DLN surface [15, 26, 27]. These were transferred to the ball as charged or uncharged fragments. The charged fragments released charges on ball and fell on the wear track as debris. Moreover the debris formed due to sliding agglomerated became larger as RH increased as seen by high resolution optical microscope (Figure 11). The average debris size at 60% RH condition was 15–17 μm, whereas it was 2–4 μm at 35% RH. Debris in the sliding interface deformed during sliding. Greater amount of energy was required for deforming larger debris than that for small scattered ones [26, 27]. This caused increase in COF, with increasing RH.

With a further increase in RH up to 80% the adhesion interaction was nullified, as dangling bonds were saturated by adsorbing the water molecule from ambient. As the ambient condition encompassed a large percentage of water, hence physisorption of water molecules led to oxidation of DLN surface [28]. Due to this oxidation of the outer most surface, the bonding strength with immediately beneath layer becomes weaker. Hence, the wearing phenomenon occurred more easily during sliding of WC ball. Moreover,

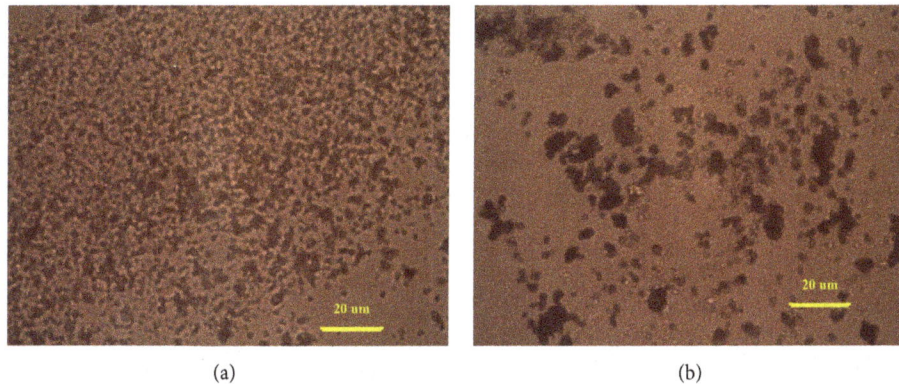

(a)   (b)

FIGURE 11: High resolution optical microscope (100X) image of debris produced by wearing in presence of air with (a) 35% RH and (b) 60% RH.

TABLE 2: wear rate, wear factor, and COF of DLN film of thickness 1.09 μm in different ambient conditions.

Ambient	Cycles/micron	Wear factor (mm^3/Nm)	Average COF during film life time
35% RH air	48200 ± 2453	$9.4 \times 10^{-8} \pm 4.78E - 09$	0.005 ± 0.001
40% RH air	65500 ± 5423	$6.9 \times 10^{-8} \pm 5.71E - 09$	0.011 ± 0.002
50% RH air	168687 ± 7534	$2.7 \times 10^{-8} \pm 1.21E - 09$	0.021 ± 0.002
60% RH air	73833 ± 5201	$6.1 \times 10^{-8} \pm 4.30E - 09$	0.055 ± 0.005
80% RH air	72000 ± 4865	$6.2 \times 10^{-8} \pm 4.19E - 09$	0.074 ± 0.005
DI Water	142100 ± 11349	$3.2 \times 10^{-8} \pm 2.56E - 09$	0.057 ± 0.004
Saline Soln.	245500 ± 12734	$1.8 \times 10^{-8} \pm 9.34E - 10$	0.021 ± 0.004

with increase in RH the debris size also increased, which in turn enhanced abrasive wear of the film as well as the COF.

In DI water ambient, the COF was less than that for air with high humidity, RH 80%, and closely matched with 60% RH air. The wear out debris contained broken pieces of DLN surface peaks, and the tribological byproducts (fragments) all floated in water. Water, in this particular case also acted as a lubricant and coolant. Thus, the COF lowered in DI water as compared to 80% RH air. It may be noted that both the COF and wear factor in DI water matched with moderate humidity ambient (RH 50%–60%).

In dilute solution of NaCl (0.9%), the COF was found to have very low value in comparison to 50% RH air, and the wear factor was also very low, even lower than that for 50% RH air. The reason behind this was, like DI water, saline water, also worked as a lubricant and coolant to reduce friction. In this case also, the debris floated leading to further reduction of friction. Moreover, the saline solution had a much higher electrical conductivity than DI water. The opposite charges that were built up at the two sliding surfaces due to friction were easily discharged via the saline solution. Hence, the adhesive force due to electrical effect reduced drastically resulting in very low wear factor, even less than that for 50% RH air.

4. Conclusion

DLN films deposited in a PACVD system from a suitable organosilicon precursor have excellent frictional and wear properties. In low humidity condition, the frictional coefficient is extremely low, and the wear factor was relatively high

(9.4×10^{-8} mm^3/Nm). The wear factor, however, dipped to a very low value (2.7×10^{-8} mm^3/Nm) at 50% RH. But the COF rose to 0.02. The wear factor was increased by a factor of 3.5 in dry air in reference to 50% RH condition. DI water and saline water (0.9% NaCl solution) acted as lubricant and coolant, to reduce wear rate. But friction was higher in DI water than in NaCl solution. The tribological properties of DLN film were extremely congenial in the ambience of dilute NaCl solution (0.9%). This type of solution is isotonic with the contents of human red blood cells. Thus DLN coating may be highly suitable for use in components for artificial hip joint and knee joint replacements. It may also be suitable as solid lubricant for machine parts.

Acknowledgments

The authors gratefully acknowledge the DST, Government of India for financial support. The XRD, FESEM, and AFM studies were carried out in the IIT, Kharagpur, IACS, Kolkata, and University of Calcutta, Kolkata, respectively.

References

[1] S. Meskinis and A. Tamuleviciene, "Structure, Properties and Applications of Diamond like Nanocomposite (SiO$_x$ Containing DLC) Films: A Review," *Materials Science*, vol. 17, pp. 358–370, 2011.

[2] C. Venkatraman, A. Goel, R. Lei, D. Kester, and C. Outten, "Electrical properties of diamond-like nanocomposite coatings," *Thin Solid Films*, vol. 308-309, no. 1–4, pp. 173–177, 1997.

[3] W. J. Yang, Y. H. Choa, T. Sekino, K. B. Shim, K. Niihara, and K. H. Auh, "Structural characteristics of diamond-like nanocomposite films grown by PECVD," *Materials Letters*, vol. 57, no. 21, pp. 3305–3310, 2003.

[4] S. Jana, S. Das, U. Gangopadhyay, P. Ghosh, and A. Mondal, "Frequency response of diamond like nanocomposite thin film based MIM capacitor and equivalent circuit modelling," *IOSR JEEE*, vol. 1, pp. 46–50, 2012.

[5] V. F. Dorfman, A. Bozhko, B. N. Pypkin et al., "Diamond-like nanocomposites: electronic transport mechanisms and some applications," *Thin Solid Films*, vol. 212, no. 1-2, pp. 274–281, 1992.

[6] K. J. Schoen, J. M. Woodall, A. Goel, and C. Venkatraman, "Electrical properties of metal-diamond-like-nanocomposite (Me-DLN) contacts to 6H SiC," *Journal of Electronic Materials*, vol. 26, no. 3, pp. 193–197, 1997.

[7] D. Neerinck, P. Persoone, M. Sercu, A. Goel, D. Kester, and D. Bray, "Diamond-like nanocomposite coatings (a-C:H/a-Si:O) for tribological applications," *Diamond and Related Materials*, vol. 7, no. 2–5, pp. 468–471, 1998.

[8] A. Pandit and N. P. Padture, "Interfacial toughness of diamond-like nanocomposite (DLN) thin films on silicon nitride substrates," *Journal of Materials Science Letters*, vol. 22, no. 18, pp. 1261–1262, 2003.

[9] W. J. Yang, K. H. Auh, C. Li, and K. Niihara, "Microstructure characteristics of diamond-like nanocomposite (DLN) film by thermally activated chemical vapor deposition," *Journal of Materials Science Letters*, vol. 19, no. 18, pp. 1649–1651, 2000.

[10] T. Das, D. Ghosh, T. K. Bhattacharyya, and T. K. Maiti, "Biocompatibility of diamond-like nanocomposite thin films," *Journal of Materials Science*, vol. 18, no. 3, pp. 493–500, 2007.

[11] A. Nath, A. Das, L. Rangan, and A. Khare, "Bacterial inhibition by Cu/Cu_2O nanocomposites prepared via laser ablation in liquids," *Science of Advanced Materials*, vol. 4, no. 1, pp. 106–109, 2012.

[12] S. J. Park, K.-R. Lee, and D.-H. Ko, "Tribochemical reaction of hydrogenated diamond-like carbon films: a clue to understand the environmental dependence," *Tribology International*, vol. 37, no. 11-12, pp. 913–921, 2004.

[13] S. I. U. Ahmed, G. Bregliozzi, and H. Haefke, "Microfrictional properties of diamond-like carbon films sliding against silicon, sapphire and steel," *Wear*, vol. 254, no. 11, pp. 1076–1083, 2003.

[14] D. Neerinck, P. Persoone, M. Sercu et al., "Diamond-like nanocomposite coatings for low-wear and low-friction applications in humid environments," *Thin Solid Films*, vol. 317, no. 1-2, pp. 402–404, 1998.

[15] T. W. Scharf, J. A. Ohlhausen, D. R. Tallant, and S. V. Prasad, "Mechanisms of friction in diamondlike nanocomposite coatings," *Journal of Applied Physics*, vol. 101, no. 6, Article ID 063521, 2007.

[16] X. Z. Ding, F. M. Zhang, X. H. Liu et al., "Ion beam assisted deposition of diamond-like nanocomposite films in an acetylene atmosphere," *Thin Solid Films*, vol. 346, no. 1, pp. 82–85, 1999.

[17] M. Park, C. W. Teng, V. Sakhrani et al., "Optical characterization of wide band gap amorphous semiconductors (a-Si:C:H): effect of hydrogen dilution," *Journal of Applied Physics*, vol. 89, no. 2, pp. 1130–1137, 2001.

[18] R. E. Shroder, R. J. Nemanich, and J. T. Glass, "Analysis of the composite structures in diamond thin films by Raman spectroscopy," *Physical Review B*, vol. 41, no. 6, pp. 3738–3745, 1990.

[19] F. Tuinstra and J.L. Koenig, "Raman spectroscopy of graphite," *Journal of Chemical Physics*, vol. 53, no. 3, pp. 1126–1130, 1970.

[20] R. O. Dillon, J. A. Woollam, and V. Katkanant, "Use of Raman scattering to investigate disorder and crystallite formation in as-deposited and annealed carbon films," *Physical Review B*, vol. 29, no. 6, pp. 3482–3489, 1984.

[21] C. Casiraghi, F. Piazza, A. C. Ferrari, D. Grambole, and J. Robertson, "Bonding in hydrogenated diamond-like carbon by Raman spectroscopy," *Diamond and Related Materials*, vol. 14, no. 3–7, pp. 1098–1102, 2005.

[22] K. Honglertkongsakul, P. W. May, and B. Paosawatyanyong, "Electrical and optical properties of diamond-like carbon films deposited by pulsed laser ablation," *Diamond and Related Materials*, vol. 19, no. 7–9, pp. 999–1002, 2010.

[23] J. Z. Wan, F. H. Pollak, and B. F. Dorfman, "Micro-Raman study of diamondlike atomic-scale composite films modified by continuous wave laser annealing," *Journal of Applied Physics*, vol. 81, no. 9, pp. 6407–6414, 1997.

[24] A. C. Ferrari and J. Robertson, "Interpretation of Raman spectra of disordered and amorphous carbon," *Physical Review B*, vol. 61, no. 20, pp. 14095–14107, 2000.

[25] S. J. Park, K.-R. Lee, and D.-H. Ko, "Tribological behavior of nano-undulated surface of diamond-like carbon films," *Diamond and Related Materials*, vol. 14, no. 8, pp. 1291–1296, 2005.

[26] E. Konca, Y.-T. Cheng, A. M. Weiner, J. M. Dasch, and A. T. Alpas, "Vacuum tribological behavior of the non-hydrogenated diamond-like carbon coatings against aluminum: effect of running-in in ambient air," *Wear*, vol. 259, no. 1–6, pp. 795–799, 2005.

[27] K. Y. Eun, K. R. Lee, E. S. Yoon, and H. Kong, "Effect of polymeric debris on the tribological behavior of diamond-like carbon films," *Surface and Coatings Technology*, vol. 86-87, no. 2, pp. 569–574, 1996.

[28] S. H. Yang, H. Kong, K. R. Lee, S. Park, and D. E. Kim, "Effect of environment on the tribological behavior of Si-incorporated diamond-like carbon films," *Wear*, vol. 252, no. 1-2, pp. 70–79, 2002.

4

Experimental Determination of Cutting Power for Turning and Material Removal Rate for Drilling of AA 6061-T6 Using Vegetable Oils as Cutting Fluid

Y. M. Shashidhara and S. R. Jayaram

Department of Mechanical Engineering, Malnad College of Engineering, Hassan, Karnataka 573201, India

Correspondence should be addressed to Y. M. Shashidhara; shashi.yms@gmail.com

Academic Editor: Shyam Bahadur

The raw and modified versions of two nonedible vegetable oils, Pongam (*Pogammia pinnata*) and Jatropha (*Jatropha curcas*), and a commercially available branded mineral oil are used as straight cutting fluids for turning AA 6061 to assess cutting forces. Minimum quantity lubrication is utilized for the supply of cutting fluids. Cutting and thrust forces are measured. Cutting power is determined for various cutting speeds, depths of cut, and feed rates. Also, drilling is performed on the material to understand the material removal rate (MRR) under these oils. The performances of vegetable oils are compared to mineral oil. A noticeable reduction in cutting forces is observed under the Jatropha family of oils compared to mineral oil. Further, better material removal rate is seen under both the vegetable oils and their versions compared to under petroleum oil for the range of thrust forces.

1. Introduction

Cutting forces are considered as important parameters in turning operation, and they dictate the power required for machining [1]. The cutting forces influence the deformation of the workpiece machined its dimensional accuracy, chip formation, tool wear, surface roughness, and machining system stability. Higher magnitudes of forces lead to distortion in workpiece, low dimensional accuracy, faster tool wear, poor surface finish, and undesirable vibrations. Cutting forces generated mainly depend on the depth of cut, cutting speed and type of cutting fluid. As the depth of cut increases, magnitudes of forces also increase. However, increase in cutting speeds reduces the magnitude of forces. Further, cutting fluids with high lubricity, high film boiling point, and quick wetting and spreading and friction reduction at extreme pressure properties reduce cutting forces significantly [2].

Metalworking fluids are extensively used in machining operations. There are several types of metalworking fluids (MWFs), which may be used to carry out such tasks [3]. The majority of the MWFs are mineral oil-based fluids. These fluids increase productivity and the quality of manufacturing

operations by cooling and lubricating during metal cutting processes [4]. The consumption of MWFs is increasing in machining industry due to their advantages. As cutting fluids are complex in their composition, they may be irritant or allergic. Even microbial toxins are generated by bacteria and fungi present, particularly in water-soluble cutting fluids [5], which are more harmful to the operators.

To overcome these challenges, various alternatives to petroleum-based MWFs are currently being explored by scientists and tribologists. Such alternatives include synthetic lubricants, solid lubricants, and vegetable-based lubricants. In general, vegetable oils are highly attractive substitutes for petroleum-based oils because they are environmentally friendly, renewable, less toxic, and readily biodegradable [6, 7]. Consequently, currently, vegetable-based oils are more potential candidates for the use in industry as MWFs. Many investigations are in progress to develop new bio-based cutting fluids based on various vegetable oils available around the world.

The performance of soybean, canola, palm, sunflower, and ground nut oils are evaluated as cutting fluids for machining different materials. Turning, drilling, reaming,

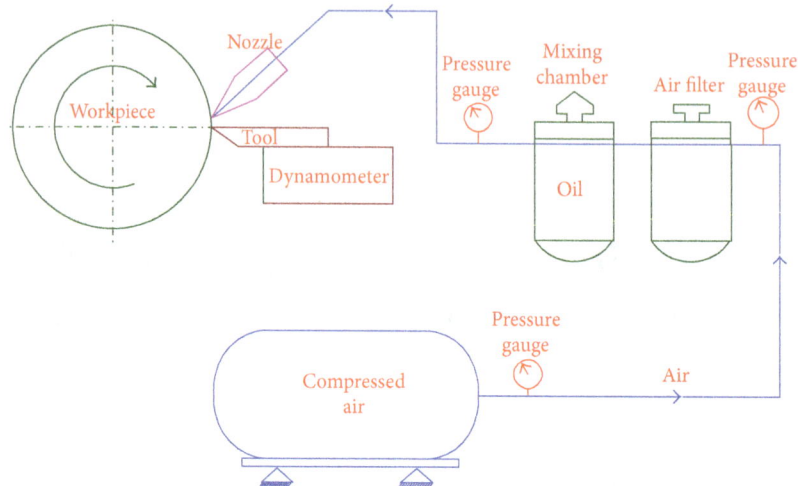

FIGURE 1: Minimum quantity lubrication circuit.

and tapping operations on austenitic stainless steel and other four materials were evaluated using vegetable-based formulations [8]. The results, on austenitic steel revealed that the cutting force under vegetable oils and esters is about 20% lower compared to reference mineral oils. Further, it is reported that about 40% increase in tool life under vegetable oils/ester modes of lubrication. Lower surface roughness, strain hardening, and thickness of the plastically deformed subsurface layer and better part accuracy are reported for a vegetable-based cutting fluid tested for reaming and tapping operations [9]. Cutting fluid formulations based on canola oil, soybean oil, and a TMP ester were made and were evaluated under tapping torque tests on 1018 cold rolled steel [10]. It is reported that all the bio-based oils exhibited about 12–14% increase in tapping torque efficiency relative to the reference soluble oil. However, Soy-based metalworking fluid had slightly higher tapping torque efficiency in the semisynthetic form compared to others.

As cutting fluids, ground nut and Shea butter oils generated lower cutting forces and coefficient friction while turning mild steel, aluminum, and copper workpieces [11]. Better surface finish is also reported under these oils. A formulated cutting fluid of canola with 8% extreme pressure additive showed comparable turning forces with mineral oils. About 61% drop in surface roughness values are reported compared to commercial mineral oil. Further, about 10% drop in tool wear is seen while turning AISI 304 L under canola oil [12]. Sunflower and canola-oil based cutting fluids showed better surface finish and produced lower cutting and feed forces while turning AISI 304 L [13]. Soybean, canola, and palm-oil based soluble cutting fluids are used for milling. About 10% lower surface roughness values reported comparable to flank wear are seen under these formulated oils compared to mineral oil [14].

The motive of the present work is to bring out the enormous potential of vegetable oils to be used in manufacturing sector as straight cutting oils or lubricants. This has gained more importance in the light of the recent restrictions made by world leaders like OSHA, HOSH, EPA, and so forth,

where they have suggested to come out with replacements for mineral oils, that are the most environmentally friendly and also are depleting. Also, there is a large consumption of cutting oils/lubricants for the manufacturing sector.

In the present study, the raw and modified versions of two nonedible vegetable oils, Pongam (*Pongammia pinnata*) and Jatropha (*Jatropha curcass*), and a commericially available branded mineral oil are used as straight cutting fluids. Turning and drilling of AA 6061 under these oils are chosen for the study. Cutting and thrust forces are measured for various cutting speeds, depth of cut, and feed. Cutting power is determined for turning the material. Further, drilling is performed on the material to assess the material removal rate (MRR) under these oils for various thrust forces. The results obtained are compared to the results under mineral oil.

2. Experiments

2.1. Oil Modification. The raw vegetable oils have certain limitations like low thermooxidative stability [15]. These problems are addressed by various methods, namely, reformulation of additives, chemical modifications, and genetic modification of the oil seed [16]. In the present work, chemical modification methods such as epoxidation [17] and transesterification [18] are used to modify the structure of two raw oils. After the modifications, their polyunsaturated C=C bonds are eliminated in the oil structure, and the thermooxidative stability, the resistance of a lubricant to molecular breakdown or molecular rearrangement at elevated temperature in the absence/presence of oxygen, is enhanced.

Pongam raw oil (PRO) is modified into Pongam methyl ester (PME) and epoxidized Pongam raw oil (EPRO). Similarly, Jatropha raw oil (JRO) is altered to Jatropha methyl ester (JME) and epoxidized Jatropha raw oil (EJRO). Further, Pongam methyl ester (PME) and Jatropha methyl ester (JME) are modified into epoxidized Pongam methyl ester (EPME) and epoxidized Jatropha methyl ester (EJME), respectively.

FIGURE 2: Experimental setup.

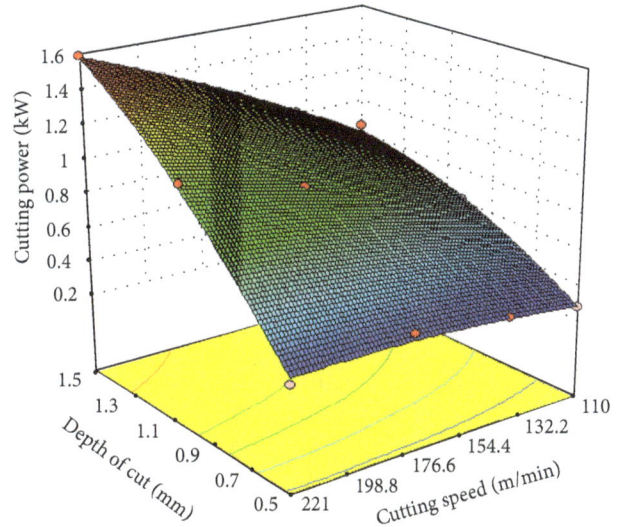

FIGURE 3: Variation of cutting power with cutting speed and depth of cut under MRO.

Turning (Figure 2) is carried out on AA 6061 using a lathe of 7.5 kW capacity. Cemented carbide tool with 0° rake angle and 14° clearance angles is utilised for cutting operation. Minimum quantity lubrication (MQL) method is adopted to supply the cutting fluid.

The MQL circuit (Figure 1) consists of an air compressor, mixing chamber, nozzle, and three pressure gauges. A nozzle of 2 mm is used for the experiments. Compressed air at 3 bar pressure is mixed with oil in the mixing chamber. The mist formed from the chamber is then fed in between cutting tool and workpiece through the nozzle.

Experiments are conducted under mineral, raw, and modified versions of the two vegetable oils as straight cutting oils for various cutting speeds (60, 90, and 140 m/min), depths of cuts (0.5, 1.0, and 1.5 mm), and feeds of 0.1 mm/rev, 0.18 mm/rev, and 0.25 mm/rev. The cutting forces are measured using Kistler lathe tool dynamometer (Figure 2). The forces are used to determine cutting power for turning.

Drilling is performed on AA 6061 sheets using HSS drill tool. Experiments are conducted for a constant spindle speed of 2625 rpm (maximum available speed). The time for drilling a hole is noted for thrust forces of 112 N, 138 N, 172 N, and 182 N under mineral, raw, and modified versions of the two vegetable oils as cutting fluids. The thrust forces are measured using a drill tool dynamometer.

3. Results and Discussion

3.1. Cutting Power. Cutting power required for machining provides the necessary input to understand the capacity of the drive. Cutting fluid with better lubrication capabilities reduces the power requirements.

Experimentally measured cutting forces for various cutting speeds, depths of cut, and feed are used to calculate the cutting power for turning AA 6061 under mineral and different vegetable oils and their versions (Table 1). Further, the results are also represented as three-dimensional pictures. The desirable conditions for turning AA 6061 under vegetable oil (raw and modified) are compared to mineral oil. An expected linear raise in cutting power is seen with an increase

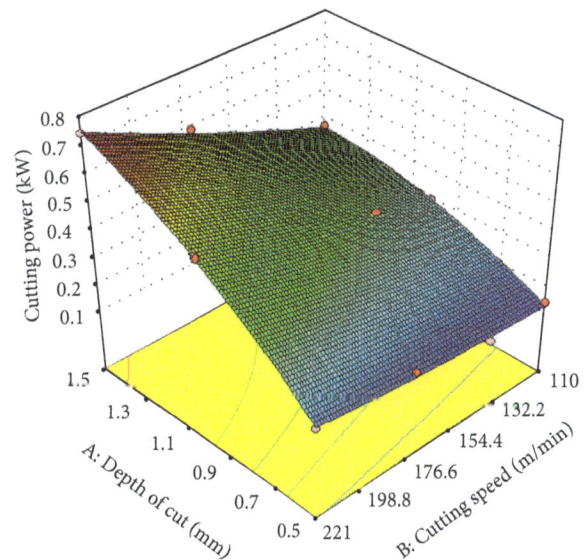

FIGURE 4: Variation of cutting power with cutting speed and depth of cut under PRO.

in cutting speeds and depths of cut [19]. Increase in cutting power is marginal for increase in cutting speed. However, it is significant with higher depths of cuts.

Figure 3 shows the cutting power distribution. Figure 3 through to Figure 6 are the plots with maximum drop in cutting power for Pongam and its versions. Figure 6 through to Figure 8 are for Jatropha and its versions.

3.1.1. Comparison of Pongam with Mineral Oil. Pongam raw oil and its modified versions show lower cutting power compared to mineral oil for the complete range of cutting speeds, depths of cut, and feed. Around 25% drop in cutting power for the feed 0.1 mm/rev and 0.5 mm depth of cut is

TABLE 1: Cutting power generated under various oils for different feed, depths of cut, and cutting speeds.

Cutting speed (m/min)	Cutting power (kW)							Cutting conditions
	MRO	PRO	EPRO	EPME	JRO	EJRO	EJME	
110	0.137	0.137	0.137	0.128	0.146	0.137	0.137	
138	0.172	0.115	0.161	0.138	0.184	0.161	0.161	$^*F = 0.1$ mm/rev
175	0.219	0.175	0.204	0.175	0.204	0.219	0.219	*DOC = 0.5 mm
221	0.257	0.202	0.239	0.202	0.220	0.257	0.276	
110	0.216	0.210	0.238	0.219	0.210	0.210	0.192	
138	0.259	0.264	0.264	0.264	0.253	0.259	0.230	$F = 0.18$ mm/rev
175	0.327	0.321	0.306	0.321	0.321	0.327	0.289	DOC = 0.5 mm
221	0.404	0.404	0.368	0.386	0.386	0.386	0.349	
110	0.284	0.293	0.284	0.274	0.274	0.265	0.265	
138	0.351	0.333	0.345	0.333	0.333	0.322	0.322	$F = 0.25$ mm/rev
175	0.438	0.394	0.423	0.408	0.408	0.408	0.394	DOC = 0.5 mm
221	0.533	0.478	0.515	0.478	0.507	0.496	0.331	
110	0.293	0.293	0.293	0.274	0.302	0.293	0.284	
138	0.363	0.356	0.333	0.333	0.356	0.356	0.333	$F = 0.1$ mm/rev
175	0.452	0.408	0.408	0.408	0.438	0.438	0.408	DOC = 1 mm
221	0.552	0.533	0.473	0.500	0.515	0.515	0.515	
110	0.485	0.467	0.549	0.458	0.485	0.458	0.430	
138	0.598	0.575	0.540	0.563	0.586	0.517	0.529	$F = 0.18$ mm/rev
175	0.715	0.686	0.671	0.700	0.730	0.642	0.642	DOC = 1 mm
221	0.846	0.828	0.791	0.846	0.846	0.791	0.791	
110	0.623	0.586	0.641	0.604	0.623	0.531	0.623	
138	0.736	0.724	0.736	0.736	0.736	0.655	0.759	$F = 0.25$ mm/rev
175	0.919	0.846	0.846	0.846	0.744	0.817	0.934	DOC = 1 mm
221	1.104	1.030	1.030	1.030	0.920	0.920	1.104	
110	0.412	0.366	0.421	0.384	0.430	0.412	0.412	
138	0.494	0.437	0.506	0.506	0.506	0.494	0.506	$F = 0.1$ mm/rev
175	0.613	0.584	0.627	0.627	0.613	0.569	0.598	DOC = 1.5 mm
221	0.736	0.736	0.809	0.901	0.736	0.736	0.736	
110	0.632	0.650	0.641	0.669	0.669	0.650	0.641	
138	0.782	0.782	0.770	0.805	0.782	0.782	0.782	$F = 0.18$ mm/rev
175	0.934	0.919	0.949	0.963	0.919	0.949	0.919	DOC = 1.5 mm
221	1.159	1.140	1.140	1.177	1.122	1.177	1.140	
110	0.852	0.834	0.760	0.870	0.779	0.806	0.639	
138	0.966	0.989	0.751	0.824	0.724	0.797	0.724	$F = 0.25$ mm/rev
175	1.197	1.197	1.416	1.126	1.109	1.211	0.898	DOC = 1.5 mm
221	1.453	1.453	1.398	1.527	1.343	1.435	1.051	

$^*F$: Feed, *DOC: Depth of Cut.

seen under PRO (Figures 3 and 4). This drop is the maximum compared to other combinations of feed and depths of cut. EPRO exhibits maximum of 20% lower cutting power for 0.1 mm/rev of feed and 1 mm depths of cut compared to MRO (Figure 5). About 28% drop in cutting power for feeds 0.1 mm/rev and 0.5 mm depth of cut is seen under EPME compared to petroleum oil (Figure 6). On the other hand, a marginal increase in cutting power is seen for higher feed and depths of cut.

3.1.2. Comparison of Jatropha with Mineral Oil. Similar to Pongam family of oils, Jatropha raw oil and its modified versions also show lower cutting power compared to mineral oil for the complete range of cutting speeds, depths of cut, and feeds. A maximum of 30% drop in cutting power is seen under JRO for the feed of 0.25 mm/rev and depth of cut 1 mm compared to mineral oil (Figure 7). Further, 10% constant drop in cutting power is observed for the other two depths of cut and feeds compared to petroleum oil.

Experimental Determination of Cutting Power for Turning and Material Removal Rate for Drilling of AA 6061-T6
Using Vegetable Oils as Cutting Fluid

35

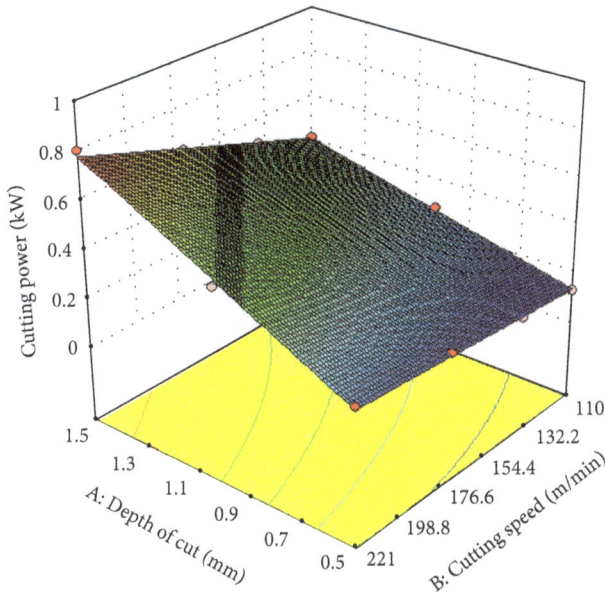

FIGURE 5: Variation of cutting power with cutting speed and depth of cut under EPRO.

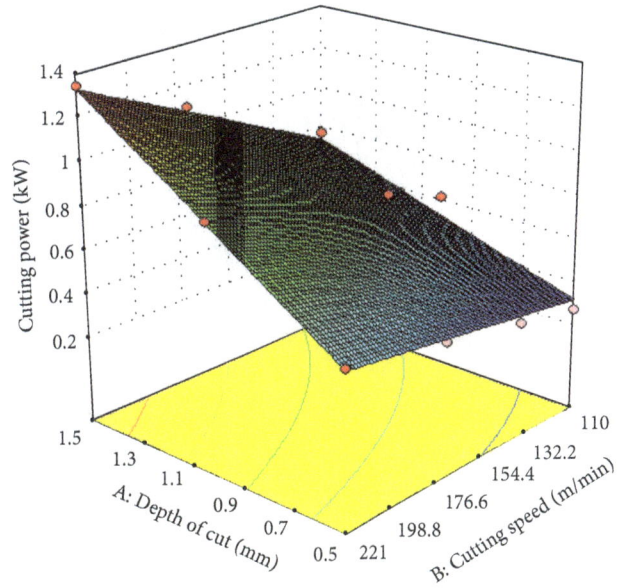

FIGURE 7: Variation of cutting power with cutting speed and depth of cut under JRO.

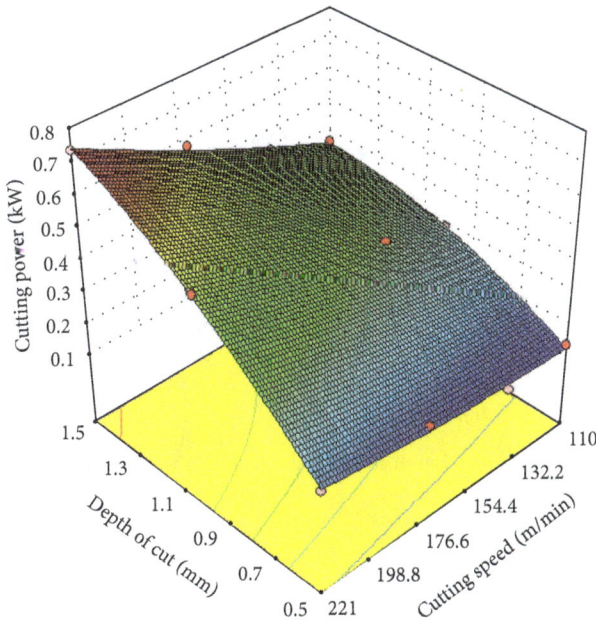

FIGURE 6: Variation of cutting power with cutting speed and depth of cut under EPME.

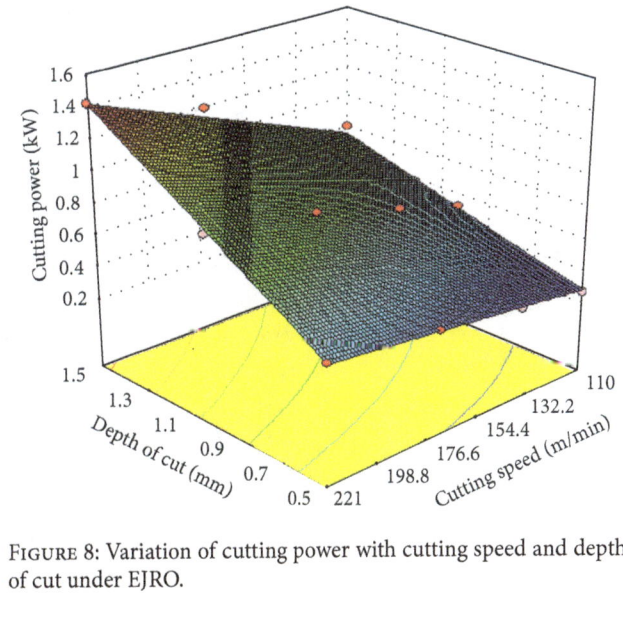

FIGURE 8: Variation of cutting power with cutting speed and depth of cut under EJRO.

About 20% reduction in cutting power under EJRO for feed 0.25 mm/rev and depth of cut 1mm compared to MRO (Figure 8). Under EJME, about 25% reduction in cutting power is seen for all the depths of cut and feed range except for feed 0.25 mm/rev and depth of cut 1.5 mm (Figure 9).

The drop in cutting power for turning under all the types of vegetable oils can be attributed to their polar nature and viscosity properties. Significant power reductions under vegetable oils can be due to the fact that, the thin surface film that develops in boundary lubrication is formed by the adsorption of polar compounds at the metal surface of the mating pair or by chemical reaction of the lubricant at the surface. Since, boundary lubrication by fatty acids is associated with the adsorption of the acid by dipolar attraction at the surface, they are capable of reducing the friction between the surfaces [20]. Jatropha versions exhibit noticeable power reductions at very high cutting speeds and depths of cut. This is due to high Oleic acid content [21] in the oil, higher viscosity index, and higher thermal conductivity of Jatropha compared to Pongam as well as mineral oil.

3.2. Drilling. In this segment, material removal rate under the two vegetable oils and their modified versions are analysed

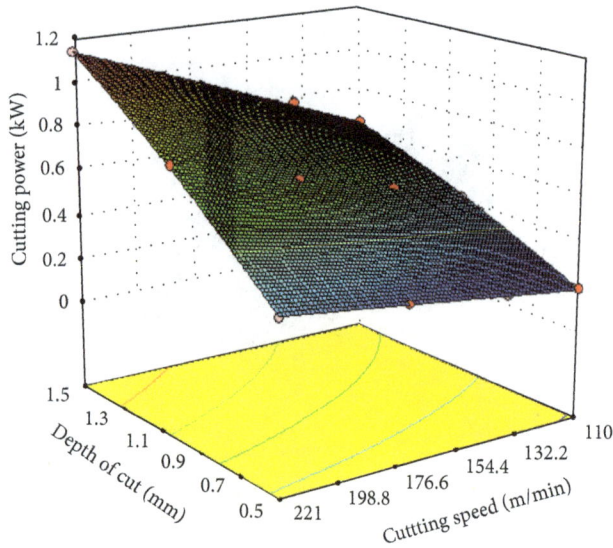

FIGURE 9: Variation of cutting power with cutting speed and depth of cut under EJME.

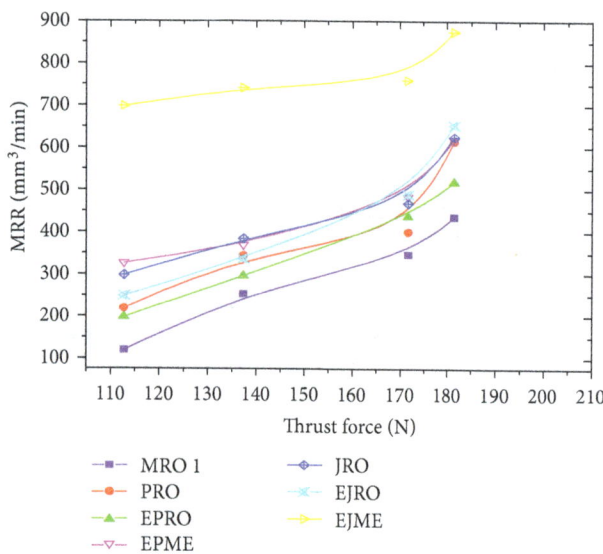

FIGURE 11: Variation of material removal rate with thrust force for Pongam family.

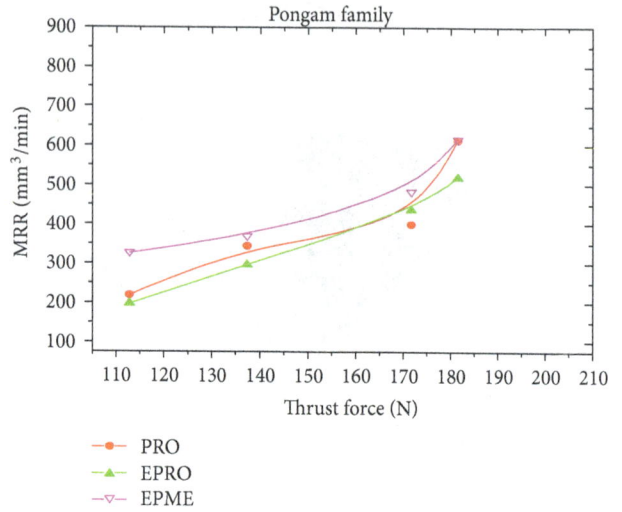

FIGURE 10: Variation of material removal rate with thrust force for various oils.

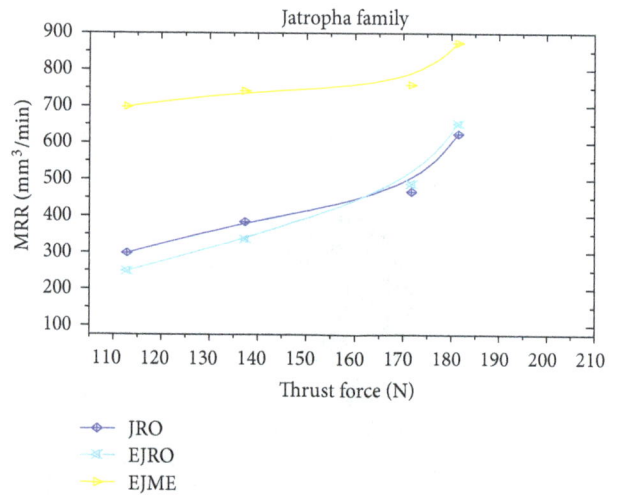

FIGURE 12: Variation of material removal rate with thrust force for Jatropha family.

and compared to mineral oil for the range of thrust force (112 N to 180 N).

3.2.1. Material Removal Rate (MRR). It is the volume of material removed/drilled out from the workpiece per unit time. It depends on the material, thrust force, feed, cutting speed, and cutting fluid. In drilling, high material removal rate and long drill life are essential to increase productivity. To achieve the desired MRR, high feed per revolution and cutting speed are required [22]. Further, a cutting fluid with good lubrication property produces higher rate of material removal.

Higher material removal is seen under both the vegetable oils and their versions compared to under mineral oil for

the range of thrust forces. This could be clearly seen from Figure 10 that, say, for the thrust force of 172 N, about 20% increase in MRR is observed under Pongam family (Figure 11) compared to under mineral oil. Similarly, under Jatropha family, about 40% increase in MRR is noticed. Interestingly, EJME offered the best MRR compared to all other oils tested with 50% increased MRR (Figure 12). It is attributed to the better lubrication capability of the oil.

The epoxidized methyl ester versions of both oils are seem to be the better cutting fluid to drilling AA 6061 as both of them show higher material removal property.

4. Conclusions

Pongam and Jatropha oils and their versions are better cutting fluids in terms of lower cutting forces and power

Experimental Determination of Cutting Power for Turning and Material Removal Rate for Drilling of AA 6061-T6
Using Vegetable Oils as Cutting Fluid

37

for turning AA 6061 compared to mineral oil. The drop in power under vegetable oils is predominant at very low and very high cutting speeds compared to mineral oil. Among the two vegetable oils, Jatropha versions show noticeable power reductions at higher cutting speeds and depths of cut. Specifically, EJME offers a better cutting fluid for turning AA 6061.

References

[1] B. Fnides, H. Aouici, and M. A. Yallese, "Cutting forces and surface roughness in hard turning of hot work steel X38CrMoV5-1 using mixed ceramic," *Mechanika*, vol. 70, no. 2, pp. 73–77, 2008.

[2] S. Kalpakjian and R. S. Steven, *Manufacturing Engineering and Technology*, Pearson Education, 4th edition.

[3] H. S. Abdalla and S. Patel, "The performance and oxidation stability of sustainable metalworking fluid derived from vegetable extracts," *Proceedings of the Institution of Mechanical Engineers B*, vol. 220, no. 12, pp. 2027–2040, 2006.

[4] J. B. Zimmerman, A. F. Clarens, K. F. Hayes, and S. J. Skerlos, "Design of hard water stable emulsifier systems for petroleum- and bio-based semi-synthetic metalworking fluids," *Environmental Science and Technology*, vol. 37, no. 23, pp. 5278–5288, 2003.

[5] A. Zeman, A. Sprengel, D. Niedermeier, and M. Späth, "Biodegradable lubricants-studies on thermo-oxidation of metal-working and hydraulic fluids by differential scanning calorimetry (DSC)," *Thermochimica Acta C*, vol. 268, pp. 9–15, 1995.

[6] T. Norrby, "Environmentally adapted lubricants—where are the opportunities?" *Industrial Lubrication and Tribology*, vol. 55, no. 6, pp. 268–274, 2003.

[7] M. T. Siniawski, N. Saniei, B. Adhikari, and L. A. Doezema, "Influence of fatty acid composition on the tribological performance of two vegetable-based lubricants," *Journal of Synthetic Lubrication*, vol. 24, no. 2, pp. 101–110, 2007.

[8] L. De Chiffre and W. Belluco, "Investigations of cutting fluid performance using different machining operations," *Lubrication Engineering*, vol. 58, no. 10, pp. 22–29, 2002.

[9] W. Belluco and L. De Chiffre, "Performance evaluation of vegetable-based oils in drilling austenitic stainless steel," *Journal of Materials Processing Technology*, vol. 148, no. 2, pp. 171–176, 2004.

[10] F. C. Andres, "Experimental comparison of vegetable and petroleum base oils in Metal working fluids using the tapping torque test," in *Proceedings of the Japan-USA Symposium on Flexible Automation*, Denver, Colo, USA, 2004.

[11] S. J. Ojolo, M. O. H. Amuda, O. Y. Ogunmola, and C. U. Ononiwu, "Experimental determination of the effect of some straight biological oils on cutting force during cylindrical turning," *Revista Materia*, vol. 13, no. 4, pp. 650–663, 2008.

[12] B. Ozcelik, E. Kuram, M. H. Cetin, and E. Demirbas, "Experimental investigations of vegetable based cutting fluids with extreme pressure during turning of AISI 304L," *Tribology International*, vol. 44, no. 12, pp. 1864–1871, 2011.

[13] H. Cetin, B. Ozcelik, E. Kuram, and E. Demirbas, "Evaluation of vegetable based cutting fluids with extreme pressure and cutting parameters in turning of AISI 304L by Taguchi method," *Journal of Cleaner Production*, vol. 19, no. 17-18, pp. 2049–2056, 2011.

[14] T. S. Lee and H. B. Choong, "An investigation on green machining: cutting process characteristics of organic metalworking fluid," *Advanced Materials Research*, vol. 230–232, pp. 809–813, 2011.

[15] B. K. Sharma, A. Adhvaryu, and S. Z. Erhan, "Friction and wear behavior of thioether hydroxy vegetable oil," *Tribology International*, vol. 42, no. 2, pp. 353–358, 2009.

[16] N. J. Fox and G. W. Stachowiak, "Vegetable oil-based lubricants-A review of oxidation," *Tribology International*, vol. 40, no. 7, pp. 1035–1046, 2007.

[17] X. Wu, X. Zhang, S. Yang, H. Chen, and D. Wang, "The study of Epoxidised Rapeseed oil used as a Potential Biodegradable Lubricant," *Journal of the American Oil Chemists' Society*, vol. 77, no. 5, pp. 561–563, 2000.

[18] R. A. Holser, "Transesterification of epoxidized soybean oil to prepare epoxy methyl esters," *Industrial Crops and Products*, vol. 27, no. 1, pp. 130–132, 2008.

[19] A. Aggarwal, H. Singh, P. Kumar, and M. Singh, "Optimizing power consumption for CNC turned parts using response surface methodology and Taguchi's technique-A comparative analysis," *Journal of Materials Processing Technology*, vol. 200, no. 1–3, pp. 373–384, 2008.

[20] S. Bhuyan, S. Sundararajan, L. Yao, E. G. Hammond, and T. Wang, "Boundary lubrication properties of lipid-based compounds evaluated using microtribological methods," *Tribology Letters*, vol. 22, no. 2, pp. 167–172, 2006.

[21] Y. M. Shashidhara and S. R. Jayaram, "Vegetable oils as a potential cutting fluid—an evolution," *Tribology International*, vol. 43, no. 5-6, pp. 1073–1081, 2010.

[22] R. Li and A. J. Shih, "Spiral point drill temperature and stress in high-throughput drilling of titanium," *International Journal of Machine Tools and Manufacture*, vol. 47, no. 12-13, pp. 2005–2017, 2007.

5

Investigations on Oil Flow Rates Projected on the Casing Walls by Splashed Lubricated Gears

G. Leprince,[1,2] C. Changenet,[1] F. Ville,[2] and P. Velex[2]

[1] *Laboratoire d'Energétique, Université de Lyon—ECAM Lyon, 40 Montée Saint-Barthélemy, 69321 Lyon Cedex 05, France*
[2] *INSA de Lyon, Université de Lyon, LaMCoS, UMR CNRS 5259, Bâtiment Jean d'Alembert, 18-20 Rue des Sciences, 69621 Villeurbanne Cedex, France*

Correspondence should be addressed to P. Velex, philippe.velex@insa-lyon.fr

Academic Editor: Ahmet Kahraman

In order to investigate the oil projected by gears rotating in an oil bath, a test rig has been set up in which the quantity of lubricant splashed at several locations on the casing walls can be measured. An oblong-shaped window of variable size is connected to a tank for flow measurements, and the system can be placed at several locations. A series of formulae have been deduced using dimensional analysis which can predict the lubricant flow rate generated by one spur gear or one disk at various places on the casing. These results have been experimentally validated over a wide range of operating conditions (rotational speed, geometry, immersion depth, etc.).

1. Introduction

Splash lubrication is traditionally used in low- to medium-speed enclosed gears such as automotive gearboxes in which the lubricant is projected by the rotation of the gears. The main disadvantages are (i) the generation of significant power losses by churning and (ii) the absence of precise control in terms of lubricant supply. Based on a number of studies [1–10], there is general agreement on the fact that losses increase with rotational speed and immersion depth. Although lubricant churning can be considered as a major source of power loss in gearboxes, splash lubrication also contributes to the regulation of gear bulk temperature since some heat is removed from the tooth faces by centrifugal fling-off as demonstrated by Blok [11]. Using a general thermal model of an automotive manual gearbox, Changenet et al. [12] have confirmed the influence of the heat exchanges between the oil sump and several rotating elements on the global thermal behaviour and emphasized, in particular, the role of the immersion depth. Höhn et al. [13] have conducted a number of experiments showing that lowering the oil level in the sump reduces churning losses but also leads to higher gear bulk temperatures. These results have been theoretically confirmed by Durand de Gevigney et al. [14].

From the above-mentioned studies, it seems possible to define an optimal lubricant level in the sump in order to reduce churning losses and, from a thermal viewpoint, ensure satisfactory gear-lubricant heat exchanges. However, as far as the authors know, the influence of the oil level on the lubricant circulation and flow has received scant attention in the open literature despite the practical importance of ensuring sufficient lubrication and cooling of sensitive elements such as bearings. Because of its intrinsically chaotic nature, splash lubrication properties are hardly predictable and a lot of the development for this lubrication technique relies on a trial-and-error approach. In automotive applications, for example, casings with transparent walls are frequently employed in order to try to visualize the oil flow for a variety of operating conditions (rotational speeds, engaged gears, etc.). Such empirical methods are expensive and cannot be used at the early stages of design thus emphasizing the interest of studies aimed at predicting the volumes and spatial distributions of projected lubricant in a gearbox. In this paper, a methodology is proposed which makes it possible to estimate the amount of lubricant projected at several positions on the casing. To this end, a specific test rig has been exploited, and some general formulae are presented based on dimensional analysis.

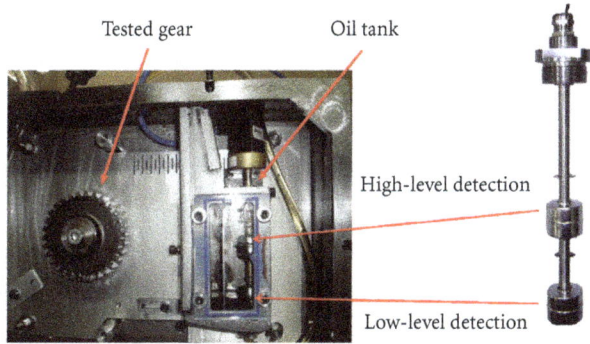

FIGURE 1: Oil tank in churning test rig.

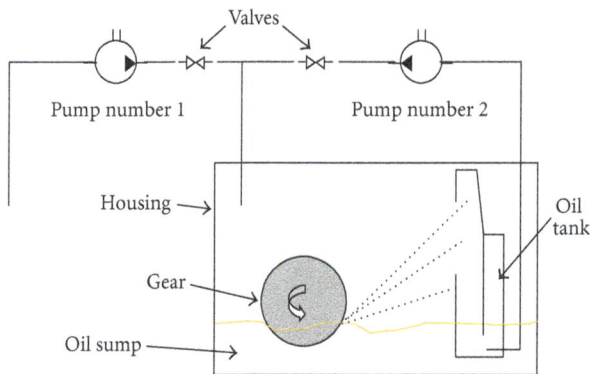

FIGURE 2: Hydraulic circuit.

2. Test Rig

The test rig developed to study churning losses and described in [7] has been modified in order to measure the oil flow rates projected on the casing walls. The pinion shaft is operated by an electric motor via a belt multiplying the rotational speed up to 7150 rpm. One of the faces of the oil sump is made of Plexiglas, and a tank has been placed in the housing to catch a part of the oil projected by the gear teeth due to centrifugal fling-off. To this end, an oblong-shaped window was drilled in the tank, and a level sensor is used to measure the time needed to fill a given volume (192 mL here) (Figure 1). In order to avoid the measurement uncertainties possibly induced by the more or less turbulent flows in the tank, a dividing wall has been introduced to separate the entry area from the measurement area in which the lubricant is far less agitated and levels can actually be read.

The hydraulic circuit in Figure 2 simultaneously enables the housing to be filled with oil and the tank to be drained when the sensor reaches the high-level detection threshold. Since measurements have to be carried out at a constant gear immersion depth, the oil in the tank is pumped and then injected into the casing (pump number 2). Finally, pump number 1 is used to fill the reservoir with lubricant at a set level.

The oil tank positioning is defined in Figure 3. Length L corresponds to the distance between the gear axis of rotation and the tank which can be varied up to a maximum of

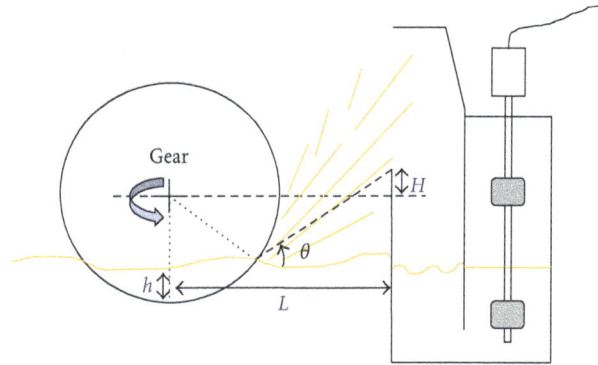

FIGURE 3: Definition of oil tank location.

130 mm. H is the height of the window from the axis of rotation which can be changed between 0 and 115 mm (the latter corresponding to the case of a completely obstructed window). The width of the window is equal to 40 mm and the rotating gear is axially centred with respect to the tank.

The results presented in this paper are limited to the disks and spur gears defined in Table 1. Experiments were conducted with two different fluids: water and a mineral oil (ν at 40°C = 45.11 Cst $-\nu$ at 100°C = 7.75 Cst $-\rho$ at 21°C = 885 kg/m^3).

3. Validity of the Measurements

Splash lubrication generates free surface flows characterised by wave trains created by the gear rotation which travel at certain speeds at the lubricant-air interface typical of transient flows. In such conditions, the number of data measurements has to be assessed in order to define a reliable experimental procedure. A number of measurements were performed for a range of operating conditions (gear geometry, rotational speed, tank location etc.) and various data acquisitions. Figure 4 presents some typical results using gear 1 partly immersed (relative immersion = h/R_p = 0.3) at 1500 rpm and for a tank location defined such that H = 10 mm and L = 110 mm. Any individual acquisition corresponds to the complete cycle for filling and emptying the tank. It can be observed that, as the number of acquisitions increases, temperature increases too because of the power losses by churning which, in turn, influence the quantity of projected lubricant by centrifugal fling-off. In order to eliminate this influence in the analysis, all the results presented in this paper correspond to measurements based on 50 data acquisitions.

In order to analyse the reliability of the test machine, measurements were performed at various rotational speeds. Table 2 presents three series of tests carried out with gear 1 for the following tank location: H = 20 mm and L = 105 mm. The results show that the maximum relative deviation in flow rate is about 8% which, considering the whole range of tests, corresponds to an average standard deviation of 4.9 mL/s. Compared with the nominal flow rates which generally lie between 50 and 200 mL/s, the standard deviation is considered as sufficiently small and the experimental results as acceptable. Finally, the uncertainty on

TABLE 1: Gear and disk data.

	Gear 1	Gear 2	Gear 3	Gear 4	Disk 1	Disk 2
Module (mm)	1.5	3	5	3	/	/
Number of teeth	102	53	30	30	/	/
Face width (mm)	14	24	24	24	25	1
Outside diameter (mm)	156	165	165	96	161	160

TABLE 2: Reliability of measurements.

N (rpm)	Flow rate (mL/s)				
	Test 1	Test 2	Test 3	Mean value	Standard deviation
500	174	168	166	169.3	4.16
1000	119	118	119	118.7	0.58
2000	86	88	93	89	3.61
3000	79	80	81	80	1.00

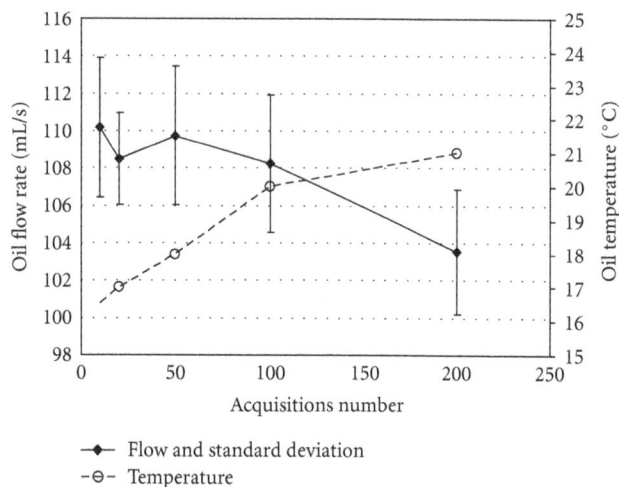

FIGURE 4: Influence of the number of acquisitions on flow rate measurement.

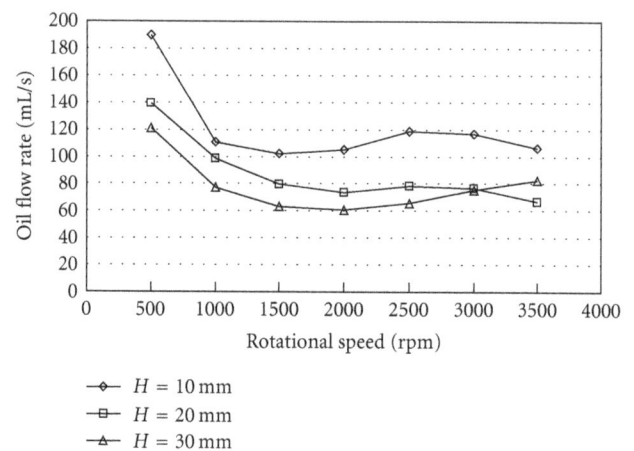

FIGURE 5: Typical evolutions of oil flow rates.

the volume measured by the two sensors was found to be around 2 mL leading to an error of 1%.

4. Experimental Results

Figure 5 shows the flow rate evolutions versus rotational speed obtained for gear 2 with a relative immersion depth of 0.25 and three different tank locations characterised by $L = 130$ mm and H between 10 and 30 mm.

As expected, Figure 5 shows that the flow rate received depends on the position on the wall and it can be noticed that a pinion rotating partly in oil does not behave as a gear pump since the flow rate does not vary linearly with the rotational speed. The curves in Figure 5 also reveal the existence of two flow regimes with a transition around 1500 rpm, above which the flow rate is nearly constant with speed (constant depending on geometry) whereas, at lower speeds, the flow decreases as speed increases. These results can be explained as follows:

(i) At low rotational speeds, near 500 rpm, a significant amount of lubricant is located at the gear periphery and moves with it so that large quantities of lubricant can be expelled and reach certain places on the walls if they are sufficiently close to the gear.

(ii) At higher rotational speeds, centrifugal forces become predominant, no bulk fluid motion occurs but the lubricant is sprayed in a large cone and only a fraction of it can reach the aperture of the measuring device.

(iii) It has also to be noticed that, below 300 rpm, no lubricant can enter the tank whereas, between 300 and 500 rpm, the intertooth spaces are filled with oil and the flow rate increases with speed.

These observations have been confirmed with other geometries. For example, Figure 6 shows the flow rate evolutions versus rotational speed obtained for disk 1 with a relative immersion depth of 0.25 and two different tank locations characterised by $L = 130$ mm and H between 20 and 40 mm. The results in this figure confirm the presence of a transition point around 1500 rpm, above which the oil flow

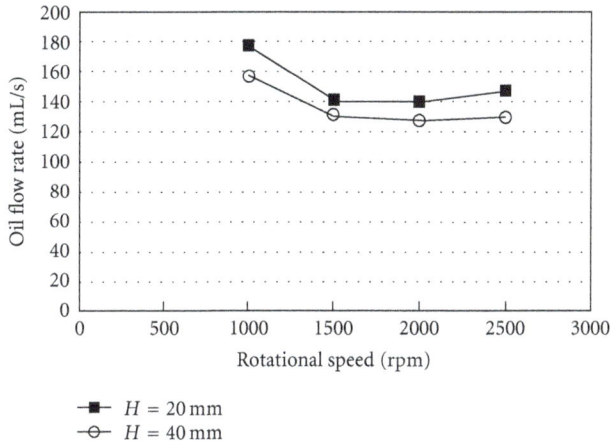

FIGURE 6: oil flow rates with a disk.

- H = 20 mm
- H = 40 mm

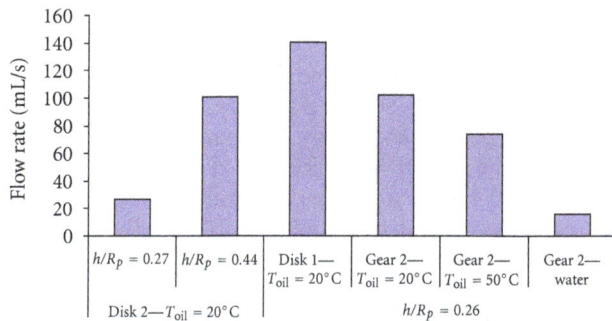

FIGURE 7: Influential parameters on flow rate.

rate becomes nearly constant versus speed. Experiments were also performed at 6000 rpm in order to verify that this result remains the same at higher rotational speeds.

In order to identify the relevant parameters with regard to the volume of oil projected by a rotating pinion on the casing walls, some specific tests have been performed at 1500 rpm for a given tank location ($L = 130$ mm and $H = 10$ mm) using disks and gears. The results are synthesised in Figure 7 from which the following conclusions have been drawn:

(i) The amount of lubricant received at one given location increases with the immersion depth.

(ii) Even disks of small face-width can generate significant flow rates suggesting that both the peripheral surface and the flanks contribute.

(iii) For equivalent geometrical parameters (radius, face width), a disk and a gear generate similar flow rates from which, it can be concluded that flow rates are largely independent of the parameters related to gear geometry, such as the module or the number of teeth.

(iv) For a given operating condition, flow rates are reduced by increasing the oil temperature or alternatively replacing oil by water suggesting that the fluid physical properties, and more specifically its viscosity, are influential.

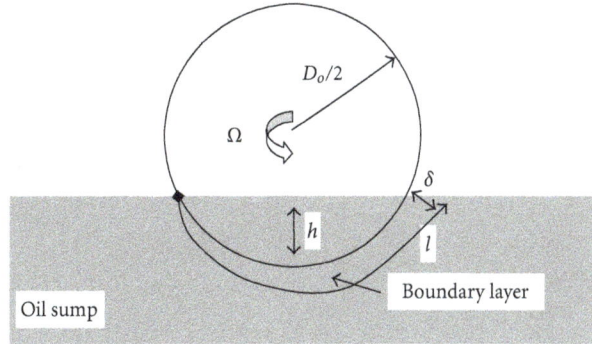

FIGURE 8: Schematic representation of the boundary layer on the side of a rotating disk.

Based on the conclusions above, it can be confirmed that the mechanism of oil projection differs from that in gear pumps. In this context, it seems interesting to investigate the hypothesis of a boundary-layer developing throughout the submerged part of a rotating disk and expelled by centrifugal forces. Theoretical flow rates can be deduced from this model and compared with the experimental evidence in order to assess the validity of this approach. Considering Figure 8, the leading edge is supposed to be located at the entry in the sump and boundary layers simultaneously develop on the two flanks and the side of the disc.

The length l of the boundary-layer on the peripheral surface of the disk depends on its outer radius and its immersion depth as:

$$l = D_o A \cos\left(\frac{D_o/2 - h}{D_o/2}\right). \tag{1}$$

Using the classic assumptions for laminar flows over flat plates, the hydrodynamic-boundary-layer thickness δ is found to be proportional to the disk arc length and the associated Reynolds number, such that

$$\delta \cong \frac{l}{\sqrt{\mathrm{Re}_l}}. \tag{2}$$

Since the flow velocity can be related to the peripheral speed of the rotating disk, an approximate order of magnitude of the flow rate generated by the side is

$$q_{\text{side}} \cong \sqrt{\frac{b^2 \nu \Omega D_o^2}{2} A \cos\left(\frac{D_o - 2h}{D_o}\right)}. \tag{3}$$

As noted before, the flanks of a rotating disk can substantially contribute to the total flow rate. Using a similar approach for every elemental slice of width dr, located at a radius r, and integrating the associated elemental flow over the total immersed surface, the flow rate associated with the immersed flanks can be expressed as

$$q_{\text{fl}} \cong \sqrt{8\Omega\nu} \int_{D_o/2-h}^{D_o/2} \sqrt{A \cos\left(\frac{D_o/2 - h}{r}\right)} r \cdot dr. \tag{4}$$

The total flow rate expelled by the disk is obtained by adding the results of (3) and (4). Based on this approach,

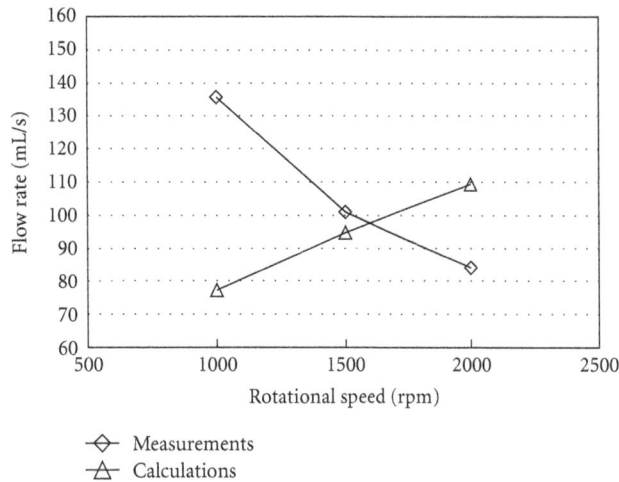

FIGURE 9: Comparison between experimental and numerical results by using boundary-layer theory.

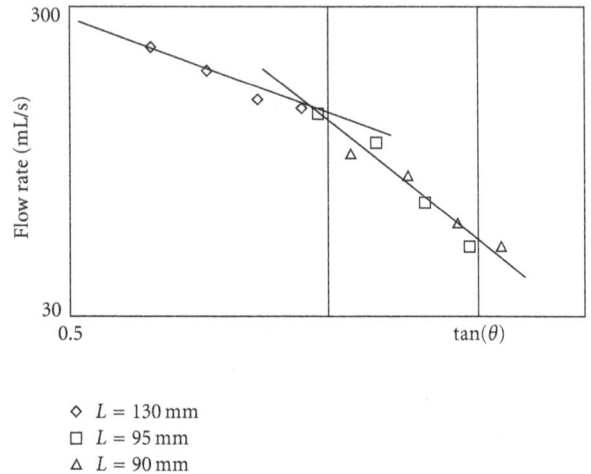

$\diamond\ L = 130\,\text{mm}$
$\square\ L = 95\,\text{mm}$
$\triangle\ L = 90\,\text{mm}$

FIGURE 10: Oil flow rate versus angle θ.

the reduction in boundary-layer thickness could explain why flow rates are reduced for lower viscosities. A comparison between the corresponding calculated values and the results from the test rig is displayed in Figure 9 for the following operating conditions: disk 2 running partly immersed in mineral oil ($T_{\text{oil}} = 20°\text{C}$) with a relative immersion depth equal to 0.44 and a tank location defined by $L = 130\,\text{mm}$ and $H = 10\,\text{mm}$. This figure clearly shows that (3) and (4) do not correctly reproduce the experimentally found variations versus speed. The actual projection mechanisms are certainly more complex than the proposed simplified approach and further studies are needed.

5. Dimensional Analysis

In an attempt to propose empirical equations valid over a range of geometry and operating conditions, dimensional analysis is used by assuming that the most influential parameters are:

(i) The geometrical parameters associated with the gear or disk: D_o, b, and h.

(ii) The geometrical parameters associated with the tank location: L and H.

(iii) The fluid parameters: ρ, μ and σ.

(iv) The dynamic parameters: Ω and g.

The oil flow rate q can therefore be sought as a function of these parameters above, that is:

$$q = f(D_o, b, h, L, H, \rho, \mu, \sigma, \Omega, g). \tag{5}$$

As noticed in the previous section, parameters such as the module or tooth number have a negligible influence and can consequently be discarded. The remaining geometrical parameters D_o, h, L, and H can be associated to form a single group of parameters defined by

$$\text{tg}(\theta) = \frac{D_o/2 - h + H}{L - (D_o/2 \sin(A \cos(1 - 2h/D_o)))}, \tag{6}$$

where $\text{tg}(\theta)$ represents the minimum initial slope of the fluid path (Figure 3) ensuring that the liquid projected by the rotating gear/disk can pass through the oblong-shaped window of the measuring device (Figure 3).

Figure 10 reveals a 2-slope linear relationship between the measured oil flow rates and $\text{tg}(\theta)$. These measurements were performed by using gear 4 with a relative immersion depth of 0.33 and different tank locations: L varies from 90 to 130 mm and H from 30 to 60 mm. It is to be noted that similar results have been obtained for other operating conditions.

From the above remark it is deduced that parameters D_o, h, L, and H are not independent, then (5) has to be transformed as follows:

$$q = f(D_o, b, \theta, \rho, \mu, \sigma, \Omega, g). \tag{7}$$

Using the results from the boundary layer approach, a dimensionless flow rate is defined by using (3) and (4), that is, the contributions from the flanks and side of a rotating element as:

$$\bar{q} = \frac{q}{q_{\text{fl}} + q_{\text{side}}}. \tag{8}$$

The three fundamental parameters D_o, Ω, and μ representing length, time, and mass are used to normalize all the other factors and, according to the theorem of Vaschy-Buckingham in dimensional analysis [15], $\bar{q}$ is found to depend on five groups of parameters as follows:

$$\bar{q} = \psi_1 \left(\frac{b}{D_o}\right)^{\psi_2} \text{tg}(\theta)^{\psi_3} \text{Re}^{\psi_4} \text{We}^{\psi_5} \text{Fr}^{\psi_6}, \tag{9}$$

where $\psi_1 \cdots \psi_6$ are constant coefficients to be adjusted from experimental results.

The influence of the Reynolds number Re can be identified from tests with a lubricant at several temperatures (20 and 50°C) while all the other parameters (geometry, rotational speed, etc.) remain constant. ψ_4 can therefore be deduced from any pair of experimental results labelled a and b, respectively, under the form:

$$\psi_4 = \frac{\ln(\overline{q_a}/\overline{q_b})}{\ln(\mathrm{Re}_a/\mathrm{Re}_b)}. \tag{10}$$

From the experimental evidence, a value of approximately 0.17 has been found for exponent ψ_4.

Tests with the pinion running partly immersed in water have been used in order to quantify the exponent of the Weber number leading to

$$\psi_5 = \frac{\ln\left(\overline{q_a}/\overline{q_b}(\mathrm{Re}_b/\mathrm{Re}_a)^{0.17}\right)}{\ln(\mathrm{We}_a/\mathrm{We}_b)}. \tag{11}$$

The experimental results show that ψ_5 is almost equal to 0 or, in other words, that surface tension is not influential and can be discarded in the dimensional analysis.

A number of complementary tests were conducted in which (a) only the rotational speed is varied and, (b) only the geometrical characteristics: $\mathrm{tg}(\theta)$ or face width are changed in order to successively determine the remaining exponents:

$$\psi_6 = \frac{\ln\left(\overline{q_a}/\overline{q_b}(\mathrm{Re}_b/\mathrm{Re}_a)^{0.17}\right)}{\ln(\mathrm{Fr}_a/\mathrm{Fr}_b)},$$

$$\psi_3 = \frac{\ln\left(\overline{q_a}/\overline{q_b}(\mathrm{Re}_b/\mathrm{Re}_a)^{0.17}(\mathrm{Fr}_b/\mathrm{Fr}_a)^{\psi_6}\right)}{\ln(\mathrm{tg}(\theta)_a/\mathrm{tg}(\theta)_b)}. \tag{12}$$

Additionally, the experimental results show that ψ_2 is almost equal to 0. The final relationships are synthesised below:
If $\mathrm{tg}(\theta) > 0.88$

(i) For $500 < N < 1500$ rpm

$$\overline{q} = 0.31\,\mathrm{tg}(\theta)^{-1.5}\mathrm{Re}^{0.17}\mathrm{Fr}^{-0.6}. \tag{13}$$

(ii) For $N > 1500$ rpm

$$\overline{q} = 0.09\,\mathrm{tg}(\theta)^{-1.5}\mathrm{Re}^{0.17}\mathrm{Fr}^{-0.3}. \tag{14}$$

If $\mathrm{tg}(\theta) < 0.88$

(i) For $500 < N < 1500$ rpm

$$\overline{q} = 0.78\mathrm{tg}(\theta)^{-1}\mathrm{Re}^{0.17}\mathrm{Fr}^{-0.6}. \tag{15}$$

(ii) For $N > 1500$ rpm

$$\overline{q} = 0.17\mathrm{tg}(\theta)^{-1}\mathrm{Re}^{0.17}\mathrm{Fr}^{-0.3}. \tag{16}$$

For rotational speeds below 300 rpm, the dimensionless flow rate is equal to zero. In the transition zone ($300 < N < 500$ rpm) a linear interpolation between this value and (13) or (15) is employed.

A typical series of comparisons between the experimental and the numerical results from (13) to (16) are displayed in Figure 11 where dimensionless flow rates are plotted against rotational speed for gear 1 and different locations of the oil tank. In this figure, the dotted lines represent

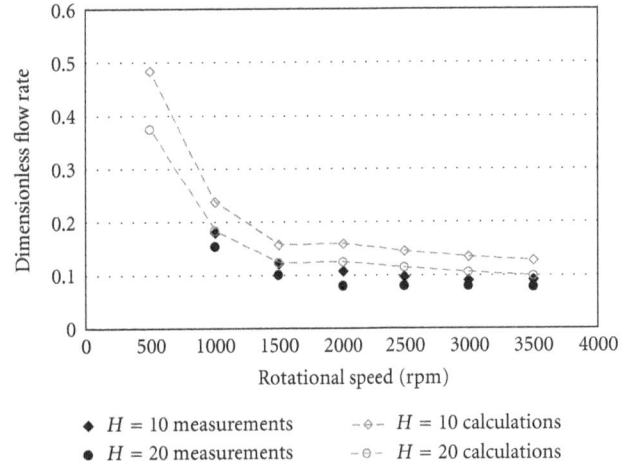

FIGURE 11: Comparison between experimental and numerical results [gear $1 - T_{\mathrm{oil}} = 20°C -h/R_p = 0.36 -L = 130$ mm].

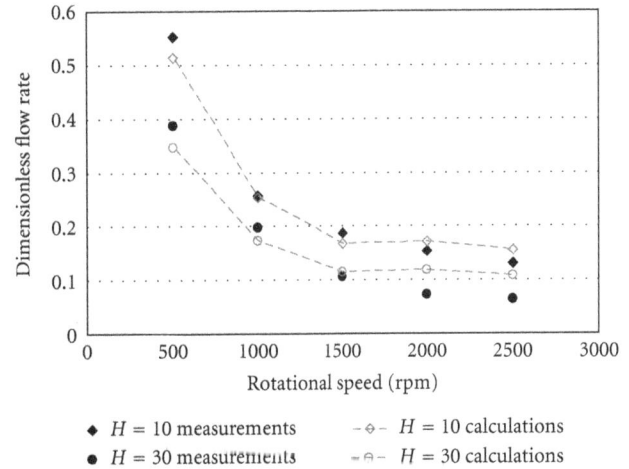

FIGURE 12: Comparison between experimental and numerical results [gear $2 - T_{\mathrm{oil}} = 50°C -h/R_p = 0.25 -L = 130$ mm].

the numerical results. The results show that the flow rate predictions are satisfactory with a larger discrepancy at the lowest rotational speeds (average error of 11%). More than 250 tests were carried out and the maximum deviation between the numerical and experimental flow rate was found to be equal to 15%. For the sake of illustration, additional curves are given in Figure 12 for different geometries and operating conditions which prove the versatility of the proposed formulae.

6. Conclusion

A test rig had been modified in order to quantify the amount of lubricant splashed at various locations on the casing walls by the rotation of a pinion partly immersed in oil. Based on a number of experimental results, a set of analytical formulae has been established using dimensional analysis in relation to an approximate theoretical approach relying on the concept of boundary layer. More than 250 different configurations have been tested and the agreement between the predictions and the experimental evidence is good. It is believed that these results can be useful at the early design stage and help understand splash lubrication characteristics particularly in automotive applications.

From a practical viewpoint, it has been demonstrated that a disk and a gear with similar geometry (outer radius and face width) generate equivalent flow rates. This conclusion is important for automotive applications as it underlines that synchronizers can also contribute to gearbox lubrication. The role of the physical properties of the lubricant is emphasized and viscosity appears as highly influential since it has been found that flow rates decrease with the operating viscosity. However, further experiments with different types of lubricants (mineral and synthetic oils) are certainly needed in order to confirm these findings or not. Experiments in this paper were limited to individual disks and spur gears; further investigations are certainly needed concerning the contribution of pinion-gear pairs along with that of helical teeth.

Nomenclature

b: Face width [m]
D_o: Outside diameter [m]
Fr: Froude number
g: Acceleration of gravity [m/s^2]
H: Height [m]
h: Immersion depth of a pinion [m]
L: Length [m]
l: Characteristic length of the boundary-layer [m]
m: Module [m]
N: Rotational speed [rpm]
q: Fluid flow rate [m^3/s]
$\bar{q}$:   Dimensionless flow rate
Re: Reynolds number
R_p: Pitch radius [m]
T: Fluid temperature [°C]
We: Weber number
δ: Thickness of the boundary-layer [m]
θ: Angle [rad]
μ: Dynamic viscosity [Pa s]
ν: Kinematic viscosity [m^2/s]
ρ: Fluid density [kg/m^3]
σ: Surface tension [N/m]
Ω: Rotational speed [rad/s].

Acknowledgments

The authors would like to gratefully thank PSA Peugeot Citroën and TOTAL for sponsoring this study. They also thank Vincent Ricol and Louis Bartolomé for their important contributions to the test rig design and construction.

References

[1] A. S. Terekhov, "Hydraulic losses in gearboxes with oil immersion," *Vestnik Mashinostroeniya*, vol. 55, no. 5, pp. 13–17, 1975.

[2] E. Lauster and M. Boos, "Zum Wärmehaushalt mechanischer Schaltgetriebe für Nutzfahrzeuge," *VDI-Berichte*, vol. 488, pp. 45–55, 1983.

[3] R. J. Boness, "Churning losses of discs and gears running partially submerged in oil," in *Proceedings of the International Power Transmission and Gearing Conference: New Technologies for Power Transmissions of the 90's*, pp. 355–359, Design Engineering Division, ASME, Chicago, Ill, USA, April 1989.

[4] B. R. Höhn, K. Michaelis, and T. Völlmer, "Thermal rating of gear drives: balance between power loss and heat dissipation," American Gear Manufacturers Association Document, 96FTM8, p. 12, 1996.

[5] P. Luke and A. V. Olver, "A study of churning losses in dip-lubricated spur gears," *Proceedings of the Institution of Mechanical Engineers G*, vol. 213, no. 5, pp. 337–346, 1999.

[6] C. Changenet and P. Velex, "A model for the prediction of churning losses in geared transmissions—preliminary results," *Journal of Mechanical Design*, vol. 129, no. 1, pp. 128–133, 2007.

[7] C. Changenet and P. Velex, "Housing influence on churning losses in geared transmissions," *Journal of Mechanical Design*, vol. 130, no. 6, Article ID 062603, 6 pages, 2008.

[8] S. Seetharaman and A. Kahraman, "Load-independent spin power losses of a spur gear pair: model formulation," *Journal of Tribology*, vol. 131, no. 2, Article ID 022201, 11 pages, 2009.

[9] G. Leprince, C. Changenet, F. Ville, P. Velex, C. Dufau, and F. Jarnias, "Influence of aerated lubricants on gear churning losses—an engineering model," *Tribology Transactions*, vol. 54, no. 6, pp. 929–938, 2011.

[10] C. Changenet, G. Leprince, F. Ville, and P. Velex, "A note on flow regimes and churning loss modeling," *Journal of Mechanical Design*, vol. 133, no. 12, Article ID 121009, 5 pages, 2011.

[11] H. Blok, "Transmission de chaleur par projection centrifuge d'huile," *Société d'Etudes de l'Industrie de l'Engrenage*, vol. 59, pp. 14–23, 1970.

[12] C. Changenet, X. Oviedo-Marlot, and P. Velex, "Power loss predictions in geared transmissions using thermal networks-applications to a six-speed manual gearbox," *Journal of Mechanical Design*, vol. 128, no. 3, pp. 618–625, 2006.

[13] B. R. Höhn, K. Michaelis, and H. P. Otto, "Influence of immersion depth of dip lubricated gears on power loss, bulk temperature and scuffing load carrying capacity," *International Journal of Mechanics and Materials in Design*, vol. 4, no. 2, pp. 145–156, 2008.

[14] J. Durand de Gevigney, C. Changenet, F. Ville, and P. Velex, "Thermal modelling of a back-to-back gearbox test machine: application to the FZG test rig," *Proceedings of the Institution of Mechanical Engineers J*, vol. 266, no. 6, pp. 501–515, 2012.

[15] S. Candel, *Mécaniques des Fluides—Cours*, Dunod, Paris, France, 2nd edition, 1995.

6

Dry Sliding Friction and Wear Studies of Fly Ash Reinforced AA-6351 Metal Matrix Composites

M. Uthayakumar,[1] **S. Thirumalai Kumaran,**[1] **and S. Aravindan**[2]

[1] *Department of Mechanical Engineering, Kalasalingam University, Krishnankoil, Tamil Nadu 626 126, India*
[2] *Department of Mechanical Engineering, Indian Institute of Technology, New Delhi 110 016, India*

Correspondence should be addressed to M. Uthayakumar; uthaykumar@gmail.com

Academic Editor: Huseyin Çimenoğlu

Fly ash particles are potentially used in metal matrix composites due to their low cost, low density, and availability in large quantities as waste by-products in thermal power plants. This study describes multifactor-based experiments that were applied to research and investigation on dry sliding wear system of stir-cast aluminum alloy 6351 with 5, 10, and 15 wt.% fly ash reinforced metal matrix composites (MMCs). The effects of parameters such as load, sliding speed, and percentage of fly ash on the sliding wear, specific wear rate, and friction coefficient were analyzed using Grey relational analysis on a pin-on-disc machine. Analysis of variance (ANOVA) was also employed to investigate which design parameters significantly affect the wear behavior of the composite. The results showed that the applied load exerted the greatest effect on the dry sliding wear followed by the sliding velocity.

1. Introduction

Metal matrix composites (MMCs) have received substantial attention due to their reputation as stronger, stiffer, lighter, and excellent wear properties over the monolithic alloys [1–3]. Though MMCs possess superior properties, they have not been widely applied due to the complexity of fabrication [4]. The conventional stir casting is an attractive processing method for fabrication, as it is relatively inexpensive and offers wide selection of materials and processing conditions. Stir casting offers better matrix particle bonding due to stirring action of particles into melts [5].

Wear is one of the most commonly encountered industrial problems, leading to frequent replacement of components, particularly abrasion. Abrasive wear occurs when hard particles or asperities penetrate a softer surface and displace material in the form of elongated chips and slivers [6]. Extensive studies on the tribological characteristics of aluminum MMCs containing various reinforcements such as silicon carbide, alumina, and short steel fiber are already done by researchers [7–9]. The variables such as composition of the matrix, particle distribution, and interface between the particles and the matrix affect the tribological behavior of metal matrix composites. These conditions include the type of countersurface, applied load, sliding speed, contact area, geometry, and environment [10]. The principle tribological parameters such as applied load [11–13], sliding speed [14, 15], and percentage of fly ash control the friction and wear performance. Fly ash is one of the residues generated in the combustion of coal. The addition of fly ash leads to the increase in wear resistance, hardness, elastic modulus and proof stress compared to unreinforced alloy. The fly ash particle size and its volume fraction also significantly affect the wear and friction properties of composites [16, 17].

The aim of the present study is to investigate the dry sliding wear of stir cast AA 6351 with 5, 10, and 15 wt.% fly ash reinforced MMCs, using a pin-on-disc type of wear machine. Furthermore, ANOVA was employed to investigate which design parameters significantly affect the wear behavior of the composite. Confirmation test was also conducted to verify the improvement of the quality characteristic using optimal levels of the design parameters.

TABLE 1: Elements of AA-6351 and its weight (%).

Elements	Al	Si	Mn	Mg
Weight (%)	97.8	1.0	0.6	0.6

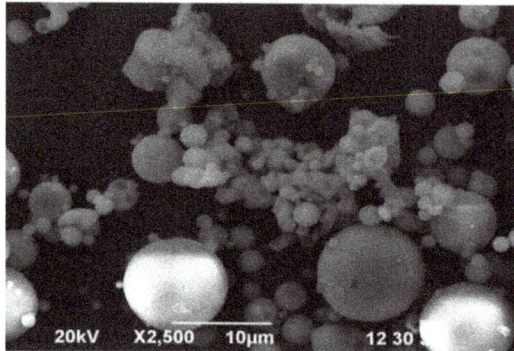

FIGURE 1: Micrograph of fly ash.

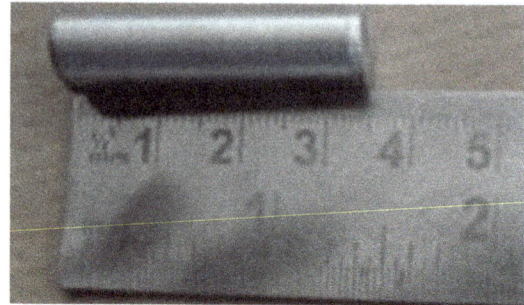

FIGURE 2: Typical specimen of AA 6351-fly ash reinforced MMC.

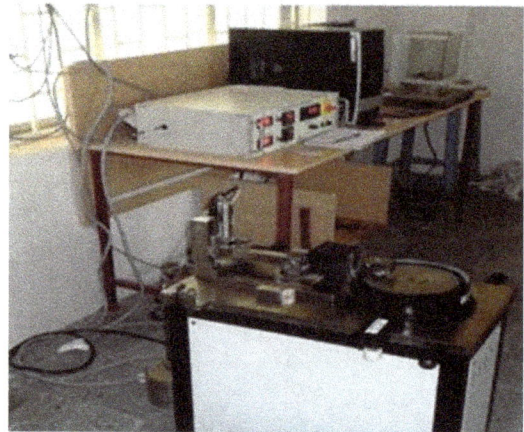

FIGURE 3: Pin-on-disc machine.

2. Materials and Experimental Procedures

2.1. Raw Material Description and Its Composition. Aluminum alloy 6351 is used in heavy engineering applications due to its strength, bearing capacity, ease of workability, and weldability. It is also used in building ship, column, chimney, rod, mould, pipe, tube, vehicle, bridge, crane, and roof [18]. The elements of aluminum alloy 6351 and the weight in % are tabulated in Table 1.

Fly ash is an industrial by-product recovered from the flue gas of coal burning electric power plants. Fly ash particles are mostly spherical in shape. The fly ash is collected from Tuticorin thermal power station, India, with an average particle size of 2–10 microns, and its typical micrograph is presented in Figure 1.

2.2. Fabrication of Composite. AA-6351 with 5, 10, and 15 wt.% fly ash reinforced MMCs are prepared by stir casting process. The melting was carried out in a resistance furnace. The fly ash particles were preheated on an open hearth furnace for one hour to remove the moisture. Scraps of AA 6351 were preheated at 450°C for 3 to 4 h before melting. The preheated aluminum scraps were first heated above the liquidus temperature to melt them completely. They were then slightly cooled below the liquidus to maintain the slurry in the semisolid state. This procedure has been adopted while stir-casting aluminum composites [19, 20]. Typical specimen is shown in Figure 2.

2.3. Wear Tests. The pin-on-disc machine was used to evaluate the friction and wear response to the sliding contact surface of the specimens as shown in Figure 3. Tests were conducted under dry conditions as per ASTM G99-95 standards. The pin was initially cleaned with acetone and weighed accurately using a digital electronic balance. The test was carried out by applying load (9.81, 19.62, and 29.43 N) and run for a constant sliding distance (3000 m) at different sliding velocities (1, 2, and 3 m/s). The disc material was made of EN-32 steel with the hardness of 65 HRC. At the end of each test, the weight loss of the specimen was noticed. The results obtained during this work have been presented in terms of sliding wear, specific wear rate, and friction coefficient. Sliding wear is related to interactions between surfaces and more specifically the removal and deformation of material on a surface as a result of mechanical action of the opposite surface. Specific wear rate is the volume loss per sliding distance and load. Friction coefficient is the ratio of the force of friction between two bodies and the force pressing them together.

2.4. Selection of Testing Parameters and Their Levels. The application of design-of-experiments (DoE) requires careful planning, prudent layout of the experiment, and expert analysis of results. Taguchi has standardized methods for each of these DoE application steps. The experiment specifies three principle wear testing conditions including applied load *(A)*, sliding speed *(B)*, and percentage of fly ash *(C)* as the process parameters. The experiments were carried out to analyze the influence of sliding wear, specific wear rate and friction coefficient on the MMCs. Control factors and their levels are shown in Table 2. This table shows that the experimental plan had three levels.

The standard Taguchi experimental plan with notation $L_9(3)^4$ was chosen based upon the degree of freedom. The degrees of freedom for the orthogonal array should be greater

TABLE 2: Control factors and their levels.

Symbols	Design parameters	Level 1	Level 2	Level 3
A	Applied load, N	9.81	19.62	29.43
B	Sliding speed, m/s	1	2	3
C	Percentage of fly ash, wt.%	5	10	15

TABLE 3: $L_9 (3)^4$ orthogonal array.

	Control factors		
Ex.	A	B	C
1	9.81	1	5
2	9.81	2	10
3	9.81	3	15
4	19.62	1	10
5	19.62	2	15
6	19.62	3	5
7	29.43	1	15
8	29.43	2	5
9	29.43	3	10

than or at least equal to those of the process parameters. Table 3 shows the $L_9(3)^4$ orthogonal array.

2.5. Experimental Details. The experimental results are transformed into a signal-to-noise (S/N) ratio. The-lower-the-better characteristics were selected for sliding wear, specific wear rate and friction coefficient due to the wear resistance of the tested samples. A pin-on-disc type of apparatus was employed to evaluate the wear characteristics at a constant sliding distance of 3000 m. Table 4 shows the experimental results of sliding wear, specific wear rate, and friction coefficient and their S/N ratios.

3. Grey Relational Analysis

3.1. Multiresponse Optimization Using Grey Relational Analysis. Taguchi method is designed to optimize single response characteristic. The-higher-the-better performance for one factor may affect the performance because another factor may demand the-lower-the-better characteristics. Hence, multiresponse optimization characteristics are complex. Here the Grey relational analysis optimization methodology for multiresponse optimization is performed with the following steps.

(i) Normalizing the experimental results.

(ii) Performing the Grey relational generating and calculating the Grey relational coefficient.

(iii) Calculating the Grey relational grade by averaging the Grey relational coefficient.

(iv) Performing statistical analysis of variance (ANOVA) for the input parameters with the Grey relational grade to find which parameter significantly affects the process.

(v) Selecting the optimal levels of process parameters.

(vi) Conducting confirmation experiment and verifying the optimal process parameters setting.

In Grey relational analysis, the complex multiple response optimizations can be simplified into the optimization of a single response Grey relational grade.

3.2. Grey Relational Generation. In the grey relational analysis, when the range of the sequence is large or the standard value is enormous, the function of factors is neglected. However, if the factors goals and directions are different, the Grey relational analysis might also produce incorrect results. Therefore, one has to preprocess the data which are related to a group of sequences, which is called "Grey relational generation."

For the-lower-the-better quality characteristics data preprocessing is calculated by

$$x_{ij} = \frac{(y_{ij})_{\max} - y_{ij}}{(y_{ij})_{\max} - (y_{ij})_{\min}}, \tag{1}$$

where y_{ij} is the ith experimental results in the jth experiment. Table 5 shows the preprocessed data results.

The grey relational coefficients of each performance characteristic are calculated in Table 6 using (2).

The grey relation coefficient is

$$\gamma(x_{0j}, x_{ij}) = \frac{\Delta_{\min} + \zeta\Delta_{\max}}{\Delta_{ij} + \zeta\Delta_{\max}} \tag{2}$$

for $i = 1, 2, \ldots, m$ $j = 1, 2, \ldots, n$. where

$$\Delta_{ij} = |x_{0j} - x_{ij}|,$$

$$\Delta_{\min} = \text{Min}\{\Delta_{ij}, i = 1, 2, \ldots, m; j = 1, 2, \ldots, n\}, \tag{3}$$

$$\Delta_{\max} = \text{Max}\{\Delta_{ij}, i = 1, 2, \ldots, m; j = 1, 2, \ldots, n\},$$

ζ is the distinguishing coefficient, $\zeta \in [0, 1]$.

3.3. Grey Relational Grade. Table 7 shows the influence of process parameters of Grey relational grade. The higher value of the Grey relational grade represents the stronger relational degree of the reference sequence and the given sequence.

The grey relational grade is obtained by

$$\alpha_j = \frac{1}{m} \sum_{i=1}^{m} \delta_{ij}, \tag{4}$$

where α_j is the Grey relational grade for the jth experiment and m is the number of performance characteristics.

As a result, optimization of the complicated multiple performance characteristics can be converted into optimization of single grey relational grade. The higher Grey relational grade represents that the experimental result is closer to the ideal normalized value.

TABLE 4: Experimental results of wear and friction coefficient and their S/N ratios.

Ex.	Experimental results			S/N ratios		
	Sliding wear (μ)	Specific wear rate $(m^2/N) \times 10^{-15}$	Friction coefficient	Sliding wear (dB)	Specific wear rate (dB)	Friction coefficient (dB)
1	146	249.4	0.254	−43.28	−47.93	11.87
2	104	237.123	0.560	−40.34	−47.49	5.02
3	82	184.3	0.244	−38.27	−45.31	12.22
4	261	237.123	0.198	−48.33	−47.49	14.03
5	298	219.07	0.214	−49.48	−46.81	13.38
6	160	181.3	0.147	−44.08	−45.16	16.60
7	369	228.93	0.383	−51.34	−47.19	8.31
8	384	290.4	0.217	−51.68	−49.25	13.25
9	414	323.48	0.224	−52.34	−50.19	12.98

TABLE 5: Preprocessed data results.

Ex.	Sliding wear (μ)	Specific wear rate $(m^2/N) \times 10^{-15}$	Friction coefficient
1	0.807	0.521	0.741
2	0.934	0.607	0.000
3	1.000	0.979	0.765
4	0.461	0.607	0.877
5	0.349	0.734	0.840
6	0.765	1.000	1.000
7	0.136	0.665	0.428
8	0.090	0.233	0.831
9	0.000	0.000	0.815

TABLE 7: Influence of process parameters of Grey relational grade.

Ex.	Grey relational grade	Order
1	0.630	3
2	0.592	6
3	0.880	2
4	0.614	5
5	0.615	4
6	0.893	1
7	0.477	8
8	0.499	7
9	0.465	9

TABLE 6: Grey relational coefficient of each performance characteristic.

Ex.	Sliding wear (μ)	Specific wear rate $(m^2/N) \times 10^{-15}$	Friction coefficient
1	0.722	0.511	0.659
2	0.883	0.560	0.333
3	1.000	0.960	0.681
4	0.481	0.560	0.802
5	0.435	0.653	0.757
6	0.680	1.000	1.000
7	0.366	0.599	0.466
8	0.355	0.395	0.748
9	0.333	0.333	0.730

TABLE 8: Response table for Grey relational grade.

Symbols	Response table			
	Level 1	Level 2	Level 3	Max−Min
A	0.701	0.708	0.481	0.227
B	0.574	0.569	0.746	0.178
C	0.674	0.557	0.657	0.117
Error	0.570	0.654	0.664	0.094
Mean value of Grey relational grade = 0.629				

4. Wear Mechanism

The typical wear mechanism of composites is presented in this section. At lower loads the sliding wear, specific wear rate, and friction coefficient decrease with increase in fly ash percentage, but as the load increases, they tend to increase with increase in fly ash percentage. The composites were also noticed to have mild wear as the sliding speed increases. The results were related and verified to the tests conducted by various researches [21, 22].

Examinations of the worn surfaces of the composites were done by scanning electron microscope. Figure 4(a) shows the worn surface of the AA-6351-flyash composite (@ 19.62 N, 3 m/s, and 5 wt.% flyash), featuring some shallow grooves that are produced when there is abrasion in the tribological pair. Figure 4(b) shows the worn surface of AA 6351-flyash composite (@ 29.43 N, 1 m/s and 15 wt.% flyash), featuring some shallow grooves and damages. Among the experiments conducted there was no seizure condition noticed.

5. Analysis of Variance

The analysis of variance (ANOVA) is used to investigate which design parameters significantly affect the quality characteristic. The traditional statistical technique can only obtain

TABLE 9: Results of analysis of variance.

Design parameters	Degree of freedom (DOF)	Sum of squares (SS)	Mean of square (MS)	F-test	Contribution (%)
Applied Load, N	2	0.100	0.050	6.25	49.71%
Sliding speed, m/s	2	0.061	0.031	3.88	30.43%
Percentage of fly ash, %	2	0.024	0.012	1.50	11.89%
Error	2	0.016	0.008		7.97%
Total	8	0.201			100.00%

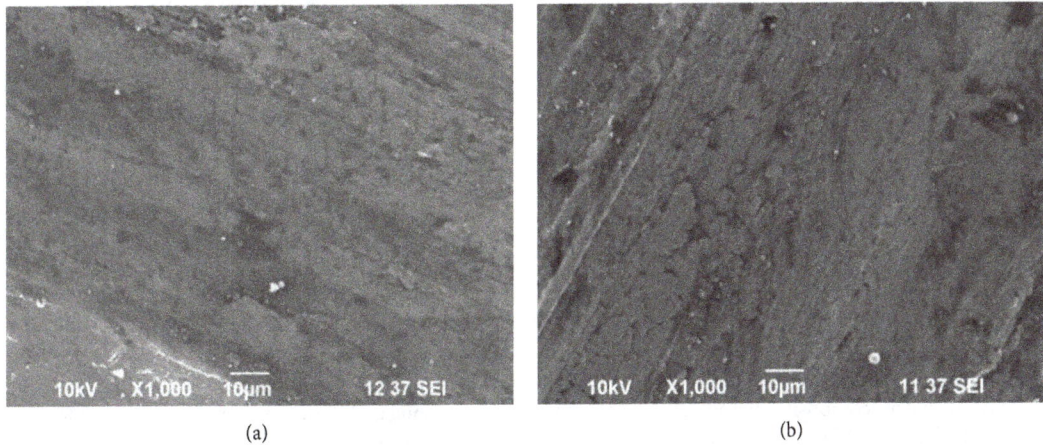

FIGURE 4: (a) SEM of worn surface with grooves. (b) SEM of worn surface with grooves and damages.

TABLE 10: Results of the confirmation experiment.

Initial design parameters	Optimal design parameters	
	Predicted	Experimental
Level	A1B1C1	A2B3C1	A2B3C1
Grey relational grade	0.630	0.870	0.893
Improvement in grade	—	0.240	0.263

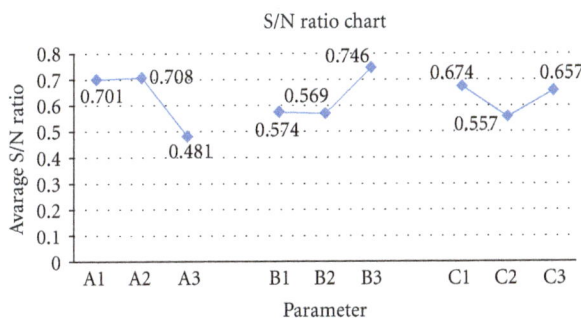

FIGURE 5: Main effects plot for S/N ratios (dB).

one parameter in a single sequence; one has to do the analysis repeatedly to obtain other factors for the experiment.

Since the experimental design is orthogonal, it is then possible to separate the effects of each process parameter at different levels. For example, the mean of grey relational grade for the applied load at level 1, 2, and 3 can be calculated by taking the average of the grey relational grade for the experiments 1–3, 4–6, and 7–9, respectively. The mean of

the grey relational grade for each level can be computed in similar manner. The mean of the grey relational grade for each level of the combining parameters is summarized in the multiresponse performance which is shown in Table 8. Figure 5 shows the main effects plot for S/N ratios (dB).

ANOVA of the response quality characteristics is shown in Table 9, and it is observed that applied load is the significant factor for minimizing the sliding wear, specific wear rate, and friction coefficient.

6. Confirmation Tests

The final step is to verify the improvement of the quality characteristic using optimal levels of the design parameters (A2B3C1). Table 10 shows a comparison of the predicted and the actual design parameters. According to Table 10, the improvement of the performance is noticed when the optimum conditions were used.

7. Conclusions

(i) Aluminum matrix reinforced with 5, 10, and 15 wt.% fly ash was successfully prepared by stir-casting process, and the wear behavior of the composites was investigated using pin-on-disc machine.

(ii) The $L_9(3)^4$ orthogonal array was adopted to investigate the effects of operating variables on the abrasive wear of various composites.

(iii) The experimental results show that the composites retain the wear resistance properties at lower loads with increase in flyash percentage. Mild wear was also observed in the composites as the sliding speed increases.

(iv) For all the trials it is observed that mild-to-severe wear exists, and it is witnessed by the microscopic results.

(v) The applied load and sliding speed are the most influencing factors, and it is observed that their contributions to wear behavior are 49.71% and 30.43%, respectively.

(vi) The optimum design parameters were predicted through Grey relational analysis (applied load = 19.62 N, sliding speed = 3 m/s, and percentage of flyash = 5 wt.%).

(vii) The confirmation experiment is conducted with the level A2B3C1 to verify the optimal design parameter, and it exhibits better wear performance.

References

[1] Y. Sahin, "Optimization of testing parameters on the wear behaviour of metal matrix composites based on the Taguchi method," *Materials Science and Engineering A*, vol. 408, no. 1-2, pp. 1–8, 2005.

[2] Y. Sahin, "Wear behaviour of aluminium alloy and its composites reinforced by SiC particles using statistical analysis," *Materials and Design*, vol. 24, no. 2, pp. 95–103, 2003.

[3] M. K. Surappa, S. V. Prasad, and P. K. Rohatgi, "Wear and abrasion of cast Al-Alumina particle composites," *Wear*, vol. 77, no. 3, pp. 295–302, 1982.

[4] C. Thiagarajan, R. Sivaramakrishnan, and S. Somasundaram, "Cylindrical grinding of SiC particles reinforced Aluminum Metal Matrix composites," *ARPN Journal of Engineering and Applied Sciences*, vol. 6, pp. 14–20, 2011.

[5] K. Kalaiselvan, N. Muruganand, and P. Siva, "Prod. and characterization of AA6061–B_4C stir cast composite," *Materials and Design*, vol. 32, pp. 4004–4009, 2011.

[6] Y. Sahin and K. Özdin, "A model for the abrasive wear behaviour of aluminium based composites," *Materials and Design*, vol. 29, no. 3, pp. 728–733, 2008.

[7] A. T. Alpas and J. Zhang, "Effect of SiC particulate reinforcement on the dry sliding wear of aluminium-silicon alloys (A356)," *Wear*, vol. 155, no. 1, pp. 83–104, 1992.

[8] O. Yilmaz and S. Buytoz, "Abrasive wear of Al_2O_3-reinforced aluminium-based MMCs," *Composites Science and Technology*, vol. 61, no. 16, pp. 2381–2392, 2001.

[9] D. Mandal, B. K. Dutta, and S. C. Panigrahi, "Dry sliding wear behavior of stir cast aluminium base short steel fiber reinforced composites," *Journal of Materials Science*, vol. 42, no. 7, pp. 2417–2425, 2007.

[10] P. K. Rohatgi, B. F. Schultz, A. Daoud, and W. W. Zhang, "Tribological performance of A206 aluminum alloy containing silica sand particles," *Tribology International*, vol. 43, no. 1-2, pp. 455–466, 2010.

[11] B. K. Prasad, S. V. Prasad, and A. A. Das, "Abrasion-induced microstructural changes and material removal mechanisms in squeeze-cast aluminium alloy-silicon carbide composites," *Journal of Materials Science*, vol. 27, no. 16, pp. 4489–4494, 1992.

[12] B. N. P. Bai, B. S. Ramasesh, and M. K. Surappa, "Dry sliding wear of A356-Al-SiCp composites," *Wear*, vol. 157, no. 2, pp. 295–304, 1992.

[13] K. J. Bhansali and R. Mehrabian, "Abrasive wear of aluminum-matrix composites," *Journal of Metals*, vol. 34, no. 9, pp. 30–34, 1982.

[14] C. S. Lee, Y. H. Kim, K. S. Han, and T. Lim, "Wear behaviour of aluminium matrix composite materials," *Journal of Materials Science*, vol. 27, no. 3, pp. 793–800, 1992.

[15] A. P. Sannino and H. J. Rack, "Dry sliding wear of discontinuously reinforced aluminum composites: review and discussion," *Wear*, vol. 189, no. 1-2, pp. 1–19, 1995.

[16] Sudarshan and M. K. Surappa, "Dry sliding wear of fly ash particle reinforced A356 Al composites," *Wear*, vol. 265, no. 3-4, pp. 349–360, 2008.

[17] Sudarshan and M. K. Surappa, "Synthesis of fly ash particle reinforced A356 Al composites and their characterization," *Materials Science and Engineering A*, vol. 480, no. 1-2, pp. 117–124, 2008.

[18] H. K. Durmuş, E. Özkaya, and C. Meriç, "The use of neural networks for the prediction of wear loss and surface roughness of AA 6351 aluminium alloy," *Materials and Design*, vol. 27, no. 2, pp. 156–159, 2006.

[19] W. Zhou and Z. M. Xu, "Casting of SiC reinforced metal matrix composites," *Journal of Materials Processing Technology*, vol. 63, no. 1–3, pp. 358–363, 1997.

[20] A. R. Ahamed, P. Asokan, and S. Aravindan, "EDM of hybrid Al-SiCp-B4Cp and Al-SiC p-Glassp MMCs," *International Journal of Advanced Manufacturing Technology*, vol. 44, no. 5-6, pp. 520–528, 2009.

[21] M. Ramachandra and K. Radhakrishna, "Effect of reinforcement of flyash on sliding wear, slurry erosive wear and corrosive behavior of aluminium matrix composite," *Wear*, vol. 262, no. 11-12, pp. 1450–1462, 2007.

[22] P. Shanmughasundaram, R. Subramanian, and G. Prabhu, "Some studies on aluminium—flyash composites fabricated by two step stir casting method," *European Journal of Scientific Research*, vol. 63, pp. 204–218, 2011.

7

A Two-Scale Approach for Lubricated Soft-Contact Modeling: An Application to Lip-Seal Geometry

Michele Scaraggi[1] and Giuseppe Carbone[2]

[1] DII, Universitá del Salento, 73100 Monteroni di Lecce, Italy
[2] DMMM, Politecnico di Bari, 70126 Bari, Italy

Correspondence should be addressed to Michele Scaraggi, michele.scaraggi@unisalento.it

Academic Editor: Michel Fillon

We consider the case of soft contacts in mixed lubrication conditions. We develop a novel, two scales contact algorithm in which the fluid- and asperity-asperity interactions are modeled within a deterministic or statistic scheme depending on the length scale at which those interactions are observed. In particular, the effects of large-scale roughness are deterministically calculated, whereas those of small-scale roughness are included by solving the corresponding homogenized problem. The contact scheme is then applied to the modeling of dynamic seals. The main advantage of the approach is the tunable compromise between the high-computing demanding characteristics of deterministic calculations and the much lower computing requirements of the homogenized solutions.

1. Introduction

Compliant contacts, most commonly known as soft-contacts, are very common in nature (e.g., cartilage lubrication, eye-eyelid contact) and technology (e.g., tires, rubber sealings, adhesives). It has long been stated that the friction and fluid leakage characteristics of wet soft-contacts are strongly related, among the other factors, to the local interactions occurring at the contact interface [1–5]. In the case of randomly rough surfaces, the basic understanding of the role played by the asperity-asperity and fluid-asperity interactions, occurring over a wide range of roughness length-scales, has been largely investigated and debated in the very recent scientific literature [6–12]. Given the (usual) fractal nature of random roughness, a number of interesting phenomena have been highlighted, as for example, the viscous-hydroplaning [6], the viscous flattening [9–15], the fluid-induced roughness anisotropic deformation [10, 11], the local [10, 11] and global [8, 16] fluid entrapment, and many others. The way to deal with random roughness contact mechanics, despite being nontrivial and suffering of a certain description fragmentation, is however well described in the current scientific literature.

On the other side, nowadays bio-inspired research [17, 18], together with the widely-spreading practice of surface engineering [19], is showing the many (mainly unexplored) opportunities offered by the physical-chemical ordered modification of surfaces in order to tailor targeted macroscopic contact characteristics, such as adhesion and friction. Bio-inspired adhesive research [20] is probably the best state of the art example of such research trend. However, investigating the combined effect of, let us say, quantized roughness and fluid action has not equally attracted the scientific community attention, apart from few experimental investigations [12, 21, 22] and basic theoretical investigations [23, 24]. This may be justified by the complexity of the numerical formulation of the problem, which is expected to not to present an analytical treatment. As a result, and to the best of authors' knowledge, the combined effect of lubricant action and single-scale (contact splitter) roughness has not been practically investigated by tribologists' community.

In this work we give our personal contribution to this research field and, in particular, we discuss a novel numerical scheme of soft mixed lubrication, which can be adopted to perform such investigations. In our model the roughness is split into two contributions. A threshold scale

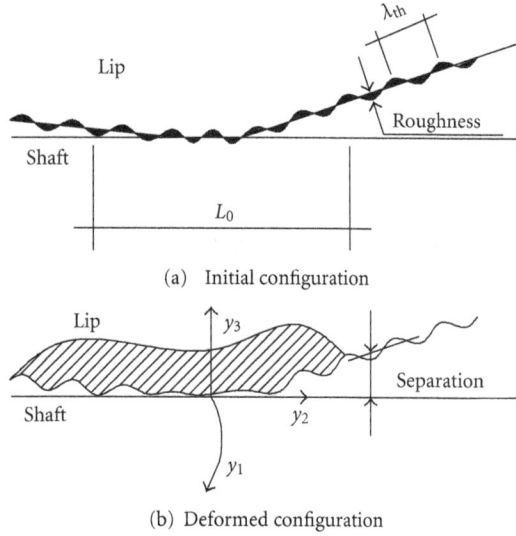

(a) Initial configuration

(b) Deformed configuration

FIGURE 1: Contact scheme: (a) reference geometry; (b) deformed geometry obtained by solving the fluid/solid contact problem.

$\zeta_{th} = L_0/\lambda_{th} > 1$ (where L_0 is the representative macroscopic size of the contact and λ_{th} is the threshold roughness length-scale) is identified. For length scales $\lambda = L_0/\zeta > \lambda_{th}$, that is $\zeta < \zeta_{th}$, the system is investigated by using a deterministic approach, whereas for $\lambda = L_0/\zeta \leq \lambda_{th}$, that is, $\zeta \geq \zeta_{th}$, the problem is treated by homogenizing the equations. This is of utmost help in performing numerical simulations, as for example in the case of microstructured or bio-inspired surfaces where the surface geometry roughness is characterized by a single scale texture (e.g., a pillars array) combined with random roughness at sub-micrometer scales. The main advantage of the proposed approach is, therefore, the strong reduction of numerical complexity. The paper is organized as follows. In Section 2 we describe the mixed lubrication model scheme (we refer to appendix for the details of numerics), whereas in Section 3 we report an example of application of the proposed model to the case of a lip sealing geometry operating in steady-sliding contact. For a detailed description on dynamic sealings modeling, the reader is referred to [25–29].

2. Problem Formulation

Here we consider a generic rough compliant solid in steady sliding contact with a rigid smooth counter surface, as shown in Figure 1(a). Due to the coupled action of asperity-fluid and asperity-asperity interactions, the compliant surface gains the actual (or deformed) configuration (Figure 1(b)). In soft-contacts, the shape difference between the deformed and initial configurations cannot be usually neglected, due to the occurrence of large deformations. Nevertheless, the small displacement assumption is usually adopted in the literature of soft-elastohydrodynamics [6, 30, 31] (soft-EHL), supported by a relevant experimental validation for ball-on-flat contact geometries. However, if one analyses the contact at much shorter length scales, that is, at the asperities

length scales, then one finds out that the asperities may be completely flattened because of the high contact and fluid pressures. Hence, neglecting the influence of large deformation would lead to strong inaccuracy in describing the evolution of the system at the micro-scales. For this reason, in this work, the small displacement assumption has been relaxed. Moreover, in some cases, for example, for dynamic sealings modeling, the precise calculation of the large tangential displacement of the rubber at the interface of the sliding contact is a must in order to capture key phenomena as the well known reverse pumping effect [26].

Consider now the schematic drawing of Figure 1(a). We assume that the whole texture belongs to the class of Reynolds roughness, that is, with $\langle |\nabla h|^2 \rangle \ll 1$ (where h is the surface height distribution). In such a case, the thin film lubrication formulation is sufficiently accurate in describing the fluid dynamics at the sliding interface at all roughness lengths. The (ensemble) average local separation, $\bar{u}(y_1, y_2)$ is calculated as:

$$\bar{u}(y_1, y_2) = u_r(x_1, x_2) + h_r(x_1, x_2) + w_3(x_1, x_2),$$
$$x_1 = y_1 - w_1(x_1, x_2), \qquad (1)$$
$$x_2 = y_2 - w_2(x_1, x_2),$$

where x_i and y_i (the subscript $i = 1$ or 2 or 3, see Figure 1(b)) are the three coordinates of the lip surface points at the initial (x_i) and at the deformed (y_i) state respectively. The quantity u_r represents the macroscopic contact shape and h_r the undeformed low-pass filtered (i.e., for $\lambda > \lambda_{th}$) surface roughness. w_i is the generic component of the displacement vector describing the average local surface deformation of the compliant body. In the classical EHL approach the displacement w_i is calculated within linear elasticity framework [31], by adopting a boundary element approach which requires $O(n^2)$ operation where n is the number of discretization points. In our case, however, we adopt a more general nonlinear rheology and the displacement components

$$w_i = f_i(\sigma_{jk}) \qquad (2)$$

are calculated by employing a classical finite-element solver, which requires $O(n^3)$ operations. Observe that the normal (locally averaged) stress $\sigma_3(\mathbf{y})$ is (see also [6])

$$\sigma_3(\mathbf{y}) = \sigma_{f3}(\mathbf{y}) + \sigma_{s3}(\mathbf{y}), \qquad (3)$$

where $\mathbf{y} = (y_1, y_2, y_3)$, $\sigma_{f3}(\mathbf{y})$ is the locally averaged fluid pressure and $\sigma_{s3}(\mathbf{y})$ the locally average solid contact pressure, the latter coming from the asperity-asperity interactions occurring at length-scales $\lambda < \lambda_{th}$. We observe that [6]:

$$\tau_1(\mathbf{y}) = \tau_{s1}(\mathbf{y}) \frac{A_c(\mathbf{y})}{A_0} + \tau_{f1}(\mathbf{y}),$$
$$\tau_2(\mathbf{y}) = \tau_{s2}(\mathbf{y}) \frac{A_c(\mathbf{y})}{A_0} + \tau_{f2}(\mathbf{y}). \qquad (4)$$

where $A_c(\mathbf{y})/A_0$ is the local normalized area of solid contact, where τ_f is the average fluid shear stress (see later for more

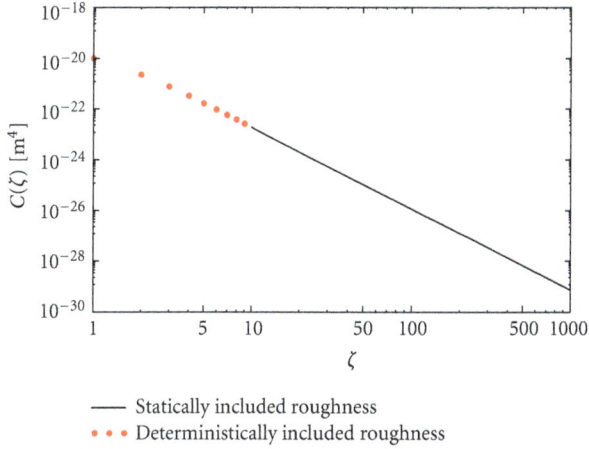

FIGURE 2: A self-affine power spectral density of a generic surface roughness. Points correspond to that spectral content included deterministically in the calculations, whereas the continuous line represents the homogenized surface roughness.

details). We assume that the solid friction shear stress τ_s is constant (and directed along the sliding direction), as it happens in the case for a rubber-inert substrate contact [6]. Clearly, the model can be easily extended to include different boundary friction conditions.

The relation between the average solid contact pressure σ_{s3} and the local average interfacial separation $\bar{u}$ is obtained from Persson's theory of contact mechanics [32, 33]. In particular, given the local elastic properties of the material and the power spectral density (PSD, see, e.g., Figure 2) $C(\mathbf{q}) = (2\pi)^{-2} \int d^2 x \langle h(\mathbf{0})h(\mathbf{x})\rangle e^{-i\mathbf{q}\cdot\mathbf{x}}$ of the surface (h is the surface roughness field, with average value $\langle h(\mathbf{x})\rangle = 0$, $q_0 = 2\pi/\lambda_0$, where λ_0 is the low frequency cut-off wavelength), the theory allows to calculate the local areal fraction of solid-solid contact $\alpha_c = A_c/A_0$ as a function of the locally averaged solid-solid contact pressure σ_{s3} as

$$\alpha_c(\mathbf{y}) = \text{erf}\left[\frac{\sqrt{2}}{E^*\left\langle[\nabla h^{\text{sto}}(\mathbf{x})]^2\right\rangle^{1/2}}\sigma_{s3}(\mathbf{y},)\right], \quad (5)$$

where the quantity $\langle[\nabla h(\mathbf{x})]^2\rangle^{1/2}$ is the root mean square gradient of the rough surface and is related to the PSD $C(\mathbf{q})$ of the rough surface through the formula $\langle[\nabla h(\mathbf{x})]^2\rangle = \int d^2 q\, q'^2 C(\mathbf{q}')$. The relation between the local interfacial separation $\bar{u}$ and the solid-solid contact pressure is instead determined, in the adhesionless case, by requiring that the change of elastic energy per unit nominal contact area dU_{el} at the interface equals the work done by the contact pressure to deform the elastic body:

$$dU_{\text{el}} = -\sigma_{s3}d\bar{u},$$
$$\bar{u} = \int_{\sigma_{s3}}^{+\infty}\frac{1}{\sigma}\frac{dU_{\text{el}}}{d\sigma}d\sigma. \quad (6)$$

In (6) the elastic energy per unit contact area U_{el} is calculated as a function of the contact pressure σ_{s3} (see [32, 33] for more

details). The final formula linking the local separation $\bar{u}$ and the solid-solid contact pressure is relatively complicated, but at large separation it simplifies and becomes [33]

$$\sigma_{s3}(\bar{u}) \approx \beta_1 E^* \exp\left(-\frac{\bar{u}}{\beta_2}\right), \quad (7)$$

where the quantities β_1 and β_2 can be easily calculated once the PSD of the rough surface is known [33], and $E^* = E/(1-\nu^2)$.

We consider now the case where a Newtonian fluid is sandwiched at the interface between the solids. We assume constant viscosity η, constant density ρ, and isothermal conditions. The adoption of a Reynolds roughness, and the assumption of a representative average interfacial separation value $\bar{u}/L_0 \ll 1$, suggests that the fluid velocity varies slowly with the coordinates y_1 and y_2 compared to the variation in the orthogonal direction y_3. In such a case, combining the equilibrium with the continuity equation, the homogenized problem formulation reads [9, 34]

$$-\frac{\partial\bar{u}(\mathbf{x},t)}{\partial t} = \nabla\cdot\left(\boldsymbol{\sigma}_{\text{eff}}'\mathbf{U}_m - \boldsymbol{\sigma}_{\text{eff}}\nabla\bar{\sigma}_{f3}\right), \quad (8)$$

where, as before, $\bar{u}$ and σ_{f3} are the locally averaged (over a length scale given by the longest wavelength surface roughness component, i.e., λ_{th}) interfacial surface and fluid pressure, respectively. The fluid flow conductivity (tensor) $\boldsymbol{\sigma}_{\text{eff}}$ and $\boldsymbol{\sigma}_{\text{eff}}'$ can be related, respectively, to the pressure flow factor (tensor) [7, 9, 35, 36] $\boldsymbol{\phi}_p = 12\eta\bar{u}^{-3}\boldsymbol{\sigma}_{\text{eff}}$ and to the shear flow factor (tensor) $\boldsymbol{\phi}_s = \bar{u}^{-1}\boldsymbol{\sigma}_{\text{eff}}'$, which we assume to be a function of the average interfacial separation $\bar{u}$ only. Here $\boldsymbol{\phi}_p$ has been calculated on the basis of the Bruggeman' effective medium theory and on the Persson's contact mechanics [34]. Within this approach, the effect of solid contacts percolation, as well as of local fluid trapping, on the fluid flow can be taken approximately into account, as recently discussed in [16]. For simplicity, we have assumed $\boldsymbol{\phi}_p = \phi_p\mathbf{I}$, and $\boldsymbol{\phi}_s = \mathbf{I}$, where $\mathbf{I}$ is the identity tensor. Moreover, by considering the steady sliding condition, the homogenized lubricant equation simplifies into:

$$6\eta U\frac{\partial\bar{u}}{\partial y_1} = \frac{\partial}{\partial y_1}\left(\bar{u}^3\phi_p\frac{\partial\sigma_{f3}}{\partial y_1}\right) + \frac{\partial}{\partial y_2}\left(\bar{u}^3\phi_p\frac{\partial\sigma_{f3}}{\partial y_2}\right), \quad (9)$$

where $U = 2U_m$ is the sliding velocity.

However, in order to include the effect of micro-cavitations occurring at large-scale roughness lengths, a JFO cavitation model (constant mixture pressure in the cavitation zone) is adopted. The Reynolds equation is then reformulated under a mass conservative equation valid throughout the cavitating/not cavitating domain:

$$\frac{6\eta U}{G}\frac{\partial\bar{u}[1 + (1-F)\phi]}{\partial y_1} = \nabla\cdot\left[F\phi_p\bar{u}^3\nabla(F\phi)\right], \quad (10)$$

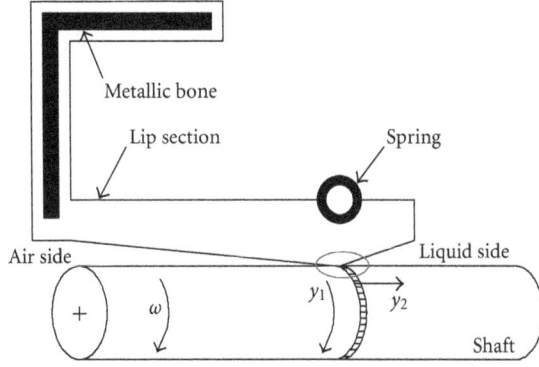

FIGURE 3: A schematic of a typical lip seal construction. A garter spring enables the radial compression of the seal lip on the shaft surface, ideally preventing fluid leakage from the high pressure side (the fluid reservoir) to the low pressure side. In the running-in stage, the seal lip is macroscopically reshaped as a consequence of the initial lip wear process, which depends on the seal-shaft designed interference and on the solids chemical and physical affinity with the actual lubricant composition. After this bulk material removal (and depending on the viscoelastic rubber properties), a successful seal reaches a steady-state configuration with a well defined roughness pattern, strongly dependent on the actual shaft micro-geometry, and with an established thin lubricant film at the interface.

where we have adopted the cavitation index F ($F = 0$ in the cavitation zones, and $F = 1$ otherwise) and the dummy variable ϕ defined as

$$\sigma_{f3} = G\phi F,$$

$$\theta = \frac{\rho - \rho_0}{\rho_0} = \phi(1 - F), \tag{11}$$

where ρ_0 is the lubricant density, and G is the representative solid shear elastic modulus. Note that (10) must be solved on the deformed configuration, which is an unknown of the problem: therefore, reformulating the fluid equation in the initial configuration may result numerically convenient. Thus (10) has been rephrased by using the mapping rule

$$x_1 = y_1 - w_1(x_1, x_2),$$

$$x_2 = y_2 - w_2(x_1, x_2). \tag{12}$$

The Jacobian $\mathbf{J} = \partial(x_1, x_2)/\partial(y_1, y_2)$ of the transformation can be calculated as the inverse $\mathbf{J} = \mathbf{g}^{-1}$ of the Jacobian $\mathbf{g} = [g_{ij}] = \partial(y_1, y_2)/\partial(x_1, x_2)$ of the transformation

$$y_1 = x_1 + w_1(x_1, x_2),$$

$$y_2 = x_2 + w_2(x_1, x_2). \tag{13}$$

Observe that the determinant of the Jacobian tensors must be necessarily larger than zero, that is, $\det \mathbf{g} = (\det \mathbf{J})^{-1} > 0$. The above transformation enables the generation of an adaptive mesh grid which follows the changing of the surface shape during the deformation process, thus without loosing spatial resolution. The detailed numerics derivation is reported in the Appendix.

The boundary conditions to be applied in the resolution of (10) depend, clearly, on the particular soft-contact problem under investigation. The dimensionless formulation of (10) is

$$\tilde{U}\frac{\partial \overline{u}[1 + (1 - F)\phi]}{\partial y_1} = \nabla \cdot \left[F\phi_p\overline{u}^3\nabla(F\phi)\right], \tag{14}$$

where $\tilde{U} = 6\mu U\lambda_0/(2Gh_{\mathrm{rms}}^2)$ and $\alpha = 2h_{\mathrm{rms}}/\lambda_0 = q_0 h_{\mathrm{rms}}/\pi$ (note that $\overline{u}$ has been made dimensionless with h_{rms}, whereas other lengths with λ_0). λ_0 corresponds in this case to the largest deterministic roughness wavelength.

3. Results

In this work we use the proposed model to investigate the lubricated contact of the lip seal schematically shown in Figure 3. Moreover, being λ_0 much smaller than the shaft diameter, it is possible to reduce the computational domain to only a small angular fraction of the lip surface. Therefore, (10) is solved with the constant pressure boundary conditions at the low pressure side $p = p_{\mathrm{env}}$ and at the high pressure side $p = p_{\mathrm{oil}}$, and with the periodicity conditions on the circumferential y_1-direction, with spatial period λ_0. The two-scale mixed lubrication model has been numerically solved, as described in the Appendix. We have opted for a fractal self-affine isotropic geometry. For any self-affine fractal surface the statistical properties are invariant under the transformation

$$\mathbf{x} \longrightarrow t\mathbf{x}; \qquad h \longrightarrow t^H h. \tag{15}$$

In such a case it can be shown that for isotropic surface the PSD is

$$C(q) = C_0\left(\frac{q}{q_0}\right)^{-2(H+1)}, \tag{16}$$

where $q_0 = 2\pi/\lambda$, $q = |\mathbf{q}|$, H is the Hurst exponent of the randomly rough profile, which is related to the fractal dimension $D_f = 3 - H$. To apply our method we need to numerically generate the surface $h_r(\mathbf{x})$ in the low frequencies spectrum, that is, for $\zeta > \zeta_{\mathrm{th}}$, see, for example, Figure 2. To this end we have utilized the spectral method described in [37] where the roughness is described by a periodic surface in the form of Fourier series

$$h_r(\mathbf{x}) = \sum_{hk=-\zeta_{\mathrm{th}}}^{\zeta_{\mathrm{th}}} a_{hk}e^{i\mathbf{q}_{kh}\cdot\mathbf{x}}, \tag{17}$$

where $\mathbf{q}_{kh} = (kq_0, hq_0)$, $\mathbf{x} = (x, y)$. Since $h_r(\mathbf{x})$ is real we must have $a_{-h,-k} = \overline{a_{hk}}$. Moreover for randomly rough surfaces the following relation must be satisfied $\langle a_{hk}a_{lm}\rangle = 0$ with $l \neq \pm h$, $m \neq \pm k$, where the symbol $\langle \cdots \rangle$ is the ensemble average operator. The PSD of Surface Equation (17) is

$$C(\mathbf{q}) = \sum_{hk=-\zeta_{\mathrm{th}}}^{\zeta_{\mathrm{th}}} \left\langle |a_{hk}|^2\right\rangle\delta(\mathbf{q} - \mathbf{q}_{hk}), \tag{18}$$

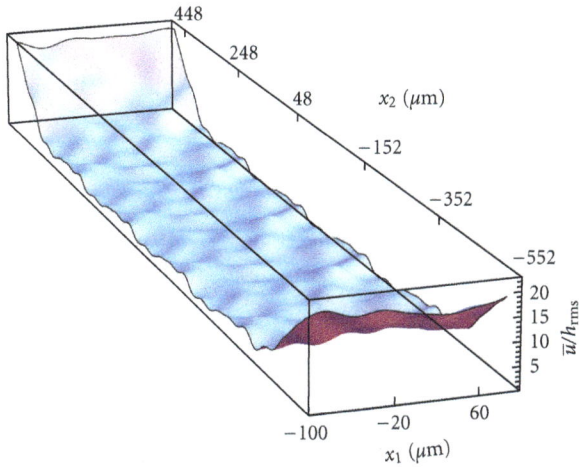

FIGURE 4: Dimensionless average interfacial separation $\bar{u}/h_{\mathrm{rms}}$ as a function of contact position, reported in the undeformed reference.

from which it follows:

$$C(\mathbf{q}_{hk}) = \left\langle |a_{hk}|^2 \right\rangle \delta(\mathbf{0}). \qquad (19)$$

For isotropic surfaces we have $C(\mathbf{q}) = C(q)$ which simply gives $C(\mathbf{q}_{hk}) = C(q_0\sqrt{h^2 + k^2})$ and assuming self-affine fractal surface (see (16)) one obtains

$$\left\langle |a_{hk}|^2 \right\rangle = \left\langle |a_{11}|^2 \right\rangle \left(\frac{h^2 + k^2}{2} \right)^{-H-1}. \qquad (20)$$

Hence the quantities $\langle |a_{hk}|^2 \rangle$ can be determined once $\langle |a_{11}|^2 \rangle$ and the Hurst exponent of the fractal surface are known. However to completely characterize the rough profile we still need the probability distribution of the quantities a_{hk}. We first observe that the condition $\langle a_{hk} a_{lm} \rangle = 0$ with $l \neq \pm h$, $m \neq \pm k$ is satisfied if the phases φ_{hk} of the complex quantities a_{hk} are random numbers uniformly distributed between 0 and 2π. We also recall the condition $a_{-h,-k} = \overline{a_{hk}}$ also implies that $|a_{-h,-k}| = |a_{h,k}|$ and that the quantities $\varphi_{-h-k} = -\varphi_{hk}$. So what we need now is only the probability distribution of $|a_{h,k}|$. Of course there are several choices and the simplest one is to assume that the probability density function of $|a_{hk}|$ is just a Dirac's delta function centered at $\langle |a_{hk}|^2 \rangle^{1/2}$, that is,

$$P(|a_{hk}|) = \delta\left(|a_{hk}| - \left\langle |a_{hk}|^2 \right\rangle^{1/2} \right). \qquad (21)$$

It can be shown that this choice guarantees also that the random profile $h(\mathbf{x})$ has a Gaussian random distribution.

3.1. Sealing at Nearly Zero Sliding Velocity.

Here we report on the model application to the case where the sliding velocity between the two mating surfaces is vanishing ($\tilde{U} \to 0$), that is, for the case of static seals. The self-affine roughness of the lip surface presents a long-distance cut-off frequency $q_0 = 2\pi/200\,\mu\text{m}^{-1}$, fractal dimension $D_f = 2.4$, $\langle h^2 \rangle^{1/2} = 10\,\mu\text{m}$, and small scale cut-off frequency $q_1 = 1000q_0$. The threshold

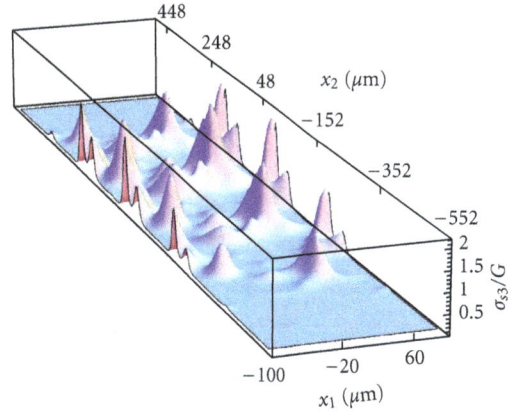

(a) Average solid contact pressure σ_{s3}

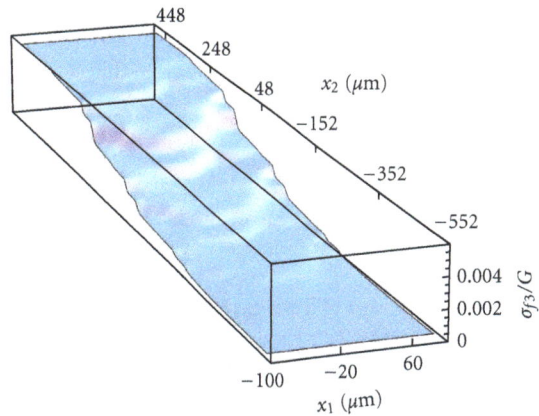

(b) Average fluid pressure σ_{f3}

FIGURE 5: Average fluid and solid contact pressure for the static linear elastic seal.

frequency adopted in the calculation is $q_{\mathrm{th}} = 2\pi/\lambda_{\mathrm{th}} = 5q_0$. In the following, the seal is assumed linear elastic. The calculation is performed at a constant rigid penetration of the lip. In Figure 4 we show the locally averaged interfacial separation field $\bar{u}$ (note that x_1 is the circumferential direction, whereas x_2 is the axial direction). Interestingly the corrugation which can be observed in the figure is just a consequence of the deterministically included roughness ($\zeta < \zeta_{\mathrm{th}}$), at larger frequency the surface appears smooth since the high frequency ($\zeta < \zeta_{\mathrm{th}}$) content of the roughness has been included through Persson's statistic model, that is, by means of homogenization. The corresponding average solid contact pressure is shown in Figure 5(a). Observe first that, due to the presence of roughness, the contact is split into many contact patches in the whole lip apparent contact area and, correspondingly, normal stresses are concentrated into contact spots. Observe also that the average solid contact pressure is smooth at the contact borders, differently, from what instead expected in the case of smooth elastic contact (e.g., in the case of the Hertzian contact). This is due to the homogenized roughness contribution, which distributes, on a wider local contact area, the pressure acting on the single deterministic asperity. The average fluid

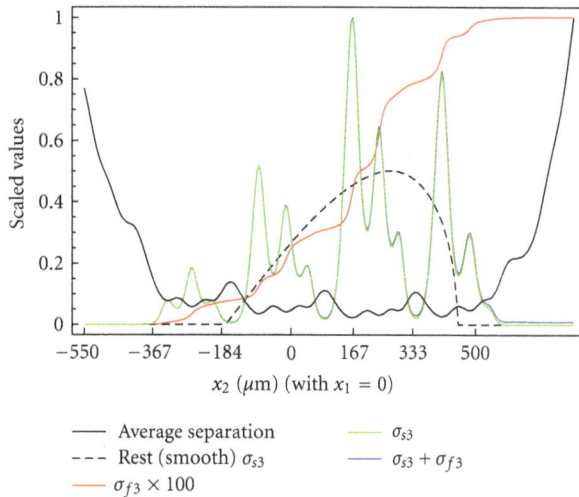

FIGURE 6: All solutions in arbitrary scaling.

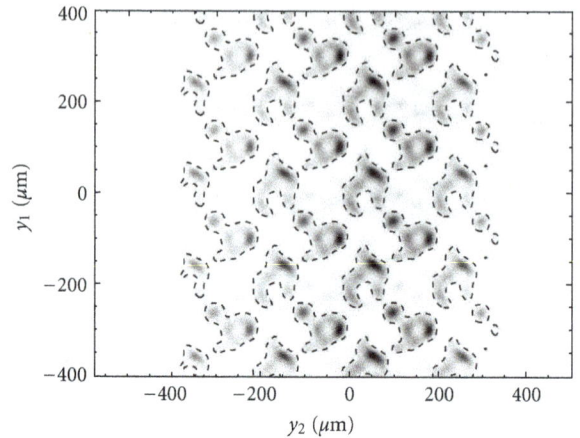

(a) Normalized solid contact area A_c/A_0

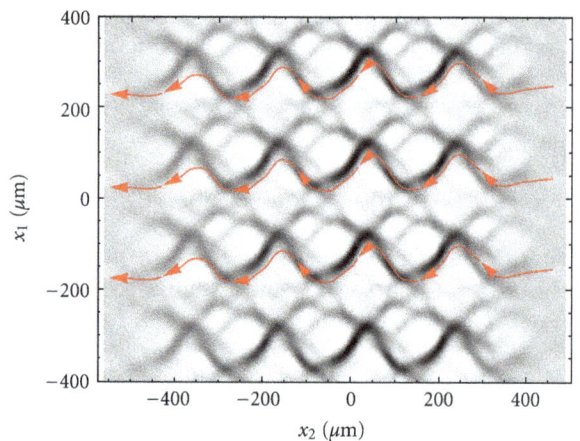

(b) Sample of fluid flux lines

FIGURE 7: Static sealing effectiveness.

pressure is shown in Figure 5(b). Observe that the fluid pressure presents steep variations (almost step-like) in those locations where the separation between the surfaces takes the smallest values. This is in perfect agreement with the critical junction theory of leakage in seals [8], which predicts that the hydraulic resistivities are only concentrated in few points where the fluid flow encounters strong restrictions. At these restrictions the pressure must present high gradients as clearly shown in Figures 5(b) and 6. In particular, in Figure 6 we report the average interfacial separation (black curve), the average solid contact pressure (green curve), the average fluid pressure (red curve), the total average normal stress (blue curve, slightly distinguished from the average solid contact pressure), and the solid contact pressure in the case of perfectly smooth contact (black dashed curve, useful for comparison), for $x_1 = 0$ (i.e., along the axial direction). Note that the largest pressure gradients occur in the very proximity of the largest values of σ_{s3}, that is, where the local average separation takes its minimum value. It is also interesting to observe the normalized solid contact area A_c/A_0 for this static contact case, see Figure 7(a). Note that, due to the high local squeezing pressure, the asperities low-frequency features (i.e., the roughness asperities described by the deterministical model) have been squeezed so much to coalesce in larger contact patches. Figure 7(b) shows a sample of the leakage flux lines calculations (red lines) superposed to the locally averaged fluid velocity intensity (in gray scale). As expected [8], the presence of asperity-asperity contacts increases the hydraulic resistivity at the interface and, in particular, largest values of fluid velocity occur at the local minimum in the average interfacial separations.

We have also carried out calculations assuming the seal obeys a Mooney-Rivlin model. The results are then compared with the elastic case. In Figure 8 we show the normalized area of real contact for the elastic (Figure 8(a)) and hyperelastic (Figure 8(b)) bulk rheology. Observe that for the linear case, the area of contact are slightly larger, as expected, then for the other case. This however, does not

strongly affect the leakage flow for the current geometry, as shown in Figure 9. However, interestingly, the nonlinear rheology is characterized by a larger value of surface area change, evaluated as $\partial w_1/\partial x_1 + \partial w_2/\partial x_2$ (see Figure 10), where we show in red colors that on part of the interaction domain there is a shrinking, that is, $\partial w_1/\partial x_1 + \partial w_2/\partial x_2 < 0$. Observe that in plane surface displacements are responsible for the frequency-shift of roughness PSD, for example, in the case of Figure 10(b) we expect a relevant modification of the homogenized power spectral density shape, which can therefore affect the local contact mechanics and flow factors. In the literature, surface displacements are not usually taken into account; however the present study shows that such a surface effect can be relevant for contact modeling.

3.2. Sealing at Non-Zero Sliding Velocity. In this section we show the case of non-zero sliding velocity. The macroscopic lip geometry is the same as before, as well as the roughness PSD. However, this time the part of roughness that is described deterministically is characterized by only one length scale, that is, just one frequency is included. The remaining roughness has been therefore included within the homogenized approach. In Figure 11 we show the cavitation

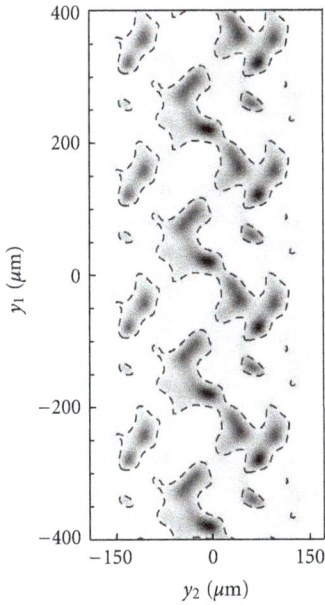

(a) Linear elastic

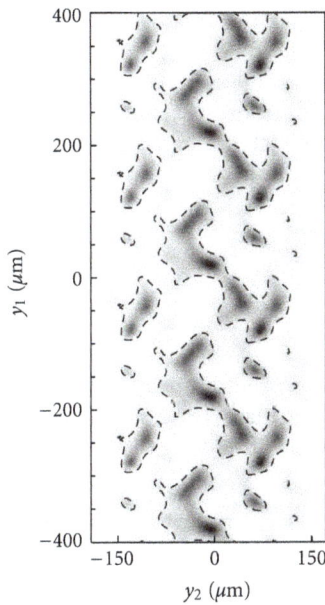

(b) Mooney-Rivlin

FIGURE 8: Normalized area of real contact for different bulk rheologies. Dashed contours represent the bearing area paths.

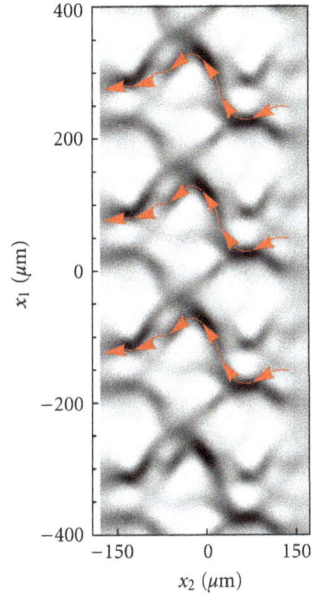

(a) Linear elastic

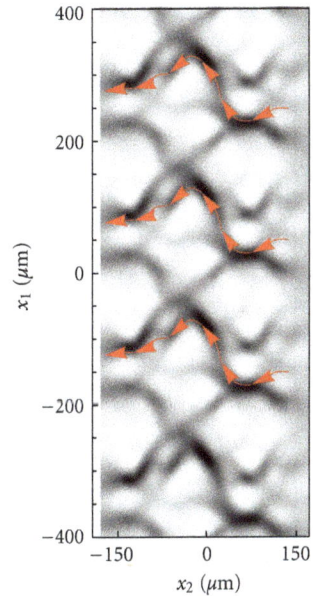

(b) Mooney-Rivlin

FIGURE 9: Sample of leakage flux lines.

areas (black areas), occurring over a circumferential portion of the lip-shaft macroscopic contact domain. Results are shown at different values of dimensionless sliding velocity. Interestingly, cavitation originates as expected in the low pressure side of the region (i.e., on the left side of the domain, whereas sliding velocity is directed from top to the bottom) and at the trailing edge of asperities. By increasing the sliding velocity, the cavitation extends from the low to the high pressure side. The cavitated areas may even coalesce at the largest sliding velocity, with the formation of cavitation

fingers shown in Figure 11(e). The adoption of our two-scale approach allows then to capture the complex solid contact and fluid-dynamics characteristics of real contact geometry, which can help engineers to have useful insights into the mixed-EHL lubrication condition occurring at the interface of soft-contacts, especially in terms of friction and leakage, with much less computation effort. Figure 12 shows the axially component of fluid flow at the interface. Red areas corresponds to the counter-gradient flow (i.e., the flow directed in the opposite direction compared to the

(a) Linear elastic

(b) Mooney-Rivlin

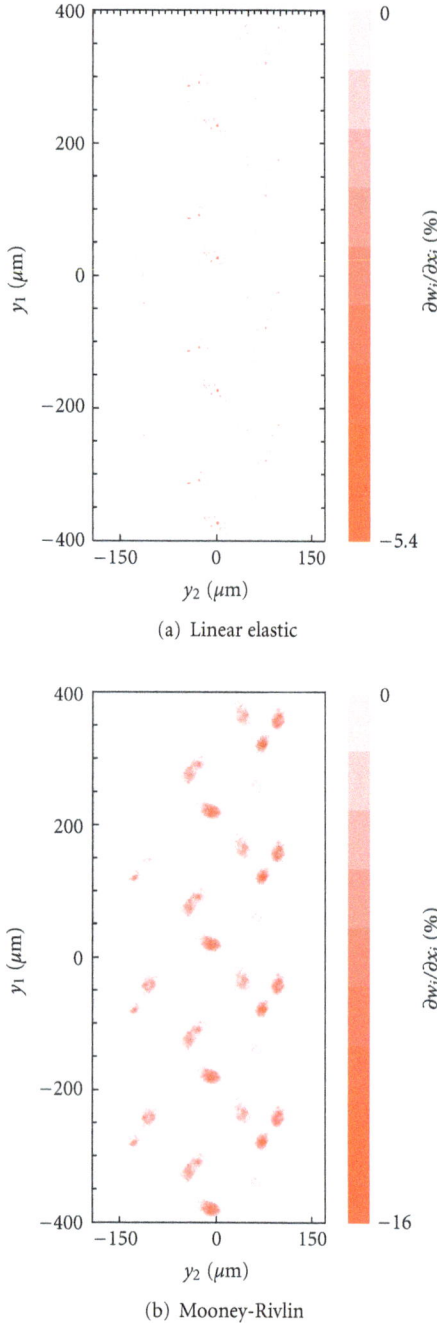

FIGURE 10: In-plane area shrinking.

externally applied fluid pressure gradient). Observe that, on the high-pressure side, due to the asperities presence, a certain fluid recirculation is observed which tends to hamper the observed net leakage (blue fingers in the low-pressure side). The recirculation is due to the effect of fluid depressurization which occurs at the divergent part of the single asperity/substrate interfaces. This depressurization determines a relevant fluid suction from the high pressure side which is partially balanced by the flow induced by the counter-pressure gradient.

4. Conclusions

We have presented a novel two-scale approach for the description of the mixed lubrication regime for real soft-contacts. We modeled the asperity-asperity and asperity-fluid interactions with a deterministic or a statistical approach (DSA) depending on length scale at which the contact region is observed. The roughness at large length scales, which mainly determines the fluid flow at the interface, is deterministically included in the model while the roughness at short wavelengths, which strongly contributes only to the friction, is included by means of a homogenization process (recently developed in [9]). We have applied the DSA to lip sealings contact mechanics modeling, and we have analyzed the mixed lubrication characteristics at nearly-zero and non-zero shaft sliding velocity. In the case of nearly-zero sliding velocity the lip seal behaves as a classical static seal, and we showed that the fluid flow at the interface is determined only at the smallest constrictions along the leakage path, in agreement with recent developments of static seals theory. In the case of non-zero sliding velocity, we showed the occurrence of micro-cavitations and cavitation fingers, whereas leakage has been shown to be associated to asperity-induced fluid suction at the high pressure side (for the given geometry). Finally, we note that DSA-based mixed lubrication models, which belong to the class of multiscale contact mechanics models, provide a high-resolution description of very complex contact problems with a reduced or, at least, tunable computational effort, opening the perspective for its application in general purpose engineering software.

Appendix

Equation (10) has been discretized with the control volume approach. In particular, (10) can be integrated in a portion of the contact area considering that the elementary area

$$dA_D = dy_1 dy_2 = \left| \frac{\partial(y_1, y_2)}{\partial(x_1, x_2)} \right| dx_1 dx_2 = |\det \mathbf{g}| dA \tag{A.1}$$

gives

$$0 = \int dA_D \frac{\tilde{U}\partial \bar{u}[1 + (1-F)\phi]}{\partial y_1} - \int dA_D \nabla \cdot \left[F\phi_p \bar{u}^3 \nabla (F\phi) \right]$$

$$= \Delta_2 [g_{22}m_1 + (-g_{21})m_2]_{x_1^+} - \Delta_2 [g_{22}m_1 + (-g_{21})m_2]_{x_1^-}$$

$$+ \Delta_1 [g_{11}m_2 + (-g_{12})m_1]_{x_2^+} - \Delta_1 [g_{11}m_2 + (-g_{12})m_1]_{x_2^-}, \tag{A.2}$$

where

$$m_1 = -F\phi_p \bar{u}^3 \left(\alpha_{11} \frac{\partial F\phi}{\partial x_1} + \alpha_{12} \frac{\partial F\phi}{\partial x_2} \right) + \tilde{U}\bar{u}[1 + (1-F)\phi],$$

$$m_2 = -F\phi_p \bar{u}^3 \left(\alpha_{21} \frac{\partial F\phi}{\partial x_1} + \alpha_{22} \frac{\partial F\phi}{\partial x_2} \right). \tag{A.3}$$

(a) $\tilde{U} = 100$

(b) $\tilde{U} = 200$

(c) $\tilde{U} = 400$

(d) $\tilde{U} = 800$

(e) $\tilde{U} = 1600$

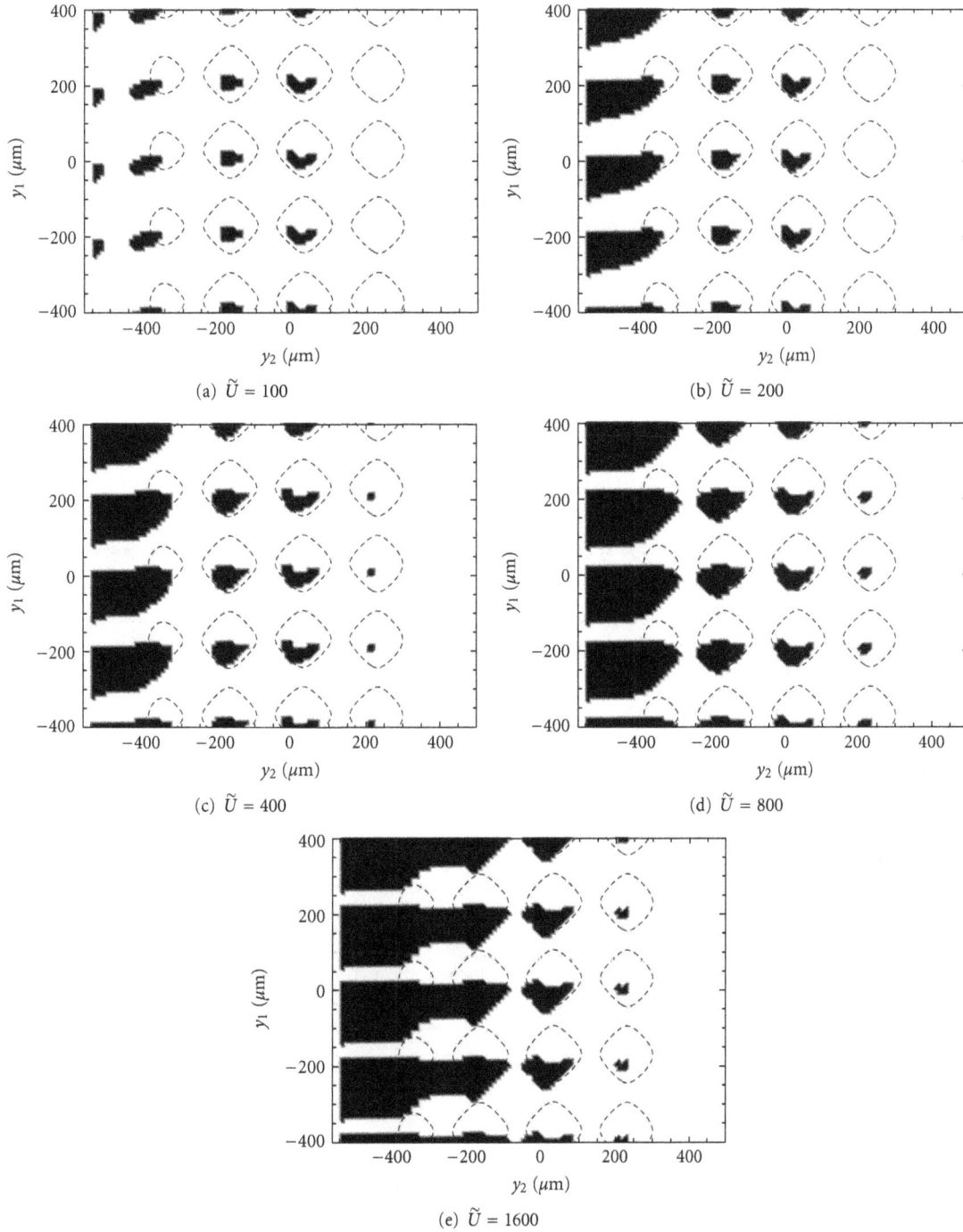

FIGURE 11: Cavitation domains (black areas) for different dimensionless sliding velocities.

The forward difference for the Couette term has been used in order to get a stable scheme, in conjunction with the adoption of the Gauss-Seidel technique.

The average fluid shear stresses in the fluid zones can be calculated as:

$$\frac{\boldsymbol{\tau}_f}{G} = -\alpha\left(-\frac{\tilde{U}}{6}\frac{1}{\overline{u}} + \frac{\overline{u}}{2}\frac{\partial p_f}{\partial y_1}\right)\mathbf{e}_1 - \alpha\left(\frac{\overline{u}}{2}\frac{\partial p_f}{\partial y_2}\right)\mathbf{e}_2 \qquad (\text{A.4})$$

whereas in the cavitation zones:

$$\frac{\boldsymbol{\tau}_f}{G} = -\alpha\left(-(1+\phi)\frac{\tilde{U}}{6}\frac{1}{\overline{u}}\right)\mathbf{e}_1. \qquad (\text{A.5})$$

The resolution scheme is summarized in Figure 13. The contact model is split into two, coupled problems, respectively, the deterministic and the homogenized problem. Given the macroscopic gap relation $\overline{u}(x_1, x_2)$, the average

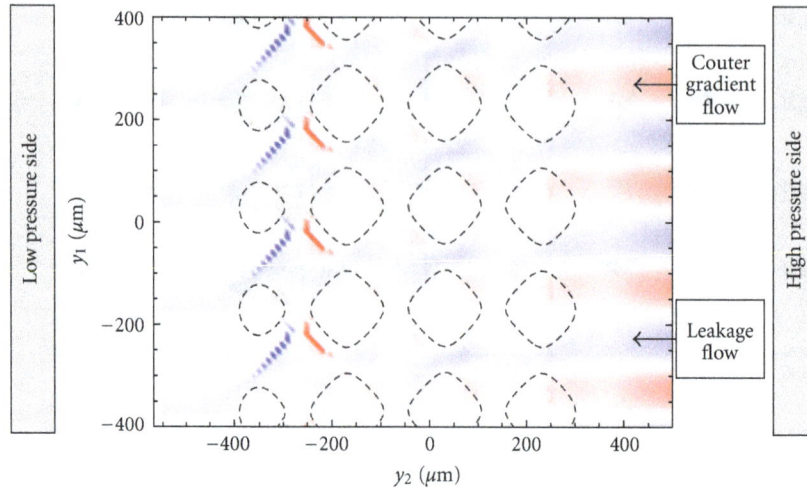

FIGURE 12: Axial average flows. Red areas correspond to the counter-gradient flows (i.e., directed from left to right). For $\tilde{U} = 800$.

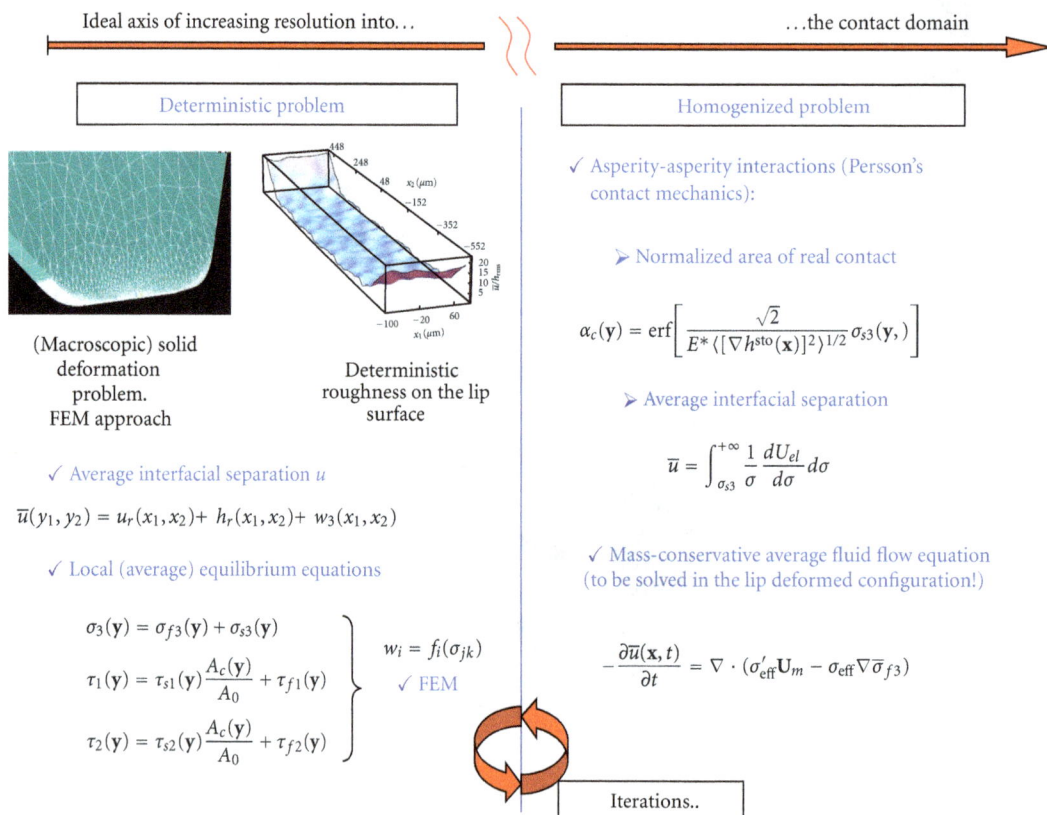

FIGURE 13: Algorithm scheme. In the two-scales approach, the solution of the contact problem is split into two, coupled problems, respectively, the deterministic and the homogenized problem. Those are solved iteratively towards convergence.

surface stress is determined by solving the homogenized part, which consists of the Persson's contact mechanics and of the homogenized fluid problem (previously discussed). The deterministic part follows, where the macroscopic deformation problem is solved as a function of the previously calculated average surface stress field. To do so, in this work we have used the Ansys finite element code; in particular, the lip geometry, similar to that described in

[27], has been meshed with tetrahedral structural elements of type 92. The macroscopic gap relation is then finally updated, determining the loop restart. The solver iterates until certain convergence criteria are satisfied. In our case, convergence is checked on the average interfacial separation field $\overline{u}(x_1, x_2)$. Under-relaxation, with relaxation factors in the range $[0.01, 0.1]$, is usually adopted to numerically damp the interfacial separation solution.

Nomenclature

η:	Fluid viscosity
ν:	Poisson's ratio
α:	Reduced root-mean-square roughness
α_c:	Local normalized area of solid contact
$\bar{u}$:	Locally averaged interfacial separation
β_1:	Parameter for the asymptotic interfacial separation law
β_2:	Parameter for the asymptotic interfacial separation law
λ_{th}:	Threshold roughness wavelength
ρ_0:	Full film density
σ'_{eff}:	Shear flow conductivity
σ_3:	Normal (locally averaged) stress
σ_{eff}:	Pressure flow conductivity
σ_{f3}:	Normal (locally averaged) fluid stress
σ_{s3}:	Normal (locally averaged) solid contact stress
τ_{fi}:	Tangential (locally averaged) fluid stress i-component
τ_i:	Tangential (locally averaged) stress i-component
τ_{si}:	Tangential (locally averaged) solid contact stress i-component
$\tilde{U}$:	Reduced sliding velocity
ζ:	Roughness magnification or wavenumber
$\zeta_{th} = L_0/\lambda_{th}$:	Threshold roughness wavenumber
A_0:	Representative area of interaction
A_c:	Area of solid contact in A_0
$C(q)$:	Total roughness power spectral density
$C^{sto}(q)$:	Power spectral density of the statistically-calculated roughness
E:	Young's modulus
E^*:	Reduced elastic modulus
F:	Cavitation index
G:	Shear elastic modulus
g_{ij}:	Mapping rule Jacobian
h_r:	Deterministic surface roughness height function
J_{ij}:	Inverse of the mapping rule Jacobian
L_0:	Representative macroscopic size of the contact
U:	Shaft sliding velocity
U_m:	Mean velocity
u_r:	Macroscopic contact shape function
U_{el}:	Locally stored elastic energy
w_i:	Generic component of the average surface displacement vector
x_i:	Initial state reference
y_i:	Actual or deformed state reference.

Acknowledgment

The authors acknowledge Regione Puglia for having supported part of this research activity through the constitution of the TRASFORMA Laboratory Network cod. 28.

References

[1] D. B. Hamilton, J. A. Walowit, and C. M. Allen, "A theory of lubrication by micro-irregularities," *Journal of Basic Engineering*, vol. 88, p. 177, 1966.

[2] K. Tønder, "Mathematical verification of the applicability of modified Reynolds equations to striated rough surfaces," *Wear*, vol. 44, no. 2, pp. 329–343, 1977.

[3] D. Dowson, "Modelling of elastohydrodynamic lubrication of real solids by real lubricants," *Meccanica*, vol. 33, no. 1, pp. 47–58, 1998.

[4] B. N. J. Persson, *Sliding Friction: Physical Principles and Applications*, Springer, 2000.

[5] K. L. Johnson, *Contact Mechanics*, Cambridge University Press, 1985.

[6] B. N. J. Persson and M. Scaraggi, "On the transition from boundary lubrication to hydrodynamic lubrication insoft contacts," *Journal of Physics Condensed Matter*, vol. 21, no. 18, Article ID 185002, 2009.

[7] B. N. J. Persson, "Fluid dynamics at the interface between contacting elastic solids with randomly rough surfaces," *Journal of Physics Condensed Matter*, vol. 22, no. 26, Article ID 265004, 2010.

[8] B. Lorenz and B. N. J. Persson, "Leak rate of seals: effective-medium theory and comparison with experiment," *European Physical Journal E*, vol. 31, no. 2, pp. 159–167, 2010.

[9] B. N. J. Persson and M. Scaraggi, "Lubricated sliding dynamics: flow factors and Stribeck curve," *European Physical Journal E*, vol. 34, p. 113, 2011.

[10] M. Scaraggi, G. Carbone, B. N. J. Persson, and D. Dini, "Lubrication in soft rough contacts: a novel homogenized approach—part I," *Soft Matter*, vol. 7, pp. 10395–10406, 2011.

[11] M. Scaraggi, B. N. J. Carbone, and D. Dini, "Lubrication in soft rough contacts: a novel homogenized approach—part II—discussion," *Soft Matter*, vol. 7, pp. 10407–10416, 2011.

[12] M. Scaraggi, G. Carbone, and D. Dini, "Experimental evidence of micro-EHL lubrication in rough soft contacts," *Tribology Letters*, vol. 43, no. 2, pp. 169–174, 2011.

[13] C. J. Hooke and C. H. Venner, "Surface roughness attenuation in line and point contacts," *Proceedings of the Institution of Mechanical Engineers, Part J*, vol. 214, no. 5, pp. 439–444, 2000.

[14] C. J. Hooke and K. Y. Li, "Rapid calculation of the pressures and clearances in rough, elastohydrodynamically lubricated contacts under pure rolling—part 1: low amplitude, sinusoidal roughness," *Proceedings of the Institution of Mechanical Engineers Part C*, vol. 220, no. 6, pp. 901–913, 2006.

[15] C. H. Venner and A. A. Lubrecht, "An engineering tool for the quantitative prediction of general roughness deformation in EHL contacts based on harmonic waviness attenuation," *Proceedings of the Institution of Mechanical Engineers, Part J: Journal of Engineering Tribology*, vol. 219, no. 5, Article ID J03804, pp. 303–312, 2005.

[16] M. Scaraggi and B. N. J. Persson, "Time-dependent fluid squeeze-out between soft elastic solids with randomly rough surfaces," *Tribology Letters*, vol. 47, no. 3, pp. 409–416, 2012.

[17] H. Gao, X. Wang, H. Yao, S. Gorb, and E. Arzt, "Mechanics of hierarchical adhesion structures of geckos," *Mechanics of Materials*, vol. 37, no. 2-3, pp. 275–285, 2005.

[18] B. Bhushan, "Bioinspired structured surfaces," *Langmuir*, vol. 28, no. 3, pp. 1698–1714, 2012.

[19] E. Stratakis, A. Ranella, and C. Fotakis, "Biomimetic micro/nanostructured functional surfaces for microfluidic and tissue engineering applications," *Biomicrofluidics*, vol. 5, no. 1, Article ID 013411, 2011.

[20] G. Carbone, E. Pierro, and S. N. Gorb, "Origin of the superior adhesive performance of mushroom-shaped microstructured surfaces," *Soft Matter*, vol. 7, no. 12, pp. 5545–5552, 2011.

[21] M. Varenberg and S. N. Gorb, "Hexagonal surface micropattern for dry and wet friction," *Advanced Materials*, vol. 21, no. 4, pp. 483–486, 2009.

[22] E. Buselli, V. Pensabene, P. Castrataro, P. Valdastri, A. Menciassi, and P. Dario, "Evaluation of friction enhancement through soft polymer micro-patterns in active capsule endoscopy," *Measurement Science and Technology*, vol. 21, no. 10, Article ID 105802, 2010.

[23] M. Scaraggi, "Lubrication of textured surfaces: a general theory for flow and shear stress factors," *Physical Review E*, vol. 86, Article ID 026314, 2012.

[24] M. Scaraggi, "Textured surface hydrodynamic lubrication: discussion," *Tribology Letters*, vol. 48, no. 3, pp. 375–391, 2012.

[25] R. F. Salant, "Modelling rotary lip seals," *Wear*, vol. 207, no. 1-2, pp. 92–99, 1997.

[26] R. F. Salant, "Theory of lubrication of elastomeric rotary shaft seals," *Proceedings of the Institution of Mechanical Engineers, Part J*, vol. 213, no. 3, pp. 189–201, 1999.

[27] M. Hajjam and D. Bonneau, "Elastohydrodynamic analysis of lip seals with microundulations," *Proceedings of the Institution of Mechanical Engineers, Part J*, vol. 218, no. 1, pp. 13–21, 2004.

[28] M. Hajjam and D. Bonneau, "Influence of the roughness model on the thermoelastohydrodynamic performances of lip seals," *Tribology International*, vol. 39, no. 3, pp. 198–205, 2006.

[29] S. R. Harp and R. F. Salant, "An average flow model of rough surface lubrication with inter-asperity cavitation," *Journal of Tribology*, vol. 123, no. 1, pp. 134–143, 2001.

[30] J. de Vicente, J. R. Stokes, and H. A. Spikes, "The frictional properties of Newtonian fluids in rolling—sliding soft-EHL contact," *Tribology Letters*, vol. 20, no. 3-4, pp. 273–286, 2005.

[31] B. J. Hamrock, *Fundamentals of Fluid Film Lubrication*, McGraw-Hill, 1994.

[32] B. N. J. Persson, "Theory of rubber friction and contact mechanics," *Journal of Chemical Physics*, vol. 115, no. 8, pp. 3840–3861, 2001.

[33] B. N. J. Persson, "Relation between interfacial separation and load: a general theory of contact mechanics," *Physical Review Letters*, vol. 99, no. 12, Article ID 125502, 2007.

[34] B. N. J. Persson, N. Prodanov, B. A. Krick et al., "Elastic contact mechanics: percolation of the contact area and fluid squeezeout," *European Physical Journal E*, vol. 35, p. 5, 2012.

[35] N. Patir and H. S. Cheng, "An average flow model for determining effects of three-dimensional roughness on partial hydrodynamic lubrication," *Journal of Lubrication Technology, Transactions of ASME*, vol. 100, no. 1, pp. 12–17, 1978.

[36] N. Patir and H. S. Cheng, "Application of average flow model to lubrication between rough sliding surfaces," *Journal of Lubrication Technology, Transactions of ASME*, vol. 101, no. 2, pp. 220–230, 1979.

[37] C. Putignano, L. Afferrante, G. Carbone, and G. P. Demelio, "A new efficient numerical method for contact mechanics of rough surfaces," *International Journal of Solids and Structures*, vol. 49, no. 2, pp. 338–343, 2012.

8

Optimization of Dry Sliding Wear Performance of Ceramic Whisker Filled Epoxy Composites Using Taguchi Approach

M. Sudheer, Ravikantha Prabhu, K. Raju, and Thirumaleshwara Bhat

Department of Mechanical Engineering, St. Joseph Engineering College, Mangalore 575 028, Karnataka, India

Correspondence should be addressed to M. Sudheer, msudheerm2002@yahoo.co.in

Academic Editor: Huseyin Çimenoğlu

This study evaluates the influence of independent parameters such as sliding velocity (A), normal load (B), filler content (C), and sliding distance (D) on wear performance of potassium-titanate-whiskers (PTW) reinforced epoxy composites using a statistical approach. The PTW were reinforced in epoxy resin to prepare whisker reinforced composites of different compositions using vacuum-assisted casting technique. Dry sliding wear tests were conducted using a standard pin on disc test setup following a well planned experimental schedule based on Taguchi's orthogonal arrays. With the signal-to-noise (S/N) ratio and analysis of variance (ANOVA) optimal combination of parameters to minimize the wear rate was determined. It was found that inclusion of PTW has greatly improved the wear resistance property of the composites. Normal load was found to be the most significant factor affecting the wear rate followed by (C), (D), and (A). Interaction effects of various control parameters were less significant on wear rate of composites.

1. Introduction

Polymer matrix composites are an important class of composite that are finding increased use in aerospace, automotive, marine, and civil infrastructure applications. In recent years, polymer composites are extensively utilized in tribological components such as cams, brakes, bearings, and gears because of their self-lubrication properties, lower friction, and better wear resistance. More and more polymer composites are now being used as sliding components, which were formerly composed of only metallic materials [1, 2]. Still, developments are underway to explore other fields of application for these materials and to tailor their properties for extreme load-bearing and environmental temperature conditions. Currently, usage of ceramic whisker-reinforced polymer composites is rapidly increasing.

Whiskers are short fiber-shaped single crystals with high perfection and very large length-to-diameter ratios. Generally whiskers possess high strength and stiffness due to their nearly perfect crystal structure [3]. Therefore whiskers are reckoned as more effective reinforcements than traditional fibers such as carbon fiber and glass fiber. Recently various inorganic whiskers such as Calcium Carbonate ($CaCO_3$),

Alumina (Al_2O_3), Silicon Carbide (SiC), Potassium Titanate (PTW, $K_2Ti_6O_{13}$), Barium Titanate ($BaTiO_3$), and so forth were prepared and employed in the manufacturing of composites with different polymer matrices.

Several researchers have observed the significant changes in the mechanical and tribological properties of polymers reinforced with different kinds of whiskers. Feng et al. [4] noticed that wear rate of PTW/PTFE composite decreased dramatically when PTW content increased from 1% to 20%. Lin et al. [5] reported that optimal content of $CaCO_3$ whisker in PEEK composites is 15% to 20% combining both mechanical and tribological properties. Zhang et al. [6] investigated the mechanical and wear properties of silicon carbide and alumina whisker-reinforced epoxy composites and observed that both whiskers significantly improved the flexural modulus and wear resistance of epoxy. However, Avella and coworkers [7] mentioned that addition of untreated SiC whisker into polypropylene lead to an enhancement of the modulus, but a decrease in the tensile strength. Wang et al. [8] revealed that ZnO whiskers have better reinforcing effect with the nylon than the ZnO particles. Jang et al. [9] proposed modifications with ceramic whiskers as an alternative to rubber toughening for improving the impact

TABLE 1: Properties of PTW.

Diameter (μm)	Length (μm)	Density (g/cc)	Tensile strength (GPa)	Tensile modulus (GPa)	Hardness (Mohs)
0.5–2.5	10–100	3.185	7	280	4

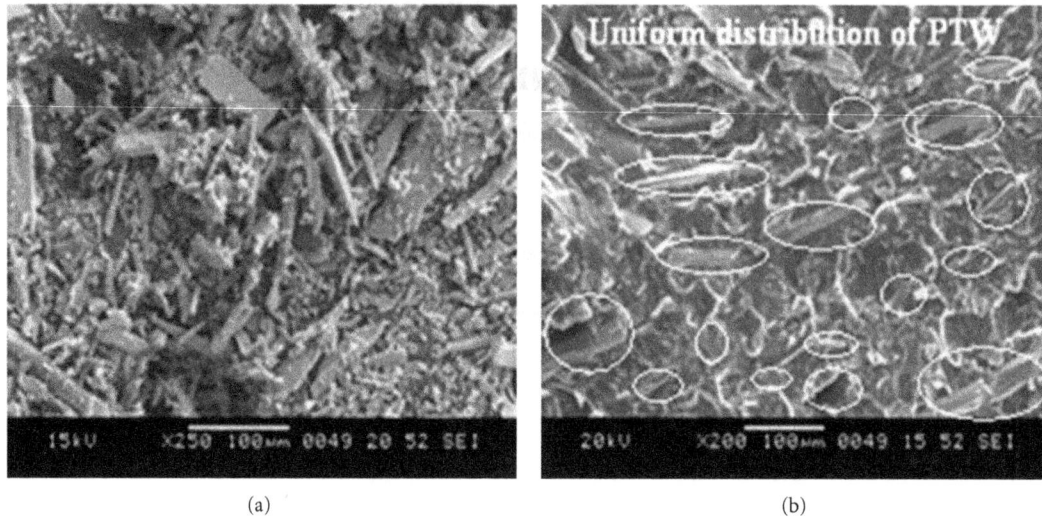

(a)   (b)

FIGURE 1: SEM picture of (a) PTW whiskers and (b) 15% PTW filled epoxy.

resistance of epoxy resins. Among the numerous inorganic fillers, potassium titanate whisker (PTW, $K_2O\cdot6TiO_2$) has been found to be a promising reinforcer for the wear resistant composites due to its unique properties, such as outstanding mechanical performance, low hardness (Mohs hardness 4), and excellent chemical stability. PTW is a kind of very fine microreinforcing material and it is suitable to reinforce the very narrow space in composites that conventional fillers are unable to do. In practice, it is an excellent fit for making products that have a complex shape, great precision, and high polished surface. The price of the PTW ranges from one-tenth to one-twentieth of the cost of SiC whiskers [10]. In this regard, PTW have been used to reinforce most of the polymers [11–18].

Design of experiment is a technique to obtain the maximum amount of conclusive information from the minimum amount of work, time, energy, money, or other limited resources. The information generally comprises the relationship between product and process parameters and the desired performance characteristic [19]. Taguchi's techniques are one of the powerful tools used in the design of experiments. Taguchi's parameter design can optimize the performance characteristics through the setting of design parameters and reduce the sensitivity of system performance to the sources of variation [20, 21]. Taguchi's experimental procedure has been successfully applied for parametric appraisal in dry sliding wear study of polymer composites [22–25]. Among the published literatures, few papers focused on tribological behavior of thermoset composites modified by ceramic whiskers. This paper discusses dry sliding wear characteristic of PTW-reinforced epoxy composites on the basis of Taguchi approach.

2. Materials and Methods

2.1. Materials. Room temperature curing epoxy resin system (LY556+HY951) supplied by Huntsman Advanced Materials India Pvt. Ltd., Bengaluru was used as the matrix system. PTW is a ceramic microfiller used as reinforcement was supplied by Hangzhou Dayangchem Co. Ltd., Hong Kong. These ceramic whiskers are of splinter shape (Figure 1(a)) and with high length/diameter ratio 20–40, and properties are listed in Table 1.

2.2. Fabrication of Composites. An open mold with cavity dimensions $225 \times 225 \times 6$ mm was fabricated to cast polymer composites. The fillers were preheated to 80°C for 2 hours to remove any moisture present and cooled to ambient temperature. The required quantities of filler were stirred gently into liquid epoxy resin, taking care to avoid the introduction of air bubbles. Resin filler mixture was then placed under the vacuum (760 Hg mm) for about 2 hours to remove any entrapped air. Hardener was then added to the resin in the ratio of 1 : 10 and then stirred to ensure complete mixing. The mixture was then poured into an open metallic mold coated with release agent and the mold was placed in a toughened glass chamber maintaining a low-vacuum level 400–450 Hg mm for about 1 hour. Specimens were allowed to cure under room temperature and released from mold after 24 hours. Cast composites plates obtained were of dimension $225 \times 225 \times 3$ mm. The plates were then postcured at 50°C for 2 hours in a hot air oven.

Composition of the test specimens was varied up to 15% of filler loading at intervals of 5%. Extreme care has been taken to avoid any undesirable filler settling effect by

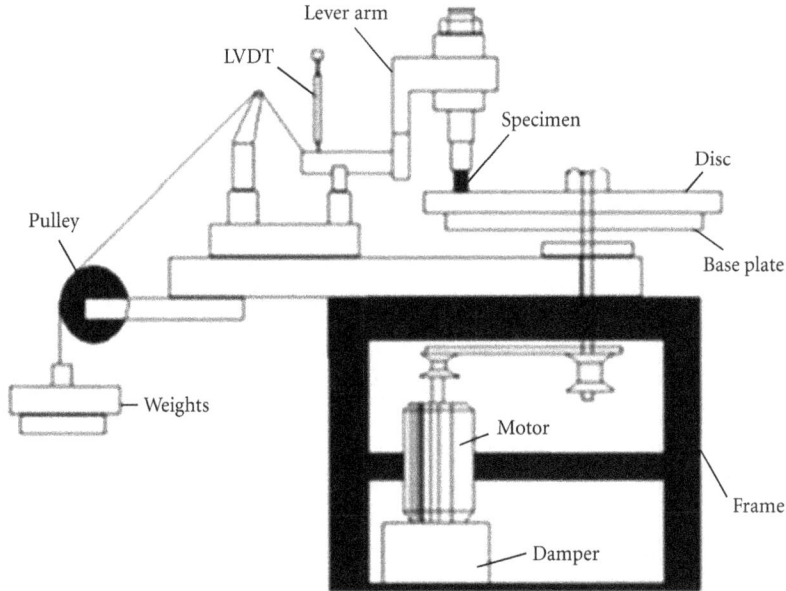

FIGURE 2: Pin on disc test setup.

casting the slurry just prior to its gelling stage, all time keeping it in a stirred condition. This was done to ensure the uniform composition of cast specimens across its volume. Higher PTW content means higher viscosity. Due to the processing difficulty, the composites with more than 15 wt% PTW were not fabricated. Wear test samples of size 10 × 10 × 3 mm were prepared from the cast composites using the diamond-tipped cutter. Figure 1(b) displays the SEM picture of 15% PTW-filled epoxy. It can be seen that the filler is uniformly distributed and has good compatibility with the epoxy matrix.

2.3. Wear Testing. The dry sliding wear tests were performed on pin on disc test setup (Ducom TR201C, Bangalore) as per ASTM G99-05 (reapproved 2010) standard [26]. This test setup is illustrated in Figure 2. Wear test samples of size 10 × 10 × 3 mm are glued to steel pins of 10 mm square cross section and 30 mm length and come in contact with (EN31 grade, 62 HRC, 1.6 μ Ra) carbon steel disc. Prior to testing, the samples were polished against fine grade sand paper (1200 grit SiC) to ensure proper contact with counterface. Test parameters are normal load: 10 N, 20 N, 30 N; sliding velocity: 0.5 m/s, 0.75 m/s, 1 m/s; and sliding distance: 500 m, 1000 m, 1500 m. The pin along with the specimen was then weighed in an electronic balance (Shimadzu Japan, AY220, 0.1 mg Accuracy). Before and after wear testing, samples were cleaned with acetone to remove wear debris. Weight loss of the test samples gives the measure of sliding wear loss. Volume loss was calculated from measured weight loss using density data of the test specimen. The specific wear rate (W_s) was calculated as per

$$W_s = \frac{V}{L \times d},\qquad(1)$$

where V is the volume loss in mm^3, L is the load in Newton, and d is the sliding distance in m.

2.4. Experimental Design. Design of experiment is the powerful analysis tool for modeling and analyzing the influence of the control factors on the performance output. The most important stage in the design of experiment lies in the selection of the control factors [19]. Four parameters, namely, sliding velocity (A), normal load (B), filler content (C), and sliding distance (D) each at three levels, are considered in this study in accordance with L$_{27}$(3^{13}) orthogonal array design. Control parameters and their levels are indicated in Table 2. Four parameters each at three levels would require $3^4 = 81$ runs in a full-factorial experiment, whereas Taguchi's factorial experiment approach reduces it to only 27 runs offering a great advantage. The plan of the experiment [23, 25] is as follows: the first column of the Taguchi orthogonal array is assigned to the sliding velocity (A), the second column to the normal load (B), the fifth column to the fiber content (C), the ninth column to sliding distance (D) and remaining columns are assigned to their interactions and experimental errors.

The experimental observations are transformed into signal-to-noise (S/N) ratio. There are several S/N ratios available depending on the type of characteristic, which can be calculated as logarithmic transformation of the loss function. For lower is the better performance characteristic S/N ratio is calculated as per

$$\frac{S}{N} = -10\log\frac{1}{n}\left(\sum y^2\right),\qquad(2)$$

where "n" is the number of observations and "y" is the observed data. "Lower is the better" (LB) characteristic, with the above S/N ratio transformation, is suitable for minimization of wear rate. A statistical analysis of variance (ANOVA) is performed to identify the control parameters that are statistically significant. With the S/N ratio and ANOVA analyses, the optimal combination of wear parameters is predicted to acceptable level of accuracy. Finally

TABLE 2: Control factors and levels used in the experiment.

Control factor	Level			Units
	I	II	III	
A: Sliding velocity	0.50	0.75	1.00	m/s
B: Normal load	10	20	30	N
C: Filler content	5	10	15	%
D: Sliding distance	500	1000	1500	m

TABLE 3: Test conditions with output results using L_{27} orthogonal array.

SI. no	Sliding velocity A (m/s)	Normal load B (N)	Filler content C (%)	Sliding distance D (m)	Specific wear rate W_s (mm^3/N-km)	S/N ratio (db)
1	0.50	10	5	500	0.132890	17.5301
2	0.50	10	10	1000	0.113821	18.8755
3	0.50	10	15	1500	0.074015	22.6136
4	0.50	20	5	1000	0.161960	15.8118
5	0.50	20	10	1500	0.132791	17.5366
6	0.50	20	15	500	0.150674	16.4392
7	0.50	30	5	1500	0.118125	18.5532
8	0.50	30	10	500	0.130081	17.7157
9	0.50	30	15	1000	0.100449	19.9611
10	0.75	10	5	1000	0.132890	17.5301
11	0.75	10	10	1500	0.113821	18.8755
12	0.75	10	15	500	0.095163	20.4307
13	0.75	20	5	1500	0.157807	16.0375
14	0.75	20	10	500	0.170732	15.3537
15	0.75	20	15	1000	0.138779	17.1535
16	0.75	30	5	500	0.149502	16.5071
17	0.75	30	10	1000	0.121951	18.2763
18	0.75	30	15	1500	0.102212	19.8100
19	1.00	10	5	1500	0.132890	17.5301
20	1.00	10	10	500	0.130081	17.7157
21	1.00	10	15	1000	0.111023	19.0917
22	1.00	20	5	500	0.224252	12.9853
23	1.00	20	10	1000	0.150407	16.4547
24	1.00	20	15	1500	0.132170	17.5773
25	1.00	30	5	1000	0.171650	15.3071
26	1.00	30	10	1500	0.130081	17.7157
27	1.00	30	15	500	0.126883	17.9319

a confirmation experiment is conducted to verify the optimal process parameters obtained from the parameter design [21].

3. Results and Discussion

3.1. Statistical Analysis of Wear Rate. The analysis was made using the software MINITAB 14 [27] specifically used for the design of experiment applications. Test conditions with output results using L_{27} orthogonal array are presented in Table 3. From Table 3, the overall mean for the S/N ratio of the wear rate was found to be 17.68 dB. Figures 3(a) and 3(b) show graphically the effect of the control factors

on specific wear rate. Process parameter settings with the highest S/N ratio always give in the optimum quality with minimum variance. The graphs show the change of the S/N ratio when the setting of the control factor was changed from one level to the other. The best wear rate was at the higher S/N values in the response graphs. From the plots it is clear that factor combination of A_1, B_1, C_3, and D_3 gives minimum specific wear rate. Thus minimum specific wear rate for the developed composites is obtained when the sliding velocity (A) and normal load (B) are at the lowest level, and filler content (C) and sliding distance (D) are at the highest level.

The effect of increasing the control variables on the specific wear rate can be observed from Figure 3(b). It is

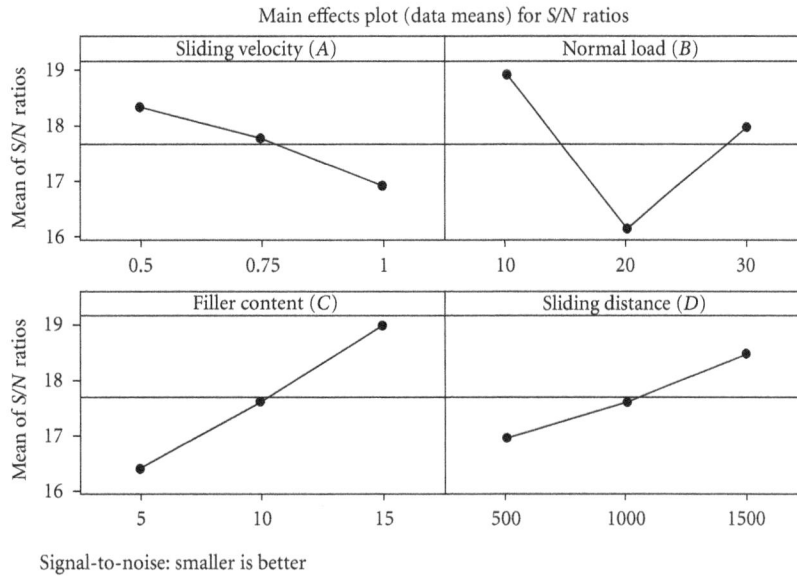

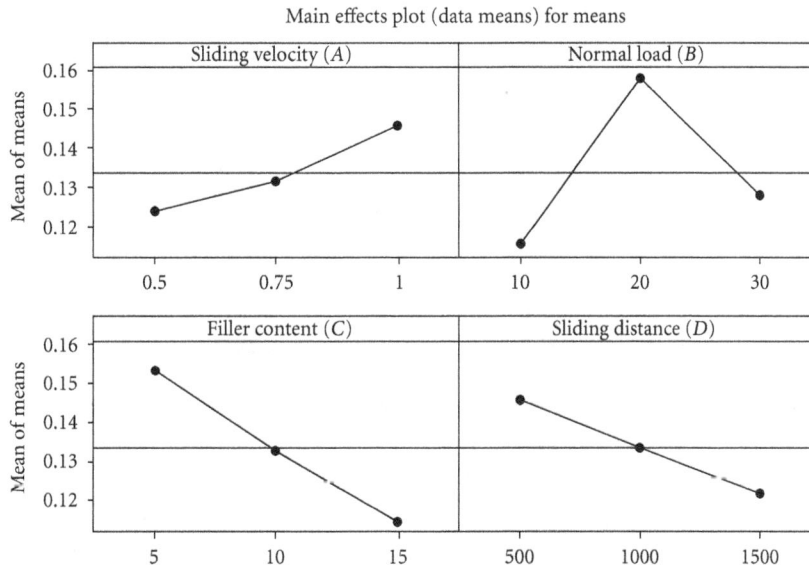

FIGURE 3: (a) Effect of control factors on wear rate (*S/N* ratio). (b) Effect of control factors on wear rate (mean).

obvious that change in the sliding conditions has a direct influence on the wear behavior of the composites. Increase in sliding velocity results in increased rubbing action at the composite and steel disc interface and consequently the wear rate increases. It is evident that wear rate has exhibited increasing and decreasing trend with the normal load. Increase in the normal load usually results in the thermal softening of the composite material and loosening of the matrix material which increases the wear rate. However, dislodging of the matrix material exposes the ceramic whiskers at the interface and these whiskers are able to take up the load applied there by reducing the wear rate at higher loads. This kind of variations in specific wear rate of polymer composites with the normal load is also reported by other researchers [17, 23]. It is observed that increase in the level of both filler content and sliding distance causes wear rate to decrease with the effect of filler content dominating the effect of sliding distance on wear rate. Inclusion of the harder ceramic phase such as PTW results in lowering the wear rate of the composite. This is because the PTW being harder phase than the epoxy matrix improves the wear resistance property of the composite. It is understandable that increase in sliding distance (time of operation) causes the transfer film formation [28] on the wear disc and this contributes to reduce the wear rate under longer duration of sliding. In an effort to identify the role of ceramic particles in the wear behavior of polymer composites, Durand et al. [29] proposed several wear mechanisms such as surface cracking, particle detachment, thin and thick transfer layer at the interface, and so forth in case of polymer composites that provide

Interaction plot (data means) for *S*/*N* ratios

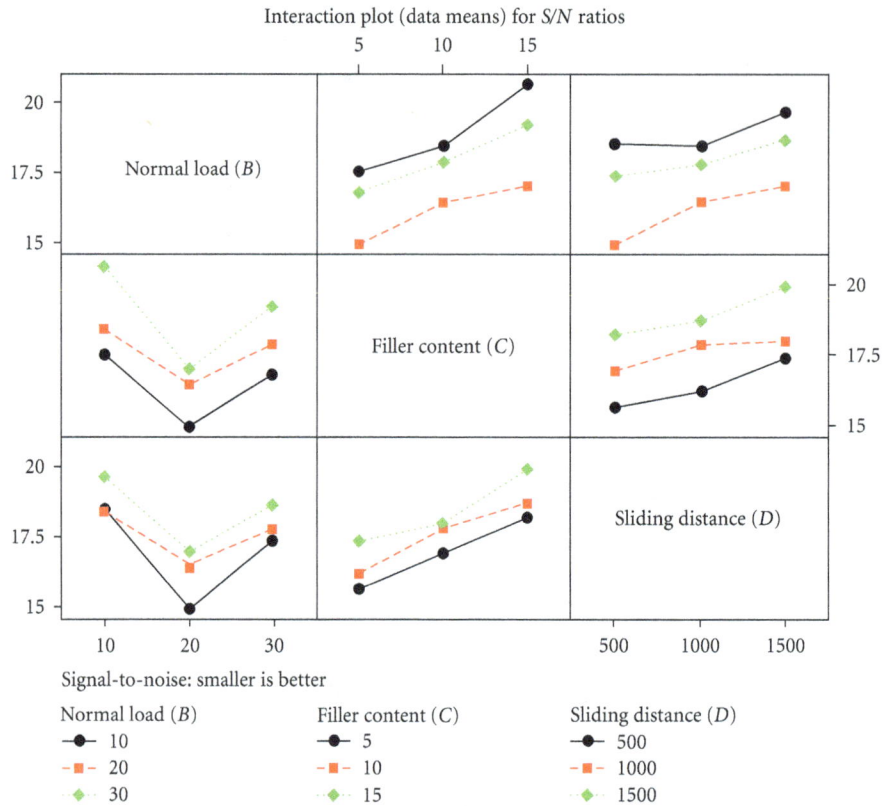

FIGURE 4: Interaction graph between *B*, *C*, and *D* for wear rate.

the effective wear protection to the matrix material. Ray and Gnanamoorthy [30] explained that three mechanisms, namely: (i) matrix material loss; (ii) filler wear; and (iii) debonding at the interface are operative in filled polymer composites and dominance of one factor one over the other control the wear behavior of composites. In the present investigation, it was noticed that PTW particles, which are brittle in nature and also have sharp edges as obvious from SEM images, easily tear the matrix and progressively get aligned in the sliding direction. These particles by virtue of their shape, size, brittleness, and hardness modify the wear performance of the composites.

The *S*/*N* ratio response is given in Table 4, from which it can be concluded that among all the factors, normal load is the most significant factor followed by filler content, sliding distance, and sliding velocity. Analysis of the results leads to the conclusion that as far as minimization of wear rate is concerned, factors *A*, *B*, *C*, and *D* have significant effect. Figure 4 illustrates the interaction effects of control parameters. It is well known that interactions do not occur when the lines on the interaction plots are parallel and strong interactions occur between parameters when the lines cross [31]. An examination of Figure 4 yields a small interaction between control parameters. In order to justify the insignificant factor and insignificant interaction a further statistical analysis (ANOVA) was carried out.

3.2. ANOVA and the Effects of Factors. ANOVA is a statistical design method used to break up the individual effects from

TABLE 4: Response table for signal to noise ratios.

Level	A	B	C	D
1	18.34	18.91	16.42	16.96
2	17.77	16.15	17.61	17.61
3	16.92	17.98	19.00	18.47
Delta	1.41	2.76	2.58	1.52
Rank	4	1	2	3

all control factors. The percentage contribution of each control factor is employed to measure the corresponding effect on the quality characteristic. Table 5 shows the results of the ANOVA with the specific wear rate. This analysis was undertaken for a level of significance of 5% that is, for level of confidence 95%. The 7th column of Table 5 indicates the order of significance among factors and interactions. From Table 5, one can observe that the normal load ($P = 0.000$) has greater static influence of 38.67%, filler content ($P = 0.000$) has an influence of 32.70%, sliding distance ($P = 0.002$) has an influence of 11.34%, and sliding velocity ($P = 0.003$) has an influence of 9.95% on specific wear rate. However, the interaction between normal load and filler content ($P = 0.200$), normal load and sliding distance ($P = 0.195$), and lastly filler content and sliding distance ($P = 0.432$) show less significance of contribution on specific wear rate.

The present analysis indicates that dry sliding wear test parameters and their interactions have both statistical and physical significance (percentage contribution > error) in

TABLE 5: ANOVA table for specific wear rate.

Source	DOF	Seq SS	Adj SS	Adj MS	F test	P value	P (%)
Sliding velocity (A)	2	9.1245	9.1245	4.5623	18.52	0.003	9.95
Normal load (B)	2	35.4780	35.4780	17.7390	71.99	0.000	38.67
Filler content (C)	2	30.0028	30.0028	15.0014	60.88	0.000	32.70
Sliding distance (D)	2	10.4056	10.4056	5.2028	21.11	0.002	11.34
$B \times C$	4	2.0655	2.0655	0.5164	2.10	0.200	2.25
$B \times D$	4	2.0972	2.0972	0.5243	2.13	0.195	2.29
$C \times D$	4	1.0942	1.0942	0.2736	1.11	0.432	1.19
Error	6	1.4784	1.4784	0.2464			1.61
Total	26	114.3166					100

S = 0.496392 R-Sq = 98.39% R-Sq (adj) = 93.02%.
DOF: degrees of freedom; Seq SS: sequential sum of squares; Adj SS: adjusted sum of squares; Adj MS: adjusted mean squares; P: percentage of contribution.

TABLE 6: Confirmation test for specific wear rate.

Level	Initial process parameters $A_1B_1C_2D_2$	Optimal process parameters Prediction $A_1B_1C_3D_3$	Experimental $A_1B_1C_3D_3$	Improvement in the result
S/N ratio (dB)	18.8755	22.3393	22.6135	3.74 dB
Specific wear rate (mm³/N-km)	0.113821	0.075656	0.074015	34.97%

the wear behavior of the Epoxy/PTW composites. However, interaction between filler content and sliding distance has statistical significance but do not have any physical significance [32–34], since error associated is more than percentage contribution of these interactions as evident from the ANOVA results.

3.3. Confirmation Experiment. The confirmation experiment is the final step in the design of experiment process. The purpose of the confirmation experiment is to validate the conclusions drawn during the analysis phase [20, 21]. The estimated S/N ratio for specific wear rate using the optimum level of parameters can be calculated with the help of the following predictive equation [21, 24, 35]:

$$\overline{\eta}_{\text{opt}} = \overline{T} + \sum_{j=1}^{k}\left(\eta_j - \overline{T}\right); \quad j = 1, 2, \ldots, k, \quad (3)$$

where $\overline{T}$ = overall experimental average of S/N ratio. η_j = mean of the S/N ratio at the optimum parameter level. k = number of main design parameters that significantly affect the wear rate of Epoxy/PTW composites. The predictive equation for the optimum wear parameters A_1, B_1, C_3, D_3 can be written as per

$$\overline{\eta}_{\text{opt}} = \overline{T} + \left(\overline{A_1} - \overline{T}\right) + \left(\overline{B_1} - \overline{T}\right) + \left(\overline{C_3} - \overline{T}\right) + \left(\overline{D_3} - \overline{T}\right)$$
$$+ \left[\left(\overline{B_1}\,\overline{C_3} - \overline{T}\right) - \left(\overline{B_1} - \overline{T}\right) - \left(\overline{C_3} - \overline{T}\right)\right]$$
$$+ \left[\left(\overline{B_1}\,\overline{D_3} - \overline{T}\right) - \left(\overline{B_1} - \overline{T}\right) - \left(\overline{D_3} - \overline{T}\right)\right]$$
$$+ \left[\left(\overline{C_3}\,\overline{D_3} - \overline{T}\right) - \left(\overline{C_3} - \overline{T}\right) - \left(\overline{D_3} - \overline{T}\right)\right], \quad (4)$$

where $\overline{A_1}$, $\overline{B_1}$, $\overline{C_3}$, and $\overline{D_3}$ = mean response for factors and interactions at designated levels. By combining the similar terms, (4) reduces to

$$\overline{\eta}_{\text{opt}} = \overline{A_1} - \overline{B_1} - \overline{C_3} - \overline{D_3} + \overline{B_1}\,\overline{C_3} + \overline{B_1}\,\overline{D_3} + \overline{C_3}\,\overline{D_3}. \quad (5)$$

The results of experimental confirmation using optimal wear parameters and comparison of the predicted wear rate with the actual wear rate using the optimal wear parameters are shown in Table 6. Good agreement seems to take place between the estimated and actual wear rate. The improvement in S/N ratio from the starting level to optimum level is 3.74 dB. The specific wear rate is reduced by 34.97%. Therefore the wear performance is greatly improved by using Taguchi method.

4. Conclusions

Taguchi's robust design method can be used to analyze the dry sliding wear behavior of the polymer matrix composites as described in this paper. The following are general conclusions that can be drawn from the work.

(i) Design of experiment approach by Taguchi method enable us to analyze successfully the wear behavior of the composite with the sliding velocity, normal load, filler content, and sliding distance as test variables. From the S/N ratio analysis, the optimal combination of wear parameters is obtained as $A_1B_1C_3D_3$ to minimize wear rate.

(ii) ANOVA results indicated that normal load is the factor which is having highest physical as well as statistical influence (38.67%) on the wear of the composites followed by filler content (32.70%), sliding distance (11.34%), and sliding velocity (9.95%). However, interactions of these factors have less significant effect on wear rate.

(iii) The confirmation tests indicated that it is possible to decrease wear rate significantly (34.97%) by using the proposed statistical technique. The experimental results confirmed the validity of Taguchi method for enhancing the wear performance and optimizing the wear parameters under dry sliding conditions.

In future, this study can be extended to learn the wear behavior of similar multiphase polymer composites.

Acknowledgments

The authors extend thanks and appreciations to Director and Principal of St. Joseph Engineering College, Mangalore for their support and encouragement during research studies.

References

[1] S. T. Peters, *Handbook of Composites*, Chapman and Hall, London, UK, 2nd edition, 1998.

[2] C. A. Happer, *Handbook of Plastics, Elastomers and Composites*, McGraw-Hill, New York, NY, USA, 4th edition, 2004.

[3] J. V. Milewski and H. S. Katz, "Whiskers," in *Handbook of Reinforcements for Plastics*, pp. 205–229, Van Nostrand Reinhold, New York, NY, USA, 1978.

[4] X. Feng, X. Diao, Y. Shi, H. Wang, S. Sun, and X. Lu, "A study on the friction and wear behavior of polytetrafluoroethylene filled with potassium titanate whiskers," *Wear*, vol. 261, no. 11-12, pp. 1208–1212, 2006.

[5] Y. Lin, C. Gao, and N. Li, "Influence of CaCO$_3$ whisker content on mechanical and tribological properties of polyetheretherketone composites," *Journal of Materials Science and Technology*, vol. 22, no. 5, pp. 584–588, 2006.

[6] Y. Zhang, C. A. Pickles, and J. Cameron, "Production and mechanical properties of silicon carbide and alumina whisker-reinforced epoxy composites," *Journal of Reinforced Plastics and Composites*, vol. 11, no. 10, pp. 1176–1186, 1992.

[7] M. Avella, E. Martuscelli, M. Raimo, R. Partch, S. G. Gangolli, and B. Pascucci, "Polypropylene reinforced with silicon carbide whiskers," *Journal of Materials Science*, vol. 32, no. 9, pp. 2411–2416, 1997.

[8] S. Wang, S. Ge, and D. Zhang, "Comparison of tribological behavior of nylon composites filled with zinc oxide particles and whiskers," *Wear*, vol. 266, no. 1-2, pp. 248–254, 2009.

[9] B. Z. Jang, J. Y. Liau, L. R. Hwang, and W. K. Shih, "Structure-property relationships in thermoplastic particulate and ceramic whisker-modified epoxy resins," *Journal of Reinforced Plastics and Composites*, vol. 8, pp. 312–333, 1989.

[10] K. Suganuma, T. Fujita, K. Niihara, and N. Suzuki, "AA6061 composite reinforced with potassium titanate whisker," *Journal of Materials Science Letters*, vol. 8, no. 7, pp. 808–810, 1989.

[11] X. Feng, H. Wang, Y. Shi, D. Chen, and X. Lu, "The effects of the size and content of potassium titanate whiskers on the properties of PTW/PTFE composites," *Materials Science and Engineering A*, vol. 448, no. 1-2, pp. 253–258, 2007.

[12] L. F. Chen, Y. P. Hong, Y. Zhang, and J. L. Qiu, "Fabrication of polymer matrix composites reinforced with controllably oriented whiskers," *Journal of Materials Science*, vol. 35, no. 21, pp. 5309–5312, 2000.

[13] S. Chen, Q. Wang, T. Wang, and X. Pei, "Preparation, damping and thermal properties of potassium titanate whiskers filled castor oil-based polyurethane/epoxy interpenetrating polymer network composites," *Materials and Design*, vol. 32, no. 2, pp. 803–807, 2011.

[14] S. C. Tjong and Y. Z. Meng, "Performance of potassium titanate whisker reinforced polyamide-6 composites," *Polymer*, vol. 39, no. 22, pp. 5461–5466, 1998.

[15] Z. Zhu, L. Xu, and G. Chen, "Effect of different whiskers on the physical and tribological properties of non-metallic friction materials," *Materials and Design*, vol. 32, no. 1, pp. 54–61, 2011.

[16] M. Kumar, B. K. Satapathy, A. Patnaik, D. K. Kolluri, and B. S. Tomar, "Hybrid composite friction materials reinforced with combination of potassium titanate whiskers and aramid fibre: assessment of fade and recovery performance," *Tribology International*, vol. 44, no. 4, pp. 359–367, 2011.

[17] G. Y. Xie, G. X. Sui, and R. Yang, "The effect of applied load on tribological behaviors of potassium titanate whiskers reinforced PEEK composites under water lubricated condition," *Tribology Letters*, vol. 38, no. 1, pp. 87–96, 2010.

[18] G. S. Zhuang, G. X. Sui, H. Meng, Z. S. Sun, and R. Yang, "Mechanical properties of potassium titanate whiskers reinforced poly(ether ether ketone) composites using different compounding processes," *Composites Science and Technology*, vol. 67, no. 6, pp. 1172–1181, 2007.

[19] C. Douglas Montgomery, *Design and Analysis of Experiments*, John Wiley and Sons, New York, NY, USA, 5th edition, 2001.

[20] P. J. Ross, *Taguchi Techniques for Quality Engineering*, McGraw-Hill, New York, NY, USA, 2nd edition, 1996.

[21] R. K. Roy, *A Primer on the Taguchi Method*, Van Nostrand Reinhold, New York, NY, USA, 1990.

[22] M. H. Cho, S. Bahadur, and A. K. Pogosian, "Friction and wear studies using Taguchi method on polyphenylene sulfide filled with a complex mixture of MoS$_2$, Al$_2$O$_3$, and other compounds," *Wear*, vol. 258, no. 11-12, pp. 1825–1835, 2005.

[23] Siddhartha, A. Patnaik, and A. D. Bhatt, "Mechanical and dry sliding wear characterization of epoxy-TiO$_2$ particulate filled functionally graded composites materials using Taguchi design of experiment," *Materials and Design*, vol. 32, no. 2, pp. 615–627, 2011.

[24] A. Satapathy and A. Patnaik, "Analysis of dry sliding wear behavior of red mud filled polyester composites using the taguchi method," *Journal of Reinforced Plastics and Composites*, vol. 29, no. 19, pp. 2883–2897, 2010.

[25] Rashmi, N. M. Renukappa, B. Suresha, R. M. Devarajaiah, and K. N. Shivakumar, "Dry sliding wear behaviour of organo-modified montmorillonite filled epoxy nanocomposites using Taguchi's techniques," *Materials and Design*, vol. 32, no. 8-9, pp. 4528–4536, 2011.

[26] American Society for Testing and Materials, "Standard test method for wear testing with a Pin-on-Disk apparatus," Tech. Rep. ASTM G99-05, American Society for Testing and Materials, 2010.

[27] Minitab User Manual, *Making Data Analysis Easier*, Minitab, Mishawaka, Ind, USA, 2001.

[28] S. Bahadur, "The development of transfer layers and their role in polymer tribology," *Wear*, vol. 245, no. 1-2, pp. 92–99, 2000.

[29] J. M. Durand, M. Vardavoulias, and M. Jeandin, "Role of reinforcing ceramic particles in the wear behaviour of polymer-based model composites," *Wear*, vol. 181–183, no. 2, pp. 833–839, 1995.

[30] D. Ray and R. Gnanamoorthy, "Friction and wear behavior of vinylester resin matrix composites filled with fly ash particles,"

Journal of Reinforced Plastics and Composites, vol. 26, no. 1, pp. 5–13, 2007.

[31] P. Déprez, P. Hivart, J. F. Coutouly, and E. Debarre, "Friction and wear studies using taguchi method: application to the characterization of carbon-silicon carbide tribological couples of automotive water pump seals," *Advances in Materials Science and Engineering*, vol. 2009, Article ID 830476, 2009.

[32] S. Basavarajappa, G. Chandramohan, and J. Paulo Davim, "Application of Taguchi techniques to study dry sliding wear behaviour of metal matrix composites," *Materials and Design*, vol. 28, no. 4, pp. 1393–1398, 2007.

[33] P. Ravindran, K. Manisekar, P. Narayanasamy, N. Selvakumar, and R. Narayanasamy, "Application of factorial techniques to study the wear of Al hybrid composites with graphite addition," *Materials and Design*, vol. 39, pp. 42–54, 2012.

[34] S. R. Chauhan, A. Kumar, I. Singh, and P. Kumar, "Effect of fly ash content on friction and dry sliding wear behavior of glass fiber reinforced polymer composites—a Taguchi approach," *Journal of Minerals & Materials Characterization & Engineering*, vol. 9, no. 4, pp. 365–387, 2010.

[35] A. Patnaik, A. Satapathy, M. Dwivedy, and S. Biswas, "Wear behavior of plant fiber (pine-bark) and cement kiln dust-reinforced polyester composites using Taguchi experimental model," *Journal of Composite Materials*, vol. 44, no. 5, pp. 559–574, 2010.

9

Effect of Strain Hardening on Elastic-Plastic Contact of a Deformable Sphere against a Rigid Flat under Full Stick Contact Condition

Biplab Chatterjee and Prasanta Sahoo

Department of Mechanical Engineering, Jadavpur University, Kolkata 700032, India

Correspondence should be addressed to Prasanta Sahoo, psjume@gmail.com

Academic Editor: Kambiz Farhang

The present study considers the effect of strain hardening on elastic-plastic contact of a deformable sphere with a rigid flat under full stick contact condition using commercial finite element software ANSYS. Different values of tangent modulus are considered to study the effect of strain hardening. It is found that under a full stick contact condition, strain hardening greatly influences the contact parameters. Comparison has also been made between perfect slip and full stick contact conditions. It is observed that the contact conditions have negligible effect on contact parameters. Studies on isotropic and kinematic hardening models reveal that the material with isotropic hardening has the higher load carrying capacity than that of kinematic hardening particularly for higher strain hardening.

1. Introduction

Surface interactions are dependent on the contacting materials and the shape of the contacting surfaces. The shape of the surface of an engineering material is a function of both its production process and the nature of the parent material. When studied carefully on a very fine scale, all solid surfaces are found to be rough. So when two such surfaces are pressed together under loading, only the peaks or the asperities of the surface are in contact and the real area of contact is only a fraction of the apparent area of contact. In such conditions the pressure in those contact spots are high. Accurate calculation of contact area and contact load are of immense importance in the field of tribology and lead to an improved understanding of friction, wear, and thermal and electrical conductance between surfaces. However, it is a difficult task as rough surfaces consist of asperities having different radius and height. The problem is simplified when Hertz [1] provides the contact analysis of two elastic solids with geometries defined by quadratic surfaces. From then, the assumption of surfaces having asperities of spherical shape is adopted to simplify the

contact problems. Greenwood and Williamson [2] used the Hertz theory and proposed an asperity-based elastic model where asperity heights follow a Gaussian distribution. Hertz assumed frictionless surfaces and his theory is restricted for perfectly elastic solids. Later on, researchers have attempted to investigate the effect of material properties beyond the Hertz restriction and the elastic plastic contact of a sphere with a flat became a fundamental problem in contact mechanics. The plastic model introduced by Abbot and Firestone [3], neglects volume conservation of the plastically deformed sphere. One of the first model of elastic-plastic contact was proposed by Chang et al. (CEB model) [4], where the sphere remains in elastic contact until a critical interference is reached, above which the volume conservation of the sphere tip is imposed. The CEB model suffers from a discontinuity in the contact load as well as in the first derivative of both the contact load and the contact area at the transition from elastic to elastic-plastic region. Later Evseev et al. [5], Chang [6], and Zhao et al. [7] have made attempts to improve the elastic-plastic contact model. An introduction of friction at the interface of contact had enabled the Hertz theory to be extended in a realistic manner. Timoshenko and

Goodier [8] stated that the results of normal loading under friction differ from the frictionless Hertzian contact problem. However, contact of spheres with the same elastic constants yields identical tangential displacements, which eliminates the possibility of interfacial slip and the Hertz theory is applicable in certain cases of frictional contact also [9]. Hence normal contact of two spheres with the same material properties exhibits the same results under full stick contact and perfect slip contact condition. This idea is used in modeling spherical contact under combined normal and tangential loading by several researchers (Mindlin [10], Bryant and Keer [11], Hamilton [12], Chang et al. [13]). These authors [10–13] assumed perfect slip contact during normal loading and used friction for tangential loading. Several researchers (Vijaywargiya and Green [14], Boucly et al. [15], Jackson et al. [16]) have also considered the elastic-plastic contact of interfering cylinders/spheres. Goodman [17] first provided the analytical solution of two dissimilar elastic spheres in normal contact under full stick (infinite friction) condition. Spence [18] solved simultaneously the dual integral equations for shear stresses and pressure distribution over the contact area and calculated the total compressive load under the full stick condition. Spence [19] extended his previous work to adhesive contact using a certain value of the friction coefficient. He found that the extent of slip region remains the same for the same elastic constants and friction coefficient.

The studies of contact mechanics of a deformable sphere and a rigid flat revealed the effect of contact conditions and material properties on the deformation and the interfacial parameters like contact load and contact area. However, accurate solutions could not be done until the finite element method was used to solve the problems. Commercial finite element software has the capability to calculate accurately the contact parameters like contact load, contact area, and pressure, and so forth removing some of the assumptions made in the earlier theories regarding asperity interaction and small deformations [20]. Kogut and Etsion [21] (KE Model) first provided an accurate result of elastic-plastic contact of a hemisphere and a rigid flat using commercial finite element software ANSYS to study the plastic zone under frictionless contact condition. They considered a wide range of material properties and sphere size and provided generalized empirical relations for contact area and contact force in terms of dimensionless contact interference for elastic, elastic-plastic, and fully plastic region. They also considered a variation of the tangent modulus up to 0.1E and found negligible effect on the contact parameters. A similar analysis has been done by Jackson and Green [22] (JG Model). The JG model observed the effect of the deformed geometry on the effective hardness and presented some empirical relations of contact area and contact load. Quicksall et al. [23] used finite element technique to model the elastic-plastic deformation of a hemisphere in contact with a rigid flat for various materials such as aluminum, bronze, copper, titanium, and malleable cast iron. They also studied the contact parameters for a generic material in which the elastic modulus and Poisson's ratio were independently varied keeping the yield strength constant. Shankar and Mayuram [24, 25] considered the effect of

material properties during transition from elastic to elastic-plastic region using the KE model. Recently, Malayalamurthi and Marappan [26] and Sahoo et al.[27, 28] concluded that the interfacial parameters like contact load, real contact area during loading are dependent on the material properties of the deformable sphere. So it can be seen that a wide range of literature is available for frictionless contact of a deformable sphere and a rigid flat. The ideal assumption of frictionless contact (perfect slip) may give an idea for interfacial interactions, but results differ from the realistic contact condition where friction is present. According to Johnson [29], "Friction can increase the total load required to produce a contact of given size by at most 5% compared with Hertz."

Two types of contact conditions are in general available in literature; perfect slip and full stick. In perfect slip, it is assumed that there is no tangential stress in the contact area. In full stick, the contacting points of the sphere and the flat, which are covered by the expanding contact zone, are prevented from further relative displacement [30]. Brizmer et al. [31] first analyzed the effect of full stick condition on the elasticity terminus of a spherical contact using finite element software ANSYS. Brizmer et al. [32] extended their study for the loading of an elastic plastic spherical contact both under full stick and perfect slip contact conditions. They found that the interfacial parameters are insensitive to the contact condition and material properties of the deformable sphere. However the contact loads and average contact pressure are slightly affected by Poisson's ratio for full stick contact condition. Zait et al. [33] performed the unloading of an elastic-plastic spherical contact under full stick contact condition. They found a substantial variation in load area curve during unloading under full stick contact condition compared to that of under perfect slip condition.

Kogut and Etsion [21] studied the effect of strain hardening with the tangential modulus up to 0.1E and concluded negligible effect of strain hardening for frictionless and non-adhesive contact. Sahoo et al. [27] used different tangent modulus (maximum 0.33E) to study the effect of strain hardening on contact parameters for frictionless and non-adhesive contact. They inferred that, with the increase in strain hardening the resistance to deformation of a material is increased and the material becomes capable of carrying higher amount of load in a smaller contact area. Kadin et al. [34] observed that secondary plastic flow is strongly affected by strain hardening and they suggested investigating the behavior with the kinematic material hardening for subsequent multiple loading-unloading. However the effect of strain hardening for the full stick contact condition is still not available in literature. Thus present investigation attempts to study the effect of strain hardening on contact parameters for full stick contact condition.

2. Theoretical Background

The contact of a deformable hemisphere and a rigid flat is shown in Figure 1 where the dashed and solid lines represent the situation before and after contact, respectively, of the sphere of radius R. The figure also shows the interference (ω) and contact radius (a) corresponding to a contact

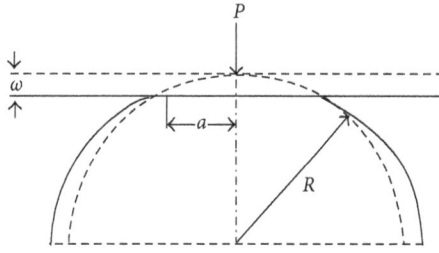

FIGURE 1: A deformable sphere pressed by a rigid flat.

load (P). As discussed earlier the present study is concentrated on the full stick contact condition and compared with results [27] for perfect slip contact condition. Tabor [35] mentioned that full stick contact condition is more realistic than that of perfect slip contact condition due to the formation of an adhesive junction at the interface. Brizmer et al. [32] provided the empirical relations for critical interference and corresponding values for critical loads and critical contact area for perfect slip and full stick contact conditions. For full stick contact condition, the contact parameters are normalized using the expressions of Brizmer et al. [32] for critical interference (ω_c), critical load (P_c), and critical contact area (A_c) given by the following:

$$\omega_c = \left(C_v \frac{\pi(1 - v^2)}{2} \left(\frac{Y}{E} \right) \right)^2 R(6.82v - 7.83(v^2 + 0.0586)),$$
(1)

$$P_c = \frac{\pi^3 Y}{6} C_v^3 \left(R(1 - v^2) \left(\frac{Y}{E} \right) \right)^2 (8.88v - 10.13(v^2 + 0.089)),$$
(2)

$$A_c = \pi \omega_c R,$$
(3)

where, $C_v = 1.234 + 1.256v$. The parameters Y, E, and v are the yield stress, Young modulus, and Poisson's ratio of the sphere material, respectively, and R is the radius of the sphere.

3. Finite Element Model

The contact of the deformable sphere with a rigid flat is modeled using finite element software (ANSYS 11). Due to the advantage of simulation of axisymmetric problems the model is reduced to a quarter circle with a straight line at its top as shown in Figure 2.

The quarter circle is divided into two different zones, for example, zone I and zone II. Zone I is within 0.1R distance from the sphere tip and zone II is the remaining region of the circle outside zone I. These two zones are significant according to their mesh density. The mesh density of zone I is high enough for the accurate calculation of the contact area of the sphere under deformation. Zone II has a coarser mesh as this zone is far away from the contact zone. The resulting mesh consists of 12986 of PLANE82 and 112 of CONTA172 elements. The rigid flat is modeled by a single,

nonflexible two-node target surface element (TARGE169). The nodes lying on the axis of symmetry of the hemisphere are restricted to move only in the radial direction. Also the nodes in the bottom of the hemisphere are restricted in the axial direction due to symmetry. The sphere size is used for this analysis is $R = 1\,\mu\text{m}$. The material properties used here are Young's Modulus (E) = 70 GPa, Poisson's Ratio (v) = 0.3, and Yield stress (σ_y) = 100 MPa. For full stick contact condition, an infinite friction condition is adopted. The ANSYS solution type is chosen as large deformation static analysis. The bilinear isotropic hardening (BISO) option in the ANSYS program is chosen to account the elastic-plastic material response for the single-asperity model. Results are compared with kinematic material hardening as well. Here displacement is applied on the target surface and the force on the hemisphere is found from the reaction solution. As this is an axisymmetric analysis the force is calculated on a full-scale basis. In the present analysis mesh configuration is validated by iteratively increasing the mesh density. The mesh density is doubled until the contact force and contact area is differed by less than 1% between the iterations. In addition to mesh convergence, the model also compares well with the Hertz elastic solution at interferences below the critical interference for perfect slip contact condition. This work uses the Lagrangian multiplier method. The tolerance of the current work is set to 1% of the element width. Computations took about 15 minutes for getting solutions up to yield inception and an hour for large deformation in a 1.6 GHz. PC.

4. Results and Discussion

Strain hardening is an increase in the strength and hardness of the metal due to a mechanical deformation in the microstructure of the metal. This is caused by the cold working of the metal. Strain hardening is expressed in terms of tangent modulus (E_t) which is the slope of the stress-strain curve. Below the proportional limit the tangent modulus is the same as the Young's modulus. Above the proportional limit the tangent modulus varies with the strain unless bilinear hardening is used. The tangent modulus is useful in describing the behavior of materials that have been stressed beyond the elastic region. When a material is plastically deformed there is no longer a linear relationship between the stress and strain as there is for elastic deformation. In elastic perfectly plastic cases, the tangent modulus tends to be zero. Very few material, exhibit elastic perfectly plastic behaviors, generally all the materials follow the multi-linear behavior with some tangent modulus. This multilinear behavior can be assumed as bilinear behavior for analysis purpose in elastic-plastic cases. In this analysis a bilinear material property, as shown in Figure 3, is assumed for the deformable hemisphere.

To study the strain hardening effect, different values of tangent modulus E_t are considered. The tangent modulus E_t is varied according to a parameter, which is known as hardening parameter (H) and defined as $H = E_t/(E - E_t)$. The value of H is taken in the range $0 \leq H \leq 0.5$ as most of the practical materials falls in this range [36]. The value of H equals to zero indicates elastic perfectly plastic

Effect of Strain Hardening on Elastic-Plastic Contact of a Deformable Sphere against a Rigid Flat under Full Stick
Contact Condition

75

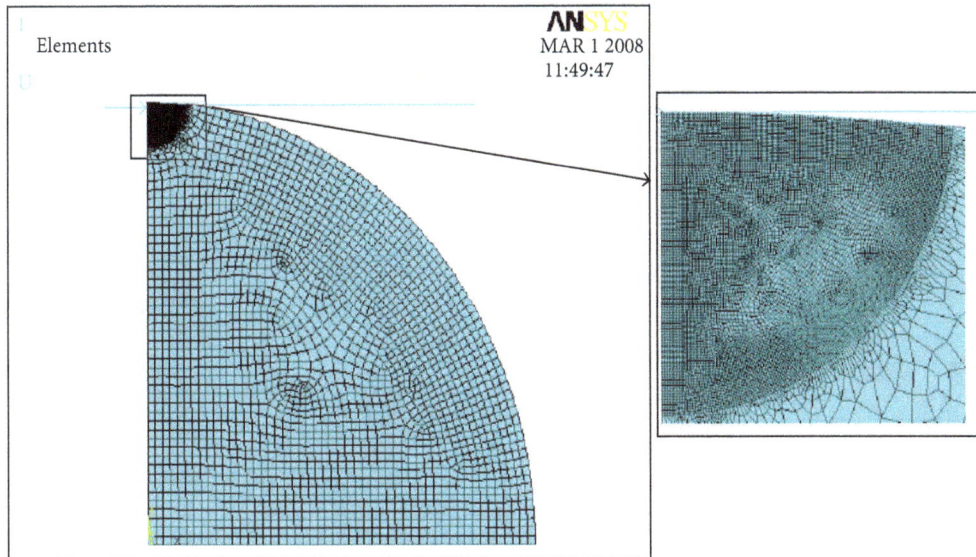

FIGURE 2: Meshed model of the hemispherical contact.

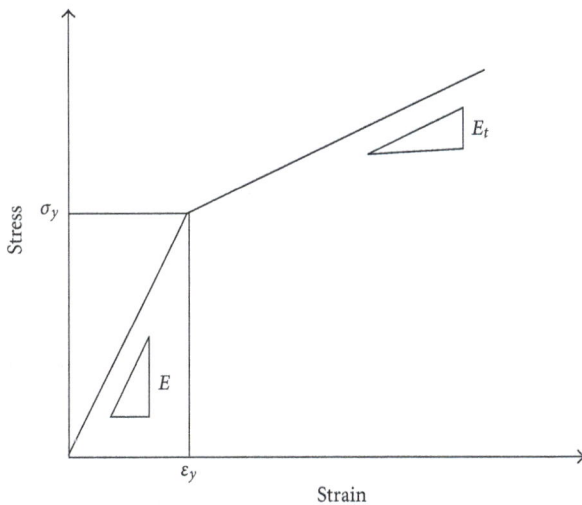

FIGURE 3: Stress-strain diagram for a material having bilinear isotropic properties.

TABLE 1: Different H and E_t values used for the study of strain hardening effect.

H	E_t in %E	E_t (GPa)
0	0.0	0.0
0.1	9.0	6.3
0.2	16.7	11.7
0.3	23.0	16.1
0.4	28.6	20.0
0.5	33.0	23.1

material behavior, which is an idealized material behavior. The hardening parameters used for this analysis and their corresponding E_t values are shown in Table 1. The wide range of values of tangent modulus is taken to make a fair idea of the effect of strain hardening in single-asperity contact analysis with the other material properties being constant.

Figure 4 presents the comparison of dimensionless contact load as a function of dimensionless interference for various models. KE model [21] and Sahoo et al. [27] consider perfect slip contact condition with elastic perfectly plastic material, whereas Brizmer et al. [32] analyze full stick contact condition assuming the material of the sphere as elastic linear isotropic hardening with a tangent modulus of 2% of the Young modulus. The maximum variation of dimensionless

contact load between KE model and Sahoo et al. is 3%. It may be noted here that Kogut and Etsion considered a large number of sphere radii in the range of $0.1 \leq R \leq 10$ (mm) as well as a large variation of material properties in the range $100 \leq (E/\sigma_y) \leq 1000$, and they also found differences in their results up to 3%. The present results with elastic perfectly plastic material under full stick contact condition agree well with the findings of Brizmer et al.

Figure 5 shows the variation of contact load at different interferences for materials having different values of tangent modulus under full stick contact condition. The plot shows a nonlinear behavior between the load and interference as the results are in the elastic-plastic and fully plastic region. The plot shows that up to a certain value of nondimensional interference ($\omega/\omega_c = 10$) the effect of strain hardening on contact parameters is negligible as in the case of perfect slip contact condition. The variation of hardening parameters shows that for a small hardening parameter ($H = 0.1$), the dimensionless contact load increases by 3–21% compared to the results of elastic perfectly plastic case with in the elastic-plastic region, that is, $10\omega_c \leq \omega \leq 100\omega_c$. While for the large hardening parameter ($H = 0.5$), the increase in dimensionless contact load is in the range 11–57% compared to that of elastic perfectly plastic case in the elastic-plastic region. In

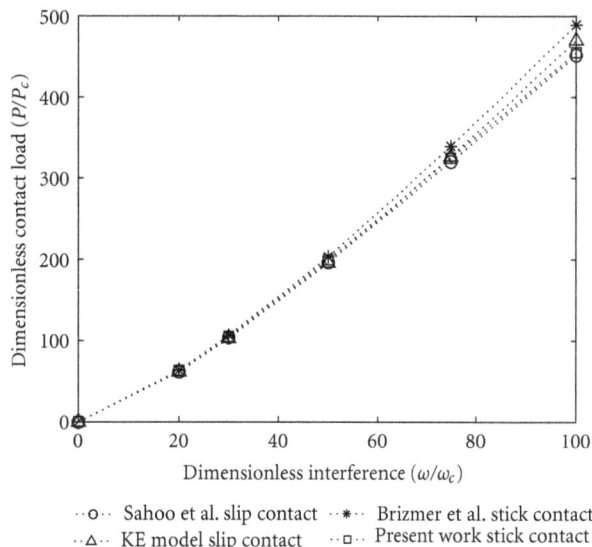

FIGURE 4: Dimensionless contact load versus dimensionless interference for various models.

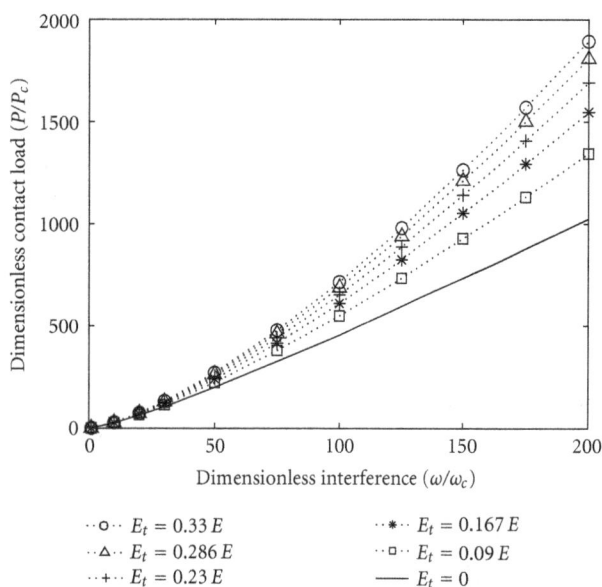

FIGURE 5: Dimensionless contact load versus dimensionless interference at stick contact condition.

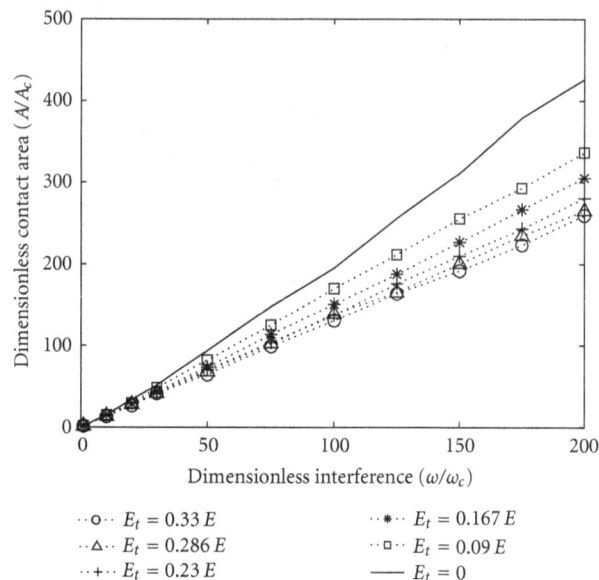

FIGURE 6: Dimensionless contact area versus dimensionless interference for stick contact condition.

fully plastic region these variations are significantly high and increase monotonically with the increase in interference.

Figure 6 represents the plot of dimensionless contact area with dimensionless interference for materials having different values of tangent modulus under full stick contact condition. For perfect slip contact condition the compressible sphere material tends to displace inward during loading up to $\omega/\omega_c = 22$ (Jackson and Green [22]). It is revealed from the finite element analysis that at $\omega/\omega_c = 20$, the dimensionless contact area for full stick contact condition is 3–12% higher than the corresponding dimensionless contact area under perfect slip contact condition. This behavior is expected as

the sphere material could not be displaced inward owing to the stick imposed by the rigid surface. The variation of hardening parameter shows that for a small hardening parameter ($H = 0.1$), the dimensionless contact area decreases by 3–13% compared to the elastic perfectly plastic case with in the elastic-plastic region, that is, $10\omega_c \leq \omega \leq 100\omega_c$. While for large hardening parameter ($H = 0.5$), the decrease in dimensionless contact area is in the range of 14–33% compared to the elastic perfectly plastic case in the elastic-plastic region.

Figure 7 shows the variation of contact area at different applied loads for materials having different values of tangent modulus. The figure shows a nonlinear behavior between the contact area and contact force. Here it is observed that the contact area is less at a particular load for a material having a higher tangent modulus value than that of a material having a lower tangent modulus. This indicates that with the increase in the effect of strain hardening, the material can support the same applied load in a smaller contact area.

The effect of kinematic bilinear hardening is also studied here. Figure 8 describes the difference between isotropic and kinematic bilinear hardening by describing the development of the yield surface with progressive yielding. In isotropic or work hardening, the yield surface is uniformly spread out from the center while in kinematic hardening the yield surface translates in stress space with constant size. Figure 9 presents the dimensionless contact load as a function of dimensionless interference for both the kinematic and isotropic bilinear hardening models. Zait et al. [37] suggested that the effect of hardening model would be most noticeable for large stress amplitudes during a single loading-unloading cycle. Thus here a comparison is made to study the effect of hardening model for three hardening parameter $H = 0$, 0.3 and 0.5 up to 200 times of critical interference during loading. It is revealed from the figure that for hardening

Effect of Strain Hardening on Elastic-Plastic Contact of a Deformable Sphere against a Rigid Flat under Full Stick Contact Condition

77

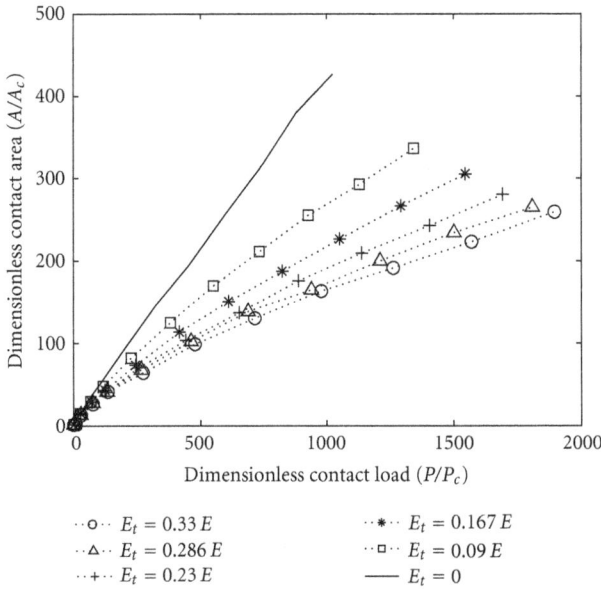

FIGURE 7: Dimensionless contact load versus dimensionless contact area for stick contact condition.

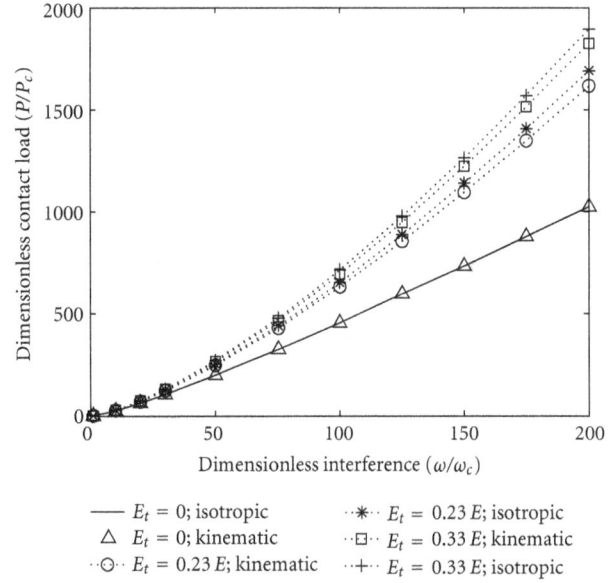

FIGURE 9: Dimensionless contact load against dimensionless interference for kinematic and isotropic material hardening.

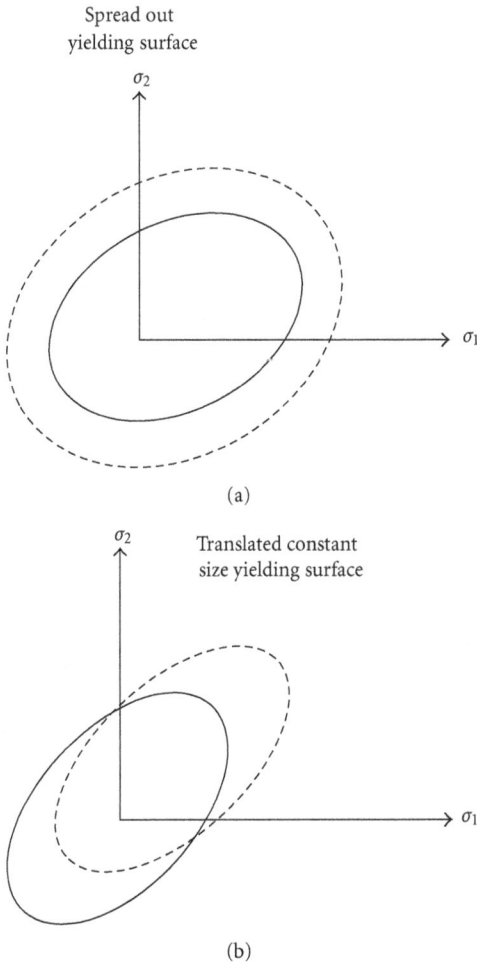

FIGURE 8: Isotropic (a) and kinematic (b) hardening models for two-dimensional stress field.

parameter of 0, that is, for elastic perfectly plastic material the dimensionless contact load for both the hardening models is exactly same even for larger interferences. Zait et al. [37] did not find any significant variation for both the hardening models. They used 2% tangent modulus (E_t/E = 2%) and studied up to ω/ω_c = 150. For hardening parameter of 0.3 and 0.5 that is; for tangent modulus of 23% and 33% of Young's modulus E, it is found from the present simulation that the dimensionless contact load is higher for isotropic hardening model than that of kinematic hardening model at larger interferences. For hardening parameter of 0.3 and 0.5, the dimensionless contact loads are higher by 1.5–4.6% and 1.4–3.9%, respectively, for isotropic hardening model than that of kinematic hardening model.

Figure 10 shows the variation of dimensionless contact area with the variation of dimensionless interference for both the kinematic and isotropic hardening models. Here also the dimensionless contact area is same for both the hardening models in case of elastic perfectly plastic material. For hardening parameter of 0.3 and 0.5, the dimensionless contact area is larger in kinematic hardening model than that for isotropic hardening model. It is revealed from the figure that at hardening parameter of 0.3 and 0.5; the dimensionless contact areas are higher by 0.4–2.9% and 0.15–1.8%, respectively, in kinematic hardening model than that of isotropic hardening model. Figure 11 shows the plots of dimensionless contact area versus dimensionless contact load for both the kinematic and isotropic hardening models. Here it is found that with isotropic hardening model, material can support a particular load with a smaller contact area compared to kinematic hardening for higher rate of hardening. The experimental results of Ovcharenko et al. [38] for full stick contact conditions have also been incorporated in Figure 11. It may be noted that the present theoretical results compare

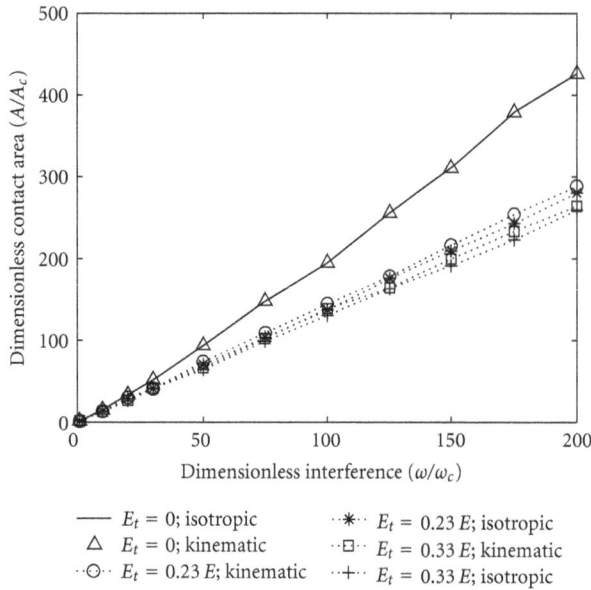

FIGURE 10: Dimensionless contact area against dimensionless interference for kinematic and isotropic material hardening.

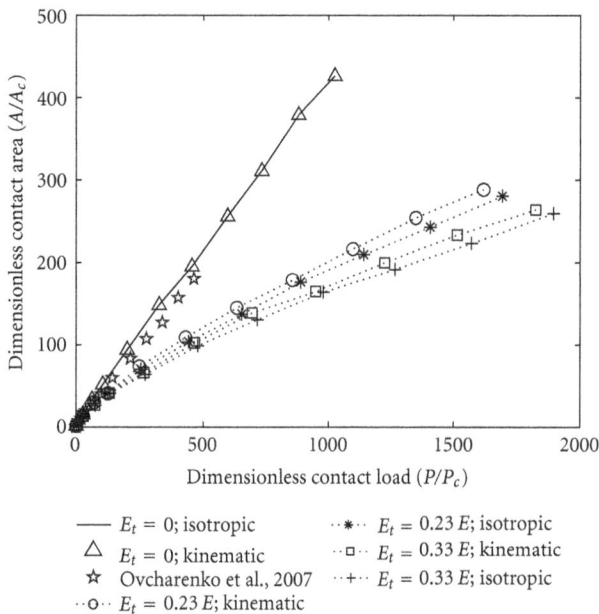

FIGURE 11: Dimensionless contact load versus dimensionless contact area for kinematic and isotropic material hardening.

well with the experimental results. It may also be noted here that any significant difference between the two hardening models is observed only when the loading is reversed [39].

5. Conclusion

The effect of strain hardening on contact parameters under the full stick contact condition for both the isotropic and kinematic hardening models is investigated. It is found that, a generalized behavior cannot be applicable under full stick

contact condition for all kind of materials as the effect of strain hardening greatly influenced the contact parameters. The dimensionless contact load increases with the increase in strain hardening while the dimensionless contact area decreases. It is also observed that the contact conditions have little effect on the interfacial parameters like contact load and contact area. With the increase in strain hardening, the resistance to deformation of a material increases and the material becomes capable of carrying a higher amount of load in a smaller contact area. The material with isotropic hardening can support higher amount of load with same contact area compared to that for material having kinematic hardening behavior.

References

[1] H. Hertz, "Über die Berührung fester elastischer Köper," Journal für die Reine und Angewandte Mathematik, vol. 92, pp. 156–171, 1882.
[2] J. A. Greenwood and J. B. P. Williamson, "Contact of nominally flat surfaces," Proceedings of the Royal Society of London. Series A, vol. 295, pp. 300–319, 1966.
[3] E. J. Abbott and F. A. Firestone, "Specifying surface quality—a method based on accurate measurement and comparison," ASME Journal of Mechanical Engineering, vol. 55, pp. 569–572, 1933.
[4] W. R. Chang, I. Etsion, and D. B. Bogy, "An elastic-plastic model for the contact of rough surfaces," ASME Journal of Tribology, vol. 109, no. 2, pp. 257–263, 1987.
[5] D. G. Evseev, B. M. Medvedev, and G. G. Grigoriyan, "Modification of the elastic-plastic model for the contact of rough surfaces," Wear, vol. 150, no. 1-2, pp. 79–88, 1991.
[6] W. R. Chang, "An elastic-plastic contact model for a rough surface with an ion-plated soft metallic coating," Wear, vol. 212, no. 2, pp. 229–237, 1997.
[7] Y. Zhao, D. M. Maietta, and L. Chang, "An asperity microcontact model incorporating the transition from elastic deformation to fully plastic flow," ASME Journal of Tribology, vol. 122, no. 1, pp. 86–93, 2000.
[8] S. P. Timoshenko and J. N. Goodier, Theory of Elasticity, McGraw-Hill, New York, NY, USA, 3rd edition, 1970.
[9] K. L. Johnson, "One hundred years of Hertz contact," Proceedings of the Institution of Mechanical Engineers, vol. 196, no. 1, pp. 363–378, 1982.
[10] R. D. Mindlin, "Compliance of elastic bodies in contact," ASME Journal of Applied Mechanics, vol. 16, pp. 259–268, 1949.
[11] M. D. Bryant and L. M. Keer, "Rough contact between elastically and geometrically identical curved bodies," ASME Journal of Applied Mechanics, vol. 49, no. 2, pp. 345–352, 1982.
[12] G. M. Hamilton, "Explicit equations for the stresses beneath a sliding spherical contact," Proceedings of the Institution of Mechanical Engineers, vol. 197, pp. 53–59, 1983.
[13] W. R. Chang, I. Etsion, and D. B. Bogy, "Static friction coefficient and model for metallic rough surfaces," Journal of Tribology, vol. 110, no. 1, pp. 57–63, 1988.
[14] R. Vijaywargiya and I. Green, "A finite element study of the deformations, forces, stress formations, and energy losses in sliding cylindrical contacts," International Journal of Non-Linear Mechanics, vol. 42, no. 7, pp. 914–927, 2007.
[15] V. Boucly, D. Nelias, and I. Green, "Modeling of the rolling and sliding contact between two asperities," Journal of Tribology, vol. 129, no. 2, pp. 235–245, 2007.

Effect of Strain Hardening on Elastic-Plastic Contact of a Deformable Sphere against a Rigid Flat under Full Stick
Contact Condition

79

[16] R. L. Jackson, R. S. Duvvuru, H. Meghani, and M. Mahajan, "An analysis of elasto-plastic sliding spherical asperity interaction," *Wear*, vol. 262, no. 1-2, pp. 210–219, 2007.

[17] L. E. Goodman, "Contact stress analysis of normally loaded rough spheres," *ASME Journal of Applied Mechanics*, vol. 29, no. 3, pp. 515–522, 1962.

[18] D. A. Spence, "Self-similar solutions to adhesive contact problems with incremental loading," *Proceedings of the Royal Society of London A*, vol. 305, pp. 55–80, 1968.

[19] D. A. Spence, "The Hertz contact problem with finite friction," *Journal of Elasticity*, vol. 5, no. 3-4, pp. 297–319, 1975.

[20] P. Sahoo, D. Adhikary, and K. Saha, "Finite element based elastic-plastic contact of fractal surfaces considering strain hardening," *Journal of Tribology and Surface Engineering*, vol. 1, no. 1-2, pp. 39–56, 2010.

[21] L. Kogut and I. Etsion, "Elastic-plastic contact analysis of a sphere and a rigid flat," *ASME Journal of Applied Mechanics*, vol. 69, no. 5, pp. 657–662, 2002.

[22] R. L. Jackson and I. Green, "A finite element study of elasto-plastic hemispherical contact against a rigid flat," *ASME Journal of Tribology*, vol. 46, no. 2, pp. 383–390, 2003.

[23] J. J. Quicksall, R. L. Jackson, and I. Green, "Elasto-plastic hemispherical contact models for various mechanical properties," *Proceedings of the Institution of Mechanical Engineers, Part J: Journal of Engineering Tribology*, vol. 218, pp. 313–322, 2004.

[24] S. Shankar and M. M. Mayuram, "A finite element based study on the elastic-plastic transition behavior in a hemisphere in contact with a rigid flat," *Journal of Tribology*, vol. 130, no. 4, Article ID 044502, 6 pages, 2008.

[25] S. Shankar and M. M. Mayuram, "Effect of strain hardening in elastic-plastic transition behavior in a hemisphere in contact with a rigid flat," *International Journal of Solids and Structures*, vol. 45, no. 10, pp. 3009–3020, 2008.

[26] R. Malayalamurthi and R. Marappan, "Elastic-plastic contact behavior of a sphere loaded against a rigid flat," *Mechanics of Advanced Materials and Structures*, vol. 15, no. 5, pp. 364–370, 2008.

[27] P. Sahoo, B. Chatterjee, and D. Adhikary, "Finite element based elastic-plastic contact behavior of a sphere against a rigid flat- effect of strain hardening," *International Journal of Engineering, Science and Technology*, vol. 2, no. 1, pp. 1–6, 2010.

[28] P. Sahoo and B. Chatterjee, "A finite element study of elastic-plastic hemispherical contact behavior against a rigid flat under varying modulus of elasticity and sphere radius," *Engineering*, vol. 2, no. 4, pp. 205–211, 2010.

[29] K. L. Johnson, *Contact Mechanics*, Cambridge University Press, Cambridge, Mass, USA, 1985.

[30] K. L. Johnson, J. J. O' Conner, and A. C. and Woodward, "The effect of the indenter elasticity on the Hertzian fracture of brittle materials," *Proceedings of the Royal Society of London A*, vol. 334, no. 1596, pp. 95–117, 1973.

[31] V. Brizmer, Y. Kligerman, and I. Etsion, "The effect of contact conditions and material properties on the elasticity terminus of a spherical contact," *International Journal of Solids and Structures*, vol. 43, no. 18-19, pp. 5736–5749, 2006.

[32] V. Brizmer, Y. Kligerman, and I. Etsion, "The effect of contact conditions and material properties on elastic-plastic spherical contact," *Journal of Mechanics of Materials and Structures*, vol. 1, no. 5, pp. 865–879, 2006.

[33] Y. Zait, Y. Kligerman, and I. Etsion, "Unloading of an elastic-plastic spherical contact under stick contact condition," *International Journal of Solids and Structures*, vol. 47, no. 7-8, pp. 990–997, 2010.

[34] Y. Kadin, Y. Kligerman, and I. Etsion, "Multiple loading-unloading of an elastic-plastic spherical contact," *International Journal of Solids and Structures*, vol. 43, no. 22-23, pp. 7119–7127, 2006.

[35] D. Tabor, "Junction growth in metallic friction: the role of combined stresses and surface contamination," *Proceedings of the Royal Society of London. Series A*, vol. 251, no. 1266, pp. 378–393, 1959.

[36] F. Wang and L. M. Keer, "Numerical simulation for three dimensional elastic-plastic contact with hardening behavior," *Journal of Tribology*, vol. 127, no. 3, pp. 494–502, 2005.

[37] Y. Zait, V. Zolotarevsky, Y. Klingerman, and I. Etsion, "Multiple normal loading-unloading cycles of a spherical contact under stick contact condition," *Journal of Tribology*, vol. 132, no. 4, Article ID 041401, 7 pages, 2010.

[38] A. Ovcharenko, G. Halperin, G. Verberne, and I. Etsion, "In situ investigation of the contact area in elastic-plastic spherical contact during loading-unloading," *Tribology Letters*, vol. 25, no. 2, pp. 153–160, 2007.

[39] V. Zolotarevskiy, Y. Kligerman, and I. Etsion, "Elastic-plastic spherical contact under cyclic tangential loading in pre-sliding," *Wear*, vol. 270, no. 11-12, pp. 888–894, 2011.

10

Tribological Properties of Metal V-Belt Type CVT Lubricant

Keiichi Narita

Lubricants Research Laboratory, Idemitsu Kosan Co., Ltd., 24-4 Anesakikaigan, Chiba, Ichihara-shi 299-0107, Japan

Correspondence should be addressed to Keiichi Narita, keiichi.narita@si.idemitsu.co.jp

Academic Editor: Philippe Velex

The priority for lubricant performance for metal V-belt-type CVT (B-CVTFs) should be the improvement of transmittable torque capacity between the belt and pulley plus excellent antishudder properties for lockup clutch used in B-CVTs. This study intends to investigate the effect of lubricant additives for improving these performances of B-CVTs. In addition, surface analysis techniques were utilized to gain a novel insight into the chemical composites and morphology of the tribofilms. As a result, it is vital for greater torque capacity to give higher boundary friction coefficient between the metal contacting interfaces, and the process of boundary lubricant film formation derived from antiwear additives used in B-CVTFs strongly impacts on the torque capacity. Moreover, it is found that a sort of lubricant formulation gave an excellent antishudder performance for wet clutch with keeping higher friction coefficient between the metals, which would result in improving the performance of B-CVTs.

1. Introduction

There are two types of transmissions used in automobiles: automatic and manual. A market share of automatic transmissions referred to as ATs has been over 97% of all transmissions installed in automobiles in Japan [1]. In particular, there is an increasing number of passenger cars that feature continuously variable transmissions (CVTs) because CVTs achieve better fuel economy. Among CVTs, the number of cars with metal push belt-type CVTs (B-CVTs) has steadily increased each year and is currently applied to cars with an engine displacement of more than 3 liters. B-CVT consists of a steel belt made up of about 400 segments and laminated rings, as shown in Figure 1. Power is transmitted by the frictional force generated between the belt and pulley. In order to improve transmission efficiency and spread the application to CVTs to larger cars, belt CVT lubricant oils (B-CVTFs) must produce a higher transmission torque capacity between the belt and pulley. Therefore, the priority for the performance of CVT fluids should be focused on the improvement of torque capacity.

There are some reports regarding the effects of lubricants on the performance of a CVT [2–4]. For example, the films generated by Zinc dithiophosphate (ZnDTP) additives in the contact regions are known to contribute to achieve a higher metal-metal friction coefficient [2]. The information

gained regarding the tribofilms of ZnDTP is helpful when considering the additive reaction from lubricants. ZnDTP tribofilms do not develop in rolling contact or if the hydrodynamic film thickness is significantly greater than the surface roughness [5]. Organic compounds around 100-nm thick were identified as the uppermost layer of the film, with P_2O_5, ZnS, FeO, FeS, and H_2O detected beneath this organic layer [6]. These latter compounds are considered to comprise a glassy phosphate, which is closely associated with antiwear performance. It has also been reported that ZnDTP films may form a pad-like structure consisting of glassy phosphate with an outer layer of Zn polyphosphate of 10 nm thickness [7, 8]. Friction-velocity characteristics of oils containing ZnDTP and analysis of their tribofilms were reported [9].

In addition to giving higher torque capacity in the belt system, B-CVTFs must be compatible with lockup clutch fitted in the torque converter. Thus, CVTFs should also give a slightly lower friction coefficient at the engagement of lockup clutch usually made of cellulose fibers, which is generally called antishudder characteristics for wet clutch system. For the improvement of antishudder, friction modifiers are needed to be added into B-CVTFs. This trend is not desirable for the friction coefficient between the metal contacting pairs such as the belt and pulley in the CVT and has let lubricant additive formulation to be complicated for CVT applications. This study intends to investigate the effect of

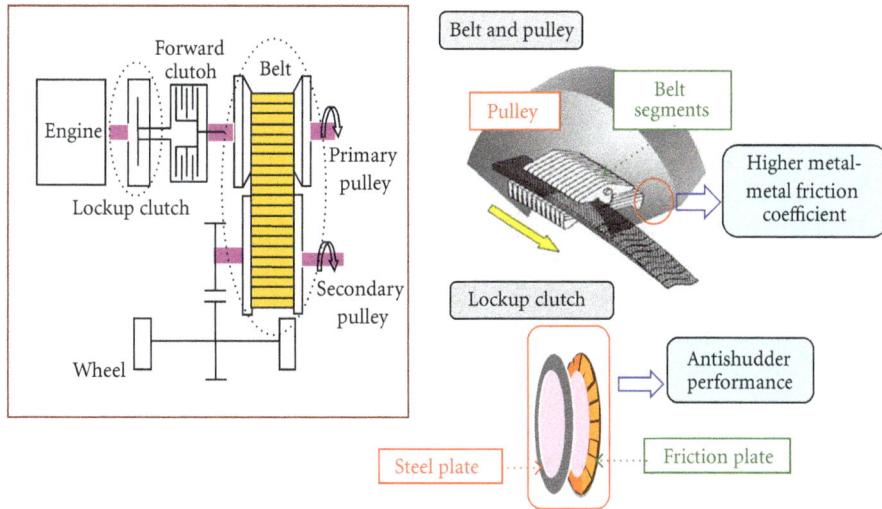

FIGURE 1: Constitution of Belt CVT unit.

lubricant additives such as antiwear additive, detergent, and friction modifier (FM) for improving these performances of B-CVTFs. Furthermore, atomic force microscopy (AFM) and X-ray photoelectron spectroscopy (XPS) were used for considering the nature of the tribofilms derived from candidate additives. The practical performance of candidate oil was evaluated by using actual belt CVT tester.

2. Experimental Methods

2.1. Block on Ring Tribometer (LFW-1). The friction characteristics between the metals were evaluated using a block on ring-type tribometer (Falex LFW-1) by varying the sliding speed from 1.0 to 0.025 m/s with a constant load of 1112N (P_{max} 0.6 GPa) and a constant oil temperature (110°C), as shown in Figure 2. The test procedure used in this study is based upon JASO M358 high-load condition [10] for assessing the torque capacity in a CVT between candidate oil and reference oil.

2.2. Belt CVT Bench Test. Parameters influencing the transmittable torque capacity were evaluated using a CVT bench tester, as shown in Figure 3. This test rig was designed to evaluate the transmission performance which is caused by the friction characteristics in essential CVT parts of the belt and pulley; it is not a whole gearbox. The belt assembly and pulley were taken out of a commercial CVT unit and set in the belt box. The bearings supporting the pulley shafts are hold in the pillow blocks. An AC motor drives the primary pulley and the drive torque is transmitted to the secondary pulley through the belt. The output from the secondary pulley is absorbed by a dynamometer. Speed and torque transducers were connected to the drive and output shafts.

There are slip phenomena due to relative motion between the belt and pulley. In particular, a large slip between the belt and pulley could cause a significant damage to the belt side and pulley sheave surface because the contacting pressure between the belt and pulley is estimated at more than 100 MPa [11]. The relative slip between the belt and pulley

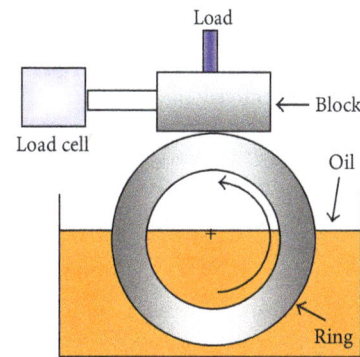

Test specimens	Ring	S10
	Block	H60
Load, N		1112
Pressure (Pmax) (GPa)		0.6
Speed (m/s)		0.025–1
Oil temperature (°C)		110

Test condition (JASO M358 high load)

FIGURE 2: Schematic of block on ring tribometer.

will be more significant with the increase of input torque, which is defined as the slip ratio between the belt and pulley, SR as shown in [12]:

$$SR = \frac{(I_L - I_N)}{I_N} \times 100(\%), \quad (1)$$

where I_L is the speed ratio at a loaded condition and I_N is the specific value at a no load.

Speed ratio is defined as

$$I = \frac{N_p}{N_s}. \quad (2)$$

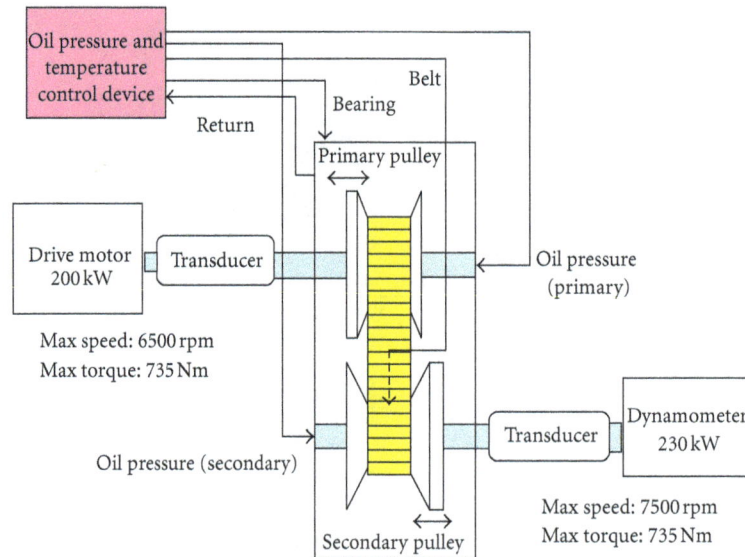

FIGURE 3: Schematic of Belt CVT bench tester.

N_p and N_s are the rotational speed on the primary and the secondary pulleys, respectively.

Under lower drive torque conditions, some of the power in the belt are passed by the bands tension not by compression in the segments, and there is a very small increase in the slip ratio. The slip ratio suddenly rises at approximately quarter load, which corresponds to the transition point, at which the compression side and the slack side change sides. When the drive torque reaches the slip limit torque, a macroslip occurs and no more torque can be transmitted. This macroslip is known to occur on the condition when the slip ratio reaches 4–6 percent [12].

Torque capacity tests were conducted by holding the primary pulley speed stable at 14 m/s and the low speed ratio of 2.3. The drive torque was raised step by step at a rate of 5 Nm every minute until a remarkable increase in the slip ratio was detected. Torque capacity of test oil was then defined as the drive torque at the moment when the slip ratio reached 3 percent so that the belt system would not be significantly damaged.

2.3. Low Velocity Friction Apparatus (LVFA). Low Velocity Friction Apparatus (LVFA) based upon JASO M349 [13] was used for antishudder properties for lockup clutch composed of friction plate and steel plate. This test method as shown in Figure 4 is widely used for evaluating friction behavior for the wet clutch system in the transmissions. Running-in operation for 30 minutes shall be conducted in accordance with the conditions at contacting pressure of 1.0 MPa, 0.6 m/s sliding speed, and 80°C oil temperature. Upon completion of running-in operations, the friction coefficient of the test fluid was measured in accordance with the conditions specified in Figure 4. Fluids showing a positive μ-V curve which means that the friction coefficient increases with sliding speed would be able to prevent uncomfortable vibration which is called shudder phenomenon. Following the initial measurement of μ-V characteristics, the durability test

TABLE 1: Compositions of test oils.

	Oil A	Oil B	Oil C
Phosphorus additive	✓	✓	✓
Calcium detergent	✓	✓	✓
Dispersant	✓	✓	✓
Friction modifier A		✓	
Friction modifier B			✓
Ca (mass %)	0.06	0.06	0.06
P (mass %)	0.06	0.06	0.06
Viscosity (mm²/s 100°C)	7.1	7.1	7.1

was repeated under the conditions specified in Figure 4, and then the μ-V performance measurement was carried out at 24-hour intervals.

2.4. Test Oils. Table 1 shows the composition of the test oils. The lubricants used were a hydrocracked mineral group III oil base blended to contain each additive. Phosphorus additive as an antiwear agent was added at the same concentration of 0.06 mass % into all tested oils. Calcium detergent and boron dispersant are blended in all samples. These additives are normally used in CVTFs in order to keep the transmission clean and are also known to play an important role in controlling μ-V characteristics for lockup clutch [14]. In order to explore effective friction modifiers (FMs), FM A (Nitrogen-type) and FM B (Ester-type) were added into Oil B and Oil C, respectively.

3. Experimental Results

3.1. Metal-Metal Friction Characteristics Tested by LFW-1. Figure 5 shows the metal-metal friction characteristics tested by LFW-1. The measurements were repeated three times for each sample, and the standard deviation of the measured data was within 0.002. First, Oil A composed of phosphorus

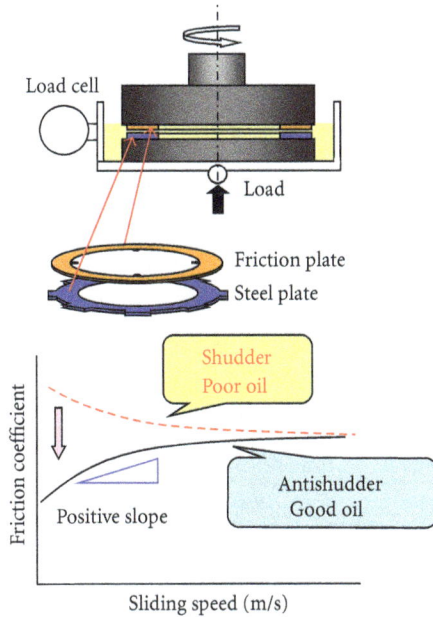

FIGURE 4: Schematic of Low Velocity Friction Apparatus (LVFA).

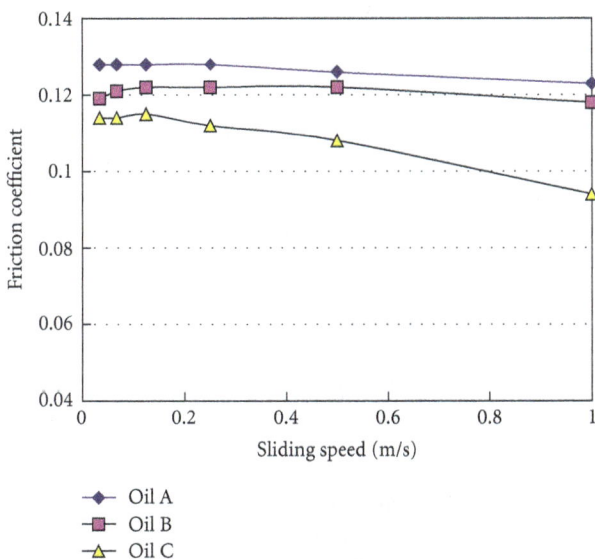

FIGURE 5: Metal-metal friction characteristic of test oils.

additive, calcium detergent, and dispersant shows the highest friction level of all tested oils. Oil B with FM A demonstrated the almost same friction level as Oil A. In the case of Oil C containing FM B, there was 24% decrease in friction coefficient at 1.0 m/s compared with Oil A. FM B gives a great impact on decreasing the metal-metal friction coefficient.

3.2. Surface Analysis on the Posttest Block. X-ray photoelectron spectroscopy (Joel JPS-9010MC XPS) and AFM (Veeco Caliber AFM) were used to investigate the potential link between the friction characteristics and the morphology of tribofilms. Before surface analysis, the test specimens were washed by immersion in hexane in order to remove residual oil and wear debris. XPS analysis identifies the elements in a tribofilm to a depth of approximately 50 nm and analyzes their chemical bonding conditions by detecting the kinetic energy of discharged photoelectrons. In this study, an X-ray beam with photon energy of Mg-Kα (10 kV) irradiated the centre of the test surface over a 1 nm diameter spot. After 15 s sputtering by argon irradiation, corresponding approximately to removal of a 5 nm thick layer from the surface, the XPS analysis was repeated to obtain information on the composition and changes in bonding conditions from the surface to the new depth.

Figure 6 shows XPS depth profile in the tribofilms on the test block from samples, Oil A, B, and C. Peaks corresponding to calcium carbonate ($CaCO_3$) and calcium phosphates ($CaHPO_4$ or $Ca_2P_2O_7$) are at 347.3 eV and 347.6 eV, respectively. In the case of Oil A, there is a sharp peak at 347 eV, as shown in Figure 6. The allowance error of the measurement was within ±0.5 eV in binding energy. Some calcium phosphates and calcium carbonate are likely to be generated on the rubbing surface. These calcium species were detected from the surface to more than 90 nm depth. The XPS profile with Oil B is quite similar to that of Oil A, which means that calcium species might be formed on the surface as well as Oil A. There is a peak at 347 eV in the case of Oil C. However, the presence of calcium species was identified from the surface to at most 40 nm and therefore the thickness of the calcium compounds from Oil C was thinner than Oil A and B.

AFM allows a real surface investigation of tribofilm morphology at the nanometer scale [15]. In this study, AFM images were recorded in contact mode with a V-shaped Si_3N_4 cantilever. To ensure consistency, the same probe was used for scanning each sample and the specified resolution in the normal direction 1 nm. It is unlikely that the AFM tip would damage the surface because this contact force level

Sample	Oil A: higher friction	Oil B: higher friction	Oil C: lower friction
Species	Calcium compounds	Calcium compounds	Calcium compounds
XPS profiling / Substrate / Uppermost			

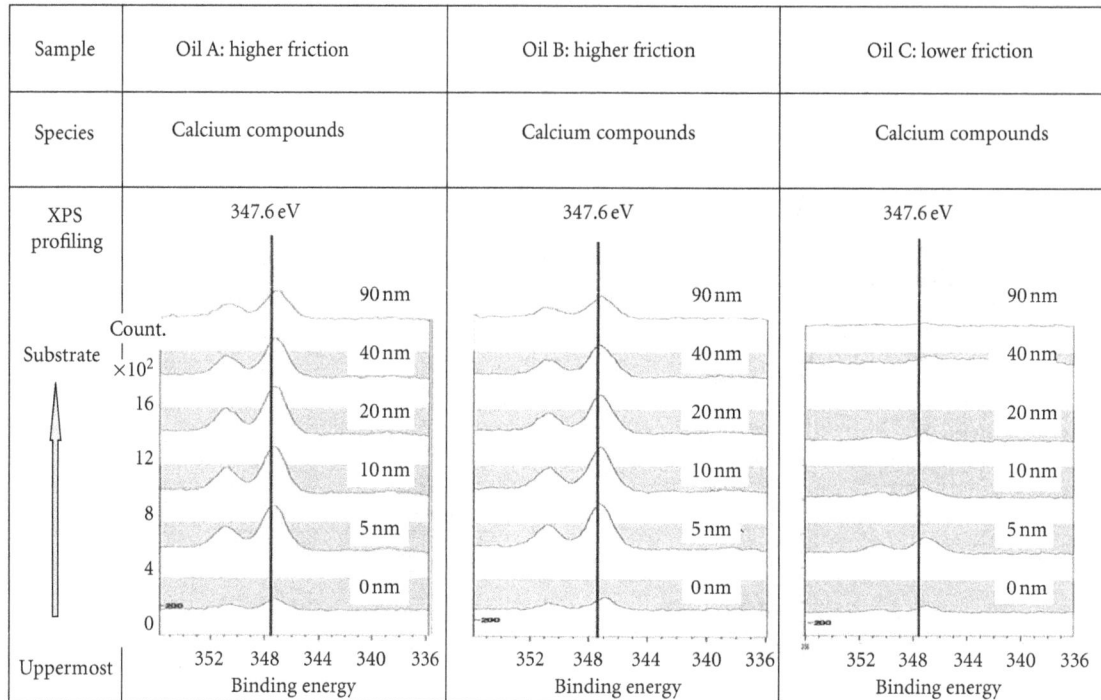

FIGURE 6: XPS spectra of tribofilm on the posttest block by LFW-1.

is extremely low compared with contacting force during the friction tests and the expected mechanical properties of the tribofilm.

Contact force mode AFM was used to characterize the nature of the tribofilm on the blocks. The central part of the wear track on the posttest block was scanned over $50\,\mu m \times 50\,\mu m$ at a rate of $200\,\mu m/s$. Figure 7 illustrates the AFM topographic images of the posttest blocks derived from the tested oils. Note that the brighter parts in the AFM images represent the higher positions on the surface. Very interestingly, the posttest surface with Oil A exhibits a clearly dense deposition structure: the depositions having $1-5\,\mu m$ elongated in the sliding direction. Furthermore, highly dense depositions can be observed on the wear scar from Oil B. These depositions might be composed of calcium carbonate and calcium phosphate from the XPS results as shown in Figure 6. However, the surface from Oil C does not possess distinct depositions. In addition, the ridges exist in the central part of wear track and the sides are polished, which might be due to the lack of the generation of calcium compound film on the surface. These results suggest that the frictional behavior strongly depends on the local morphology of the tribofilms derived from the lubricants' additives.

3.3. *Transmittable Torque Capacity of Test Oils.* Figure 8 shows the torque capacity of test oils, which were conducted varying the secondary pulley clamping force from 19.3 to 36.8 kN at a constant primary pulley speed of 14 m/s and a low speed ratio of 2.3. The torque capacity of Oil A at pulley force 36.8 N was normalized as a torque capacity of 1. The torque capacity of tested oils increased in proportion with the pulley force, and the repeatability of measured values was

within ±3%. The torque for Oil A was the greatest of all tested oils and 17% greater than that of Oil C. The torque for Oil B was almost same level as Oil A at higher load condition. The difference in the torque capacity of the test oils corresponds with the results by LFW-1, as shown in Figure 5, which justifies the use of LFW-1 tribometer as a test method for assessing torque capacity of B-CVT. It is vital for greater torque capacity to give higher boundary friction coefficient between the metal contacting interfaces.

Higher torque capacity fluids will bring out a potential for improving the transmission efficiency in the CVT units. Oil A gave 17% greater torque capacity than Oil C at 36.8 kN pulley clamping force. When the secondary pulley force was reduced from 36.8 to 25 kN (by 32%) for Oil A, Oil A obtained the same torque capacity level as Oil C with a pulley force of 36.8 kN. Therefore, applying higher torque capacity fluid to CVT units could contribute to reduce the maximum required pulley clamping force. It would be possible to have lower oil pump load, which results in the reduction of total power loss in the CVT unit.

3.4. *Antishudder Performance for Lockup Clutch.* Figure 9 shows antishudder performance of test oils evaluated at 80°C by using LVFA. The friction coefficient Oil A is almost constant at a lower speed condition between 0.006 and 0.3 m/s, which implies a negative impact on antishudder property. Oil B and C with friction modifiers demonstrated a positive μ-V curve at less than 0.3 m/s, thus excellent antishudder performance. Kugimiya [14] studied the effects of each typical additive applied to ATFs on μ-V characteristics. Most of the friction modifiers (alcohol, amine, acid, and amide) reduce friction coefficients at lower sliding speeds.

Oil A: higher friction

(a)

Oil B: higher friction

(b)

Oil C: lower friction

(c)

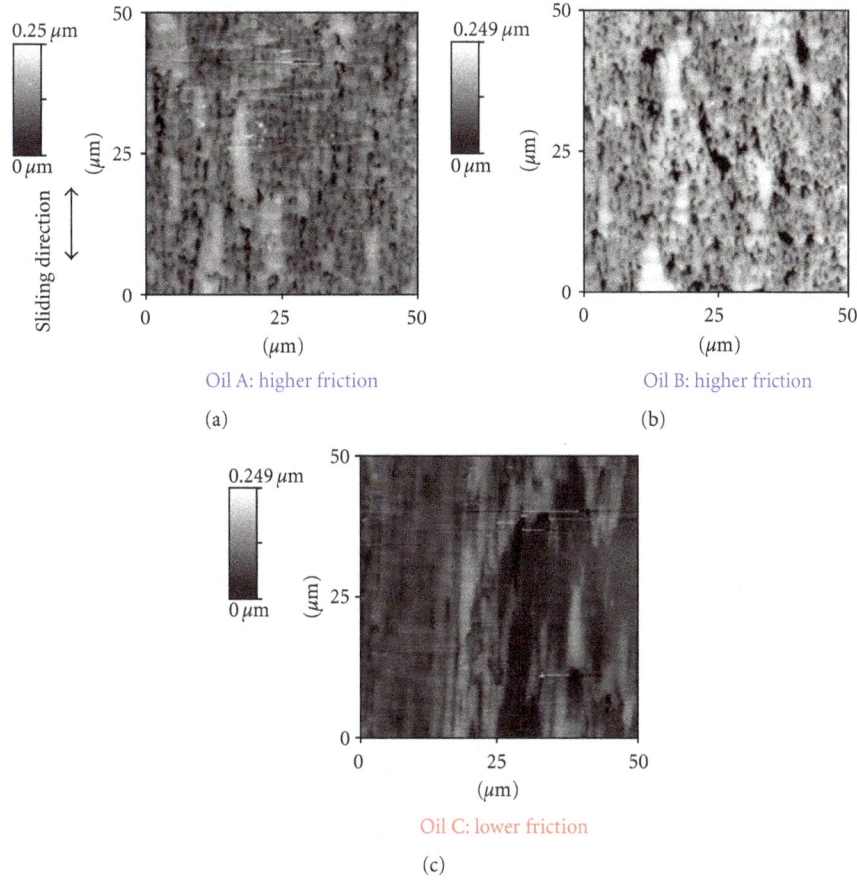

FIGURE 7: AFM image on the posttest block by LFW-1.

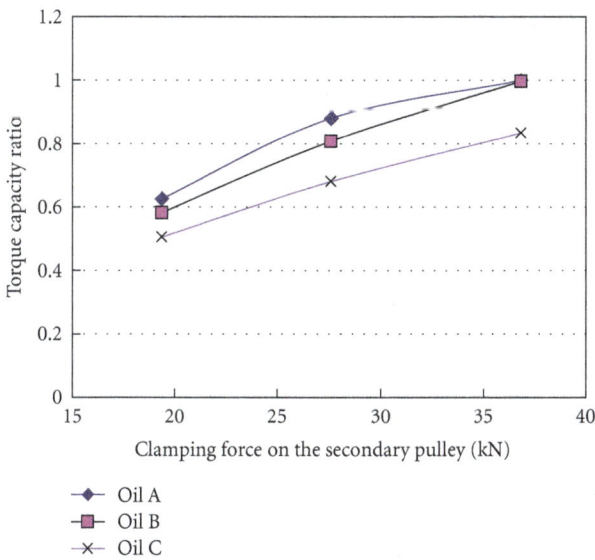

FIGURE 8: Transmittable torque capacity of test oils using belt CVT bench tester.

Dispersant and detergent can increase friction coefficients at higher sliding speed. From these effects, Oil B and C showed a positive μ-V curve in the LVFA.

Next, the antishudder durability of test oils is shown in Figure 10. An approximate expression representing μ-V characteristics at 40°C was obtained from the friction calculated by (3), and the reference for judgment of antishudder service life shall be $d\mu/dV(0.3)$:

$$\frac{d\mu}{dV(0.3)} = \frac{(\mu_{0.3} - \mu_{0.006})}{(0.3 - 0.006)}, \tag{3}$$

where $\mu_{0.3}$ and $\mu_{0.006}$ are friction coefficients at sliding speed of 0.3 and 0.006 m/s, respectively.

In this study, antishudder service life was defined when the value of $d\mu/dV(0.3)$ reaches zero.

The result as shown in Figure 10 includes the antishudder life of JASO reference transmission fluid TIII evaluated by the same test apparatus, and the $d\mu/dV(0.3)$ for TIII was below zero at 168 hours durability. On the other hand, the $d\mu/dV(0.3)$ for Oil B still maintained a positive value at more than 240 hours, indicating longer antishudder service life than TIII.

In commercial use of CVT fluids, two different lubricants would not permit optimizing the performances according their respective purposes. This means that the CVT fluid must be compatible with the components in the whole CVT unit such as the belt and pulley, wet clutch, gears, and bearings. From these results, this additive formulation such as Oil

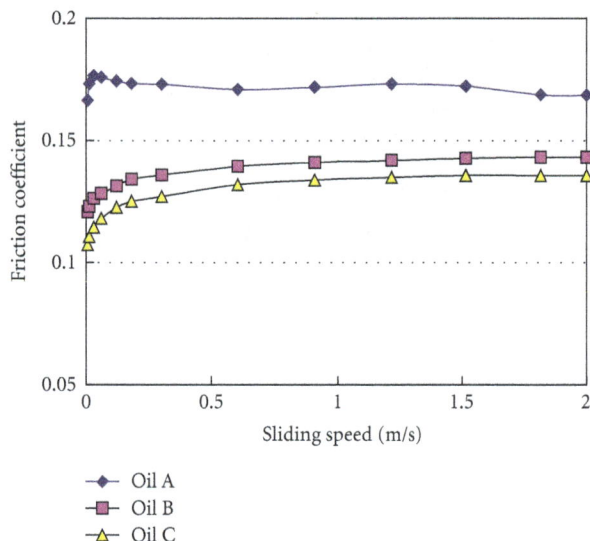

FIGURE 9: Antishudder performance of test oils by LVFA. (Oil temp. 80°C, after running-in).

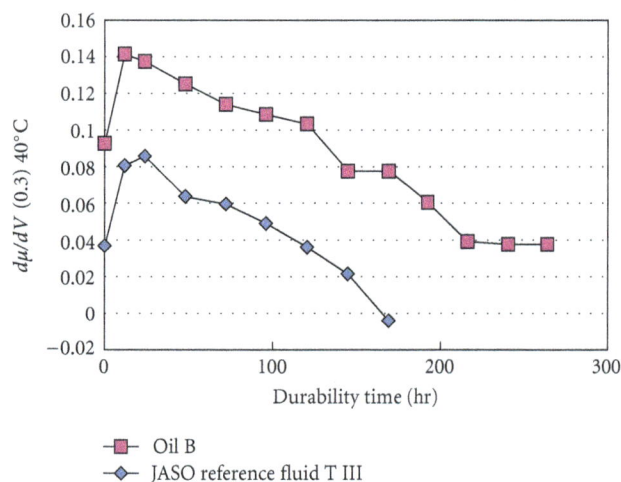

FIGURE 10: Antishudder durability of test oils.

B would attribute for improving antishudder performance keeping metal-metal friction coefficient.

4. Discussion

Oil B including phosphorus additive, calcium detergent, dispersant, and FM A showed a higher metal-metal friction coefficient and good antishudder performance. Here, the reason for the higher friction between the metals obtained by Oil B is discussed. From the results of XPS analysis as shown in Figure 6, much thicker calcium tribofilm species were identified on the wear scar with Oil B than the case of Oil C. The films derived from Oil B shown in Figure 7 therefore exhibit dense depositions in the sliding direction while the surface with Oil C was not uniform and the side parts were worn out. Taylor and Spikes [16] showed that the ZnDTP reaction film appears to inhibit lubricant entrainment into

the contact, thereby leading to a reduced elastohydrodynamic lubrication (EHL) film thickness compared with ZnDTP-free lubricants. It is assumed that the thicker and dense tribofilm derived from Oil B plays a role preventing wear and inhibiting the entrainment of lubricant into the interface, resulting in higher friction.

Furthermore, the difference in the behaviors of tribofilm formation and friction characteristics between FM A in Oil B and FM B in Oil C in the LFW-1 and LVFA is discussed. Both FM A and B could play a role in decreasing friction coefficient under a lower speed around 0.06 m/s and lower contacting pressure of 1 MPa in the LVFA. However, FM A and B gave a different impact on the metal-metal friction at a higher pressure of 0.6 GPa in the LFW-1. FM A in Oil B did not disturb the formation of calcium tribofilm species on the rubbing surface, as shown in Figure 7 AFM topography. FM B in Oil C might more strongly adsorb the surface leading to preventing from forming calcium tribofilm species, consequently with a lower friction. The different effects of FM A and B on the LFW-1 and LVFA test are likely to be influenced by tribological conditions such as contacting pressure and sliding speed between the two tests.

5. Conclusions

Belt CVT lubrication oils (B-CVTFs) must be able to produce a greater transmittable torque capacity between the belt and pulley and an excellent antishudder property for lockup clutch system in CVT units. The effect of lubricant additives was investigated for improving these performances of B-CVTs. In addition, surface analysis techniques were utilized to gain a novel insight into the chemical composites and morphology of the tribofilms. The findings may be summarized as follows.

(1) The lubricant additive formulation composed of calcium detergent and phosphorus antiwear agent demonstrated a higher metal-metal friction coefficient, and the process of boundary lubricant film formation derived from these additives used in B-CVTFs strongly impacts on the tribological property.

(2) A sort of lubricant formulation is found to give an excellent antishudder performance for wet clutch with keeping higher friction coefficient between the metals, which would result in improving the performance of B-CVTs.

References

[1] Japan Automobile Dealers Association, *Annual Report of the Number of New Cars Registered*, 31st edition, 2008.

[2] H. Mitsui, "Trends and requirements of fluids for metal pushing belt type CVTs," *Journal of the Japanese Society of Tribologists*, vol. 45, no. 6, pp. 13–18, 2000.

[3] K. Narita and M. Priest, "Metal-metal friction characteristics and the transmission efficiency of a metal V-belt-type continuously variable transmission," *Proceedings of the Institution of Mechanical Engineers, Part J*, vol. 221, no. 1, pp. 11–26, 2007.

[4] T. Ishikawa, Y. Murakami, R. Yauchibara, and A. Sano, "The effect of belt drive CVT fluid on the friction coefficient between metal components," SAE Paper 972921, 1997.

[5] H. Fujita and H. A. Spikes, "The formation of zinc dithiophosphate antiwear films," *Proceedings of the Institution of Mechanical Engineers, Part J*, vol. 218, no. 4, pp. 265–277, 2004.

[6] S. Bec, A. Tonck, J. M. Georges, R. C. Coy, J. C. Bell, and G. W. Roper, "Relationship between mechanical properties and structures of zinc dithiophosphate anti-wear films," *Proceedings of the Royal Society A*, vol. 455, no. 1992, pp. 4181–4203, 1999.

[7] Z. Yin, M. Kasrai, M. Fuller, G. M. Bancroft, K. Fyfe, and K. H. Tan, "Application of soft X-ray absorption spectroscopy in chemical characterization of antiwear films generated by ZDDP Part I: the effects of physical parameters," *Wear*, vol. 202, no. 2, pp. 172–191, 1997.

[8] J. M. Martin, C. Grossiord, T. L. Mogne, S. Bec, and A. Tonck, "The two-layer structure of Zndtp tribofilms, part I: AES, XPS and XANES analyses," *Tribology International*, vol. 34, no. 8, pp. 523–530, 2001.

[9] M. Muraki and A. Oshima, "Friction-velocity characteristics of oils containing zinc dialkyldithiophosphates and analysis of their tribofilms," *Journal of Japanese Society of Tribologists*, vol. 56, no. 8, pp. 523–529, 2011.

[10] JASO M358, "Standard test method for metal on metal friction characteristics of belt CVT fluids," 2005.

[11] J. D. Micklem, D. K. Longmore, and C. R. Burrows, "Modelling of the steel pushing V-belt continuously variable transmission," *Proceedings of the Institution of Mechanical Engineers, Part C*, vol. 208, no. 1, pp. 13–27, 1994.

[12] C. Morgan and R. Fewkes, "Development of a belt CVT fluid test performance using the VT20/25E," SAE Paper 2002-01-2819, 2002.

[13] JASO M349, "Road vehicles—test method for anti-shudder performance of automatic transmission fluids," 1998.

[14] T. Kugimiya, "Effects of additives for ATF on μ-V characteristics," in *Proceedings of International Tribology Conference*, vol. 2, pp. 1355–1360, Nagasaki, Japan, 2000.

[15] A. J. Pidduck and G. C. Smith, "Scanning probe microscopy of automotive anti wear films," *Wear*, vol. 212, no. 2, pp. 254–264, 1997.

[16] L. J. Taylor and H. A. Spikes, "Friction-enhancing properties of ZDDP antiwear additive: part I—friction and morphology of ZDDP reaction films," *Tribology Transactions*, vol. 46, no. 3, pp. 303–309, 2003.

11

Effect of Nitrogen Implantation on Metal Transfer during Sliding Wear under Ambient Conditions

Luke Autry and Harris Marcus

Department of Chemical, Materials and Biomolecular Engineering, Institute of Materials Science, University of CT, Storrs, CT 06269, USA

Correspondence should be addressed to Luke Autry; autch97@gmail.com

Academic Editor: Patrick De Baets

Nitrogen implantation in Interstitial-Free steel was evaluated for its impact on metal transfer and 1100 Al rider wear. It was determined that nitrogen implantation reduced metal transfer in a trend that increased with dose; the Archard wear coefficient reductions of two orders of magnitude were achieved using a dose of 2e17 ions/cm^2, 100 kV. Cold-rolling the steel and making volumetric wear measurements of the Al-rider determined that the hardness of the harder material had little impact on volumetric wear or friction. Nitrogen implantation had chemically affected the tribological process studied in two ways: directly reducing the rider wear and reducing the fraction of rider wear that ended up sticking to the ISF steel surface. The structure of the nitrogen in the ISF steel did not affect the tribological behavior because no differences in friction/wear measurements were detected after postimplantation heat treating to decompose the as-implanted ε-Fe$_3$N to γ-Fe$_4$N. The fraction of rider-wear sticking to the steel depended primarily on the near-surface nitrogen content. Covariance analysis of the debris oxygen and nitrogen contents indicated that nitrogen implantation enhanced the tribo-oxidation process with reference to the unimplanted material. As a result, the reduction in metal transfer was likely related to the observed tribo-oxidation in addition to the introduction of nitride wear elements into the debris. The primary Al rider wear mechanism was stick-slip, and implantation reduced the friction and friction noise associated with that wear mechanism. Calculations based on the Tabor junction growth formula indicate that the mitigation of the stick-slip mechanism resulted from a reduced adhesive strength at the interface during the sticking phase.

1. Introduction

Metal transfer is an insidious process occurring during sliding of metallic contacts that can result in galling [1, 2] as defined by ASTM [3]. The traditional model of metal transfer usually starts with adhesive wear events between interacting asperities. Fractured metal from the cohesively weaker material transfers to the stronger material. As a result, small particles or "transfer elements" of transferred material are stuck on the harder materials surface. The individual fragments of the transferred material resulting from an isolated tribo-interaction are known as "transfer elements." A transfer element acts like an additional asperity that can continue to interact with the softer material and build-up into a multi-particle debris through a process known as "Mutual Material Transfer" [4, 5]. Metal transfer results in protrusions

of the softer material on the harder materials surface. On subsequent passes on the wear track this results in a change in wear mechanism; the softer material ends up also sliding against work hardened and oxidized protrusions of the same material resulting in a change of friction coefficient. For an excellent example of this phenomenon, see the wear studies based on sheet metal drawing [1, 6]. In the worst case, the protrusions continue to build up on subsequent passes until the surface has "galled." Qualitative galling tests rely on the accumulation of material from one surface to another during continuous contact sliding [7, 8], and the galling criterion is subjective. Typically, if the surface roughens because of material transfer, galling occurs at that particular load for the test. Quantitative tests are based on measuring the coefficient of friction [9, 10] where the galling load is the load at which the coefficient of friction rapidly increases. At the heart of galling is metal transfer, and this why there is a need to study the influence of different surface treatments on its severity.

In the past 30 years, researchers have studied ion implantation for the modification of tribosurfaces [11, 12]. In the case of pure iron [13, 14] and some types of steels [15] nitrogen ion implantation hardens the surface via nitride formation. In some cases, 3-4x relative improvements in the microhardness of the surface of steels using a Knoop indenter were observed [16]. The modification of the surface results in improved abrasive [15–17] and adhesive [18, 19] wear resistance. The improvement in the abrasive wear resistance is obviously because of the surface hardening via implantation. However, for adhesive wear regimes the mechanism for improvement is not so obvious. The environment and any films/layers formed on either surface affect adhesive wear resistance. In addition, the hardness properties of the implanted surface and its counter material may be important when phenomena such as junction growth and deformation-induced surface, roughening [20] are considered. These various factors work together to influence the adhesive wear resistance of a surface and as a result it makes fundamentally studying the adhesive wear resistance of an implanted surface difficult.

The Archard wear equation is a good starting point for discussing adhesive wear for full plastic contact. The volume of material removed per unit sliding length is

$$\frac{V}{x} = k\frac{W}{H},\qquad(1)$$

where W is the applied load and H is the hardness of the softer material. Note that hardness is the load divided by the projected area on the indented surface of the indentation. The constant k is often interpreted as the probability of two touching asperities producing a wear event. Fixing constant k shows that increasing the load or reducing the softer materials hardness increases the wear rate due to the increase in real contact area, $A_{\text{real}} \approx (W/H)$. In this case, indentation of the softer material by the asperities of the harder surface is occurring. The indentation area between the harder material and softer material is equal to the projected area of the indented interface onto the flat surface. The plasticity index (The equivalent roughness parameters used for this analysis determined by optical profilometry were $R_a = 0.095\ \mu m$ and $\rho = 25\ \mu m$. The hardness of the 1100 Al was 441 Mpa (HK). The Young Modulus/Poisson Ratio for Fe and Al were taken from their elemental values.) in this case is 6.5 indicating that full plastic contact is occurring at the surface and that the W/H approximation for the real contact area is appropriate.

The validity of using (1) to analyze wear data depends on the type of wear that occurs. For pure hard on soft adhesive wear (1) is a suitable empirical relationship. However, as noted by Rigney [21], the hardness of the harder surface could also affect the wear regime. Examples of this are found with steel-on-steel wear studies where the hardness of the disk varied by changing the tempering conditions [22]. Borland and Bian [23] created an alternative to the Archard wear equation by defining a variable known as the "index of wear intensity," which was a function of the "equivalent hardness"—H_{eq}. In that study, a linear combination of the hard and soft material hardness values defined H_{eq}. The index of wear intensity, I,

was used in an exponential relationship to relate wear rate to harder materials hardness:

$$\frac{V}{x} = a\exp^{(bI)},\qquad(2)$$

where I is the index wear intensity given by $I = (Ws)/H_{\text{eq}}$. s is the sliding velocity. a and b are fitted parameters. The advantage of modeling wear rate by using an exponential relationship is that mild and severe wear regimes can be captured by one relationship. The linear relationship given by (1) breaks down at a mean contact pressure of $\approx H_{\text{soft}}/3$ and the wear rate exponentially increases with load. The equivalent hardness used by Borland has the disadvantage that, as the hardness of the harder material becomes much larger than that of the softer material, the H_{eq} becomes $\approx CH_{\text{Hard}}$ and can underpredict the wear rate. This study proposes the following alternative expression for H_{Eq} while using (1) for the basic wear equation:

$$H_{\text{eq}} = \frac{H_{\text{hard}}H_{\text{soft}}}{H_{\text{hard}} + H_{\text{soft}}}\ \ \text{or}\ \ \frac{1}{H_{\text{eq}}} = \frac{1}{H_{\text{hard}}} + \frac{1}{H_{\text{soft}}}.\qquad(3)$$

The use of this form for H_{eq} is justified because in the extreme case $H_{\text{hard}} \rightarrow \infty$, $H_{\text{eq}} \rightarrow H_{\text{soft}}$ describes the case of a rigid indenter penetrating into soft material. Therefore, $A_{\text{real}} \approx W/H_{\text{eq}} = W/H_{\text{soft}}$, which was used in the derivation of (1). When $H_{\text{hard}} \rightarrow H_{\text{soft}}$, $H_{\text{eq}} \rightarrow \{H_{\text{soft}}/2\}$ resulting in both asperities bilaterally deforming with no indentation of either surface occurring. This gives the second limiting case of $A_{\text{real}} \approx W/H_{\text{eq}} = \{2W/H\}$, which is close to that predicted by a plane stress model of a square asperity against a flat surface without shear stresses: $(3W/H)$. In the presence of shear stresses, the real contact area is larger than $\{3W/H\}$ (This argument is based on using the Von Mises Criterion in plane stress: $\sigma_{11}^2 + 3\sigma_{12} = \sigma_y^2$. Note that in the expression $H \approx 3\sigma_y$ and $\sigma_{11} = W/A$ were used. In addition, shear stresses were absent: $\sigma_{12} = 0$.) as predicted by the Tabor Junction Growth Model, and hence the accuracy of the W/H approximation for A_{real} should get worse. For intermediate H_{hard} ranges, $\infty > H_{\text{hard}} > H_{\text{soft}}$, a combination of indentation and junction growth can be thought to occur. Unfortunately, no publication of a full plastic, asperity contact analysis where both surfaces are yielding is available to date, but it is likely that the real contact area will be between W/H and $3W/H$.

In a system in which adhesive wear dominates, material hardness is important in that it influences the contact area. In this paper, (3) modifies the Archard Wear equation (1) by using (3) for H:

$$\frac{V}{X} = k\frac{W}{H_{\text{eq}}} = a + \frac{b}{H_{\text{hard}}}.\qquad(4)$$

The empirical coefficients a and b are determined by V/X versus H_{hard} regression analysis for a given set of testing conditions. Notice that, since W is a constant, it was lumped into a ($= kW/H_{\text{soft}}$) and b ($= kW$). The wear coefficient determined by b/W or aH_{soft}/W is equal to an averaged Archard wear coefficient over the series of fit data. The study will evaluate V/X versus H_{hard} for a series of work-hardened, unimplanted ISF steel samples in order to determine how the wear rate changes with harder material hardness. If the results of the unimplanted tests validate (4), then using (4) to adjust the hardening effect of ion implantation is valid. This paper reports the rider wear rates (V/X) and percentage of rider wear ending up as metal transfer for unimplanted and implanted conditions. With the aforementioned regression analysis and using surface analysis techniques this paper will show the effect of chemical modification by ion implantation.

The reduction in adhesion between the surfaces is estimated by measuring friction coefficient. Assuming that adhesive wear dominates, which is a reasonable assumption if the surface roughnesses are controlled, the Tabor Junction Growth formula [24] relates shear strength at the interface and bulk yield strength to the maximum friction coefficient during sliding:

$$\mu = \frac{1}{2\sqrt{(s_o/s_i)^2 - 1}}, \qquad (5)$$

where s_o is the yield shear strength of the bulk and s_i is the shear strength of the interface. Rearranging (5) allows the interfacial shear strength to be deduced. This model is valid when considering the stick-slip mechanism of adhesive wear where junctions are dynamically interacting in the following order: welding, transmitting shear stress, growing, and eventually delaminating or separating. Equation (5) applies during the junction growth phase, which is when most of the shear stresses transmit across the interface. Since the measured friction coefficient is a composite of all the types of asperity interactions, friction coefficients are by nature noisy. This paper reports the average friction coefficient μ_{ave} and standard deviation of the noise. The μ_{ave} plus the standard deviation of the noise is used to calculate the interfacial shear strength through (5) because during the junction growth phase the shear strength at the interface is a maximum.

2. Methodology

The implantations were carried out using two separate ion implantation laboratories: the University of Michigan Ion Beam Laboratory (MIBL) and the University of Albany (SUNY) Ion Beam Laboratory. The ion implanter used at the University of Michigan was a 400 kV piecemeal implanter with the *ion source* manufactured by Danfysik, the *ion accelerator* manufactured by National Electrostatics Corp., and the *implantation stage* manufactured by High Volatge Engineering Europa. The implanter used at the SUNY Ion Beam Laboratory was a 400-Kilovolt Varian (Extrion) 400-10A Implanter. Nitrogen gas was used as the implanter ion source material. The current was kept between 50 and 75 μA. At those currents, the sample temperature was not expected to exceed 200°C; this prediction was based on a heat balance using the beam energy as the heat input and blackbody radiation as the heat output. Several dose/energy combinations were used for these experiments. Ion doses ranged from 5e16 ions/cm^2 to 2e17 ions/cm^2. Two acceleration energies were used: 100 kV and 50 kV. In addition, precold rolling the

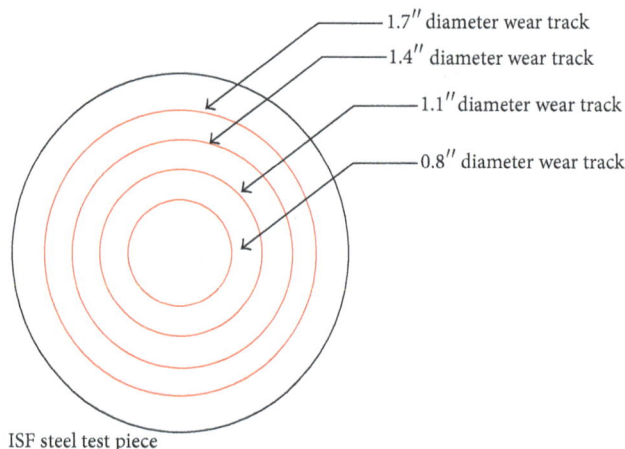

ISF steel test piece

FIGURE 1: An example of the wear tracks used for the unimplanted samples. Note the nitrogen-implanted sample had an extra wear track.

ISF steel prior to implantation was performed on two samples in order to provide extra hardening.

The samples used for this experiment were Interstitial Free Steel (ISF) with a chemistry of (in Wt%) 0.0014% C, 0.12% Mn, 0.034% Al, 0.038% Ti, and the balance Fe. The measured Knoop hardness (HK) of this material was 92 ± 6 (gf/mm^2). The 1100 Al had chemistry of (in Wt%) 0.05% Cu, 0.35% Fe, 0.25% Si, and the balance Al with a HK of 48 ± 4. The ISF Steel was ground with 600 grit and chemically polished using a 7% (by Vol.) HF solution with a balance of H$_2$O$_2$ (30% vol. concentration). The solution was kept at temperature 25–20°C using packed ice. The resulting finish (RMS) of the samples (as checked by an optical profilometer) was 0.01–0.08 μm when inspected over 50 μm lengths. The 1100 Al samples were received in the form of 1/8″ Dia. balls with an HK of 38 ± 4 (gf/mm^2). Prior to testing, the samples were chemically polished by a solution composed of 75% (by vol.) phosphoric acid, 5% nitric acid, and 20% H$_2$O. The temperature of the solution was kept 200–210°F during the process. The resulting RMS of the samples was 0.05–0.11 μm when inspected over 50 μm lengths.

The single pass wear testing was performed using a Falex multispecimen wear tester. The speed setting was 1 RPM, and the load setting was 3 lbs. (Because the 1100 Al pin was tested on different diameter tracks, the pin speed varied by at most a factor of 3.4. Pin speed variations between the five wear tracks were not a factor in friction measurements, as the coefficient of friction did not measurably vary from track to track. This was likely due to the test speed being kept at 1 RPM. This implies that the speed variations did not significantly affect the working hardening rate of the 1100 Al. Al-Kα EDS mapping did not reveal and differences surface area coverage between different tracks.) A 3 lbs. load was chosen because it maximized the amount of metal transfer without tearing out the ball bearing from its mount. A pneumatic pressure sensor controlled the load to within 0.1 lbs. For the unimplanted samples, the traced tracks on the sample were at the following diameters: 1.7″, 1.4″, 1.1″, and 0.8″ giving an

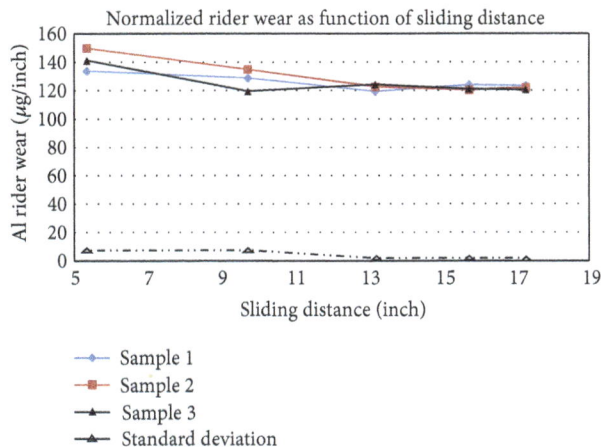

FIGURE 2: Plot of 1100 Al rider wear versus siding distance for three unimplanted ISF steel.

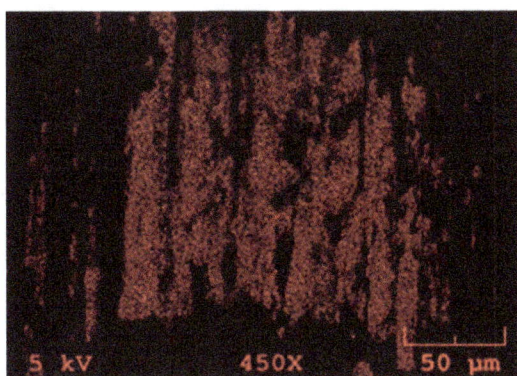

FIGURE 3: Al-Kα EDS map for the 3 lbs single-pass wear test.

effective sliding distance of 15.7″. The N-implanted samples had five tracks (1.7″, 1.4″, 1.1″, 0.8″, and 0.5″ in diameter) giving an effective sliding distance of 17.3″. An schematic of the wear tracks is provided in Figure 1. During sliding the friction coefficient was measured using a 10 lbs load cell with an error of 0.01 lbs offset 3″ from center of rotation. After sliding, the 1100 Al rider and ISF Steel plates were reweighed. The metal transfer and rider wear rates were determined by the weight difference from before and after the wear tests using a Mettler Toledo micro balance with an accuracy of ±4 μg. Using a large load and long track lengths as described above the balance error impacts on the weight changes were minimized. Other sources of error in the wear experiment come from the repeatability of the wear experiment due to errors in load/speed control and variability in the surface conditions such as roughness/chemistry differences. Wearing gloves during handling and storing the samples in dry desiccators under 0.5 atm pressure controlled the surface chemistry variation.

Due to the quantitative nature of this study, the reproducibility error of the single-pass wear experiments needed to be established. Using three unimplanted ISF steel samples with three 1100 Al rider counter materials, the weight change

of each 1100 Al rider as a function of sliding distance was monitored in order to establish the reproducibility error of the wear tests. Figure 2 illustrates the results from the wear tests. At the smallest sliding distance, the reproducibility of the normalized (μg/inch) rider wear is the most poor as the standard deviation was the highest. As the sliding distance increased the reproducibility improved, and the standard deviation changed very little after a 13″ sliding distance. The two major sources of error at small sliding distances are local ISF steel roughness and chemical variations. In addition, the decrease in the average normalized rider wear from 142 μg/inch (at a 5.3″ sliding distance) to 127 μg/inch (at a 9.7″ sliding distance) indicates that Al rider run-in effects were present in the early stages of sliding. Because of run-in, the Al rider developed a flat, which saw repeated contact as the wear tracks were changed. It should be noted that despite developing a flat the real contact area stays the same because the ISF steel surface does not experience run-in and is sufficiently rough to create full-plastic contact condition: plasticity index >1. Due to material constraints, such reproducibity tests were not possible at all testing conditions, and we assumed for this study that the same reproducibility error (2 μg/inch) applies to all other tests.

The average height of the Al area coverage on the ISF steel surface was determined by evaluating the coverage area of five Al-Kα EDS maps (see Figure 3 as an example) for each testing condition. Dividing the volumetric aluminum transfer to the ISF steel by the coverage area gives the average debris height. Adobe Photoshop determined the average red-pixel intensity on the wear track ($I_{wear, ave}$), which is on a scale of 1 to 256. In addition, the software also determined the total number of red pixels (N_{wear}). The average red-pixel intensity ($I_{ref, ave}$) and total number of red pixels (N_{ref}) were determined for an 1100 Al reference sample. The coverage area was determined by multiplying the area of the EDS map by $\{I_{wear, ave} N_{wear}\}/\{I_{ref, ave} N_{ref}\}$.

Prior to testing the ion implanted samples unimplanted samples with varying degrees of cold work provided by flat-rolling were tested to determine how the hardness of the ISF steel affects metal transfer and to also provide a baseline

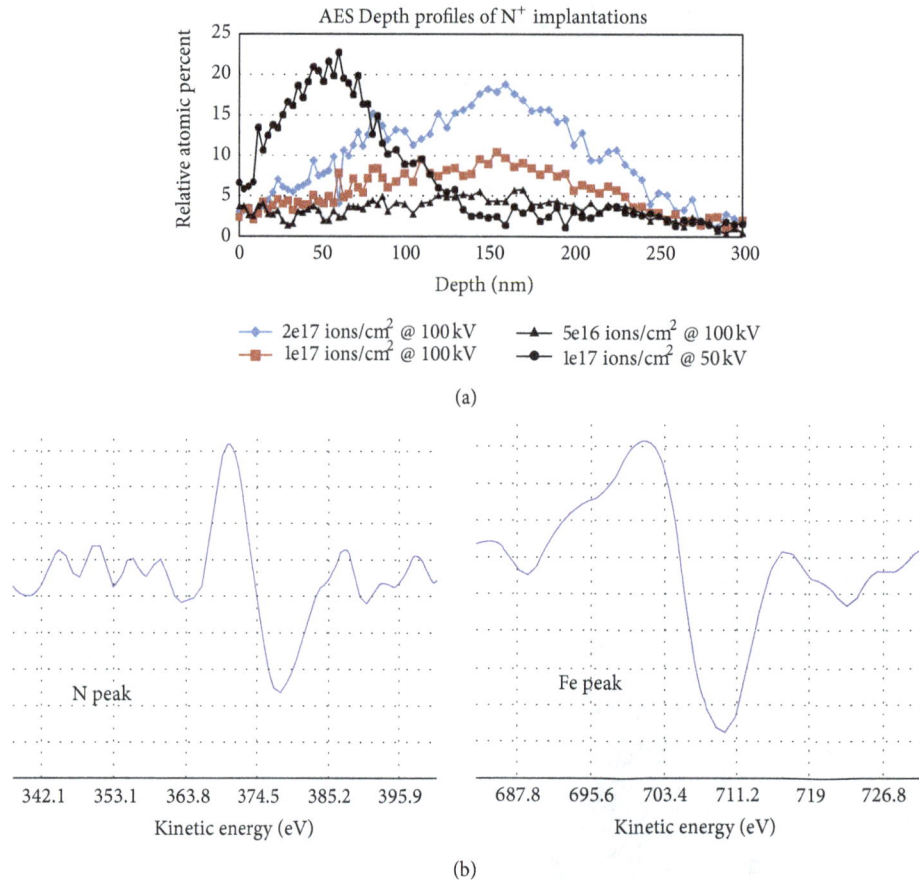

FIGURE 4: AES N depth profile and sample Auger peaks for four implantation conditions. Sample differentiated spectrums are provided in (b).

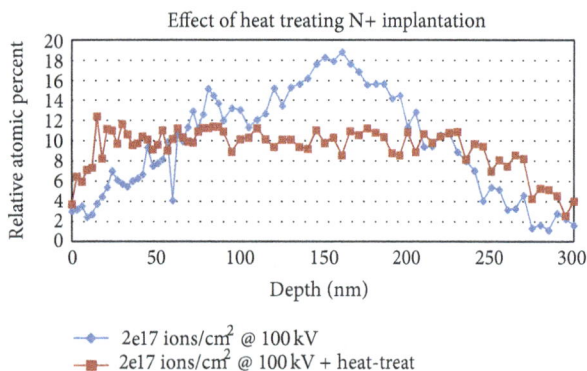

FIGURE 5: Depth profile of heat-treated 2e17 ions/cm^2, 100 kV compared to nonheat-treated sample implanted under same conditions.

that the N-implanted samples could be compared to. The cold-rolled samples were rolled prior to grinding/chemical polishing and implantation if applicable.

The implantations were characterized using Auger Electron Spectroscopy (AES) with a PHI-590 Auger spectrometer with Ar ion beam depth profiling. The energy of electron probe was 3 kV, and the Ar ion beam energy was 1 kV.

The depth versus sputtering time was estimated using the following relationship:

$$d = \frac{kCt}{\rho}, \qquad (6)$$

where k is the sputtering coefficient determined by a SRIM simulation, C is the ion current density, and ρ is the atomic packing density. The current was controlled 100–80 nA, which when rastered over a 1.5 mm × 1.5 mm area yields a sputtering rate of 2-3 Ang/s.

A Bruker D5005 diffractometer using a Cu X-ray tube performed Glancing Angle X-Ray Diffraction (GAXRD) using a 1.5 deg. angle of incidence. A 1.5 deg. angle of incidence gave a 150 nm layer sensitivity 77–84% using kinematical diffraction theory (This calculation uses the formula derived by Rafaja [25] where it is assumed that the reflection intensity from a layer dz is given by $dI = I_o \exp\{-\mu z((1/\sin(\theta_{\text{incident}})) + (1/\sin(\theta_{\text{exit}})))\}dz$. I_o is the reflection intensity of the top layer and μ is the mass absorption coefficient. The evaluated integration limits were [0, 150 nm] for the implantation layer and [0, 400 nm], where 400 nm reflects the maximum depth for a detectable reflection signal as determined by Beer's Law. The layer sensitivity is defined as $(I/I_o)_{[0,150]}/(I/I_o)_{[0,400]} * 100$). A Topcon-90 SEM with EDS capability analyzed the wear

Glancing angle X-ray diffraction
results for nitrogen implantations

(a) —— 2e17 ions/cm^2; 100 kV
(b) ······ 1e17 ions/cm^2; 100 kV
(c) – – – 5e16 ions/cm^2; 100 kV
(d) ·–·– 2e17 ions/cm^2; 100 kV; heat-treat
(e) – – – 1e17 ions/cm^2; 50 kV

(a)

Glancing angle X-ray diffraction
results for unimplaned samples

(b)

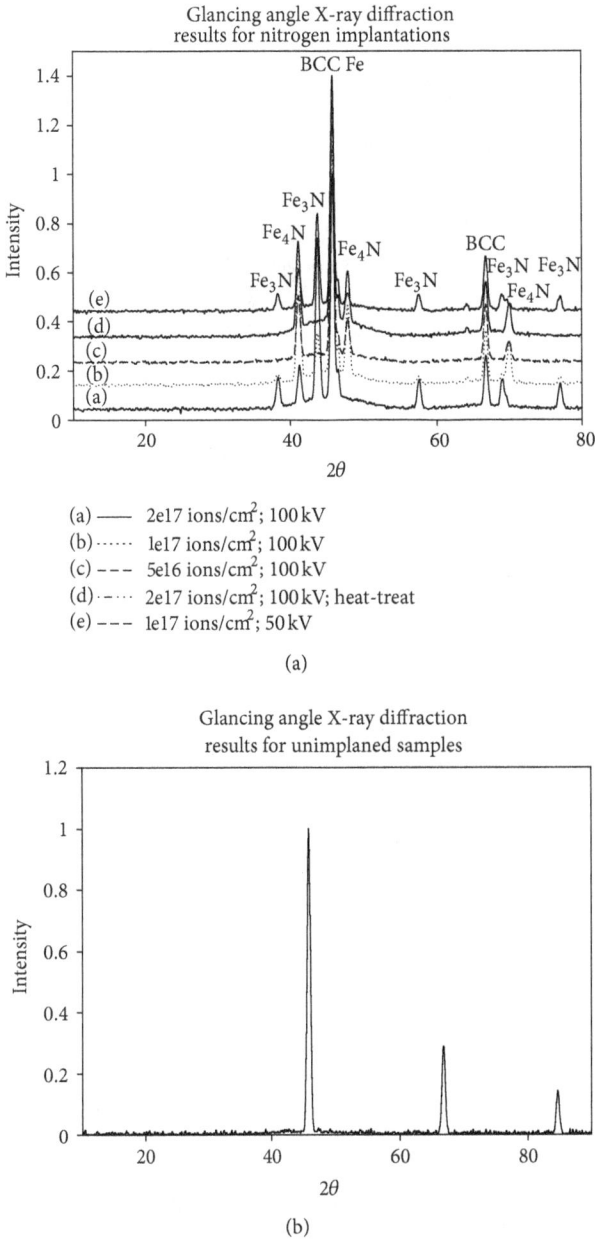

FIGURE 6: (a) Diffraction patterns of implanted surfaces using Cu Kα at a 1.5 deg. angle of incidence. (b) Diffraction patterns of unimplanted sample.

samples for debris morphology and chemistry. The electron acceleration used varied and is noted in this document. White light interferometry via a Zygo New-View 5000 determined the surface roughness.

3. Results

3.1. Implant Characterization. The AES depth profiles for the nitrogen implantations are in Figure 4. The N KLL peak at 380 eV and Fe LVV (III) peak at 708 eV were monitored during the depth profiling, and the peak intensities were determined by using the peak-to-peak height method with

a differentiated spectrum. The relative atomic percent of nitrogen was evaluated by

$$X_N = \frac{I_N/S_N}{(I_N/S_N) + (I_{Fe}/S_{Fe})}, \tag{7}$$

where I is the peak intensity and S is the peak sensitivity factor (determined from the *Handbook of Auger Electron Spectroscopy*, 3rd edition, Physical Electronics). As expected, the data shows that the peak concentration shifts towards the surface as the implantation energy reduces from 100 keV to 50 keV, and the peak concentration increases with ion dose. The fact that the depth profiles still retain their Gaussian shape indicates that the samples were kept sufficiently cool during the implantation to limit diffusion. The depth profile of the heat-treated (200°C, 1 hr in a stainless steel-canned argon environment) 2e17 ions/cm^2; 100 kV sample is given in Figure 5. The depth profile shows a leveling of the as-implanted profile. AES analysis of the oxygen content revealed higher surface oxygen levels than the nonheat-treated ones.

Figure 6(a) provides the GAXRD patterns for the implants, and Figure 6(b) has the reference unimplanted diffraction pattern. The pattern reveals the presence of γ'-Fe$_4$N (cP5, Perovskite prototype) and ε-Fe$_3$N (hP4, NiAs prototype) in varying proportions depending on the implantation conditions. As one would expect, the proportion of Fe$_3$N peaks increased as the dose increased for constant ion energy. Annealing (200°C, 1 hr) the samples after implantation decreased the proportion of Fe$_3$N while increasing the Fe$_4$N.

Other studies [13, 26, 27] reported this decomposition of Fe$_3$N to Fe$_4$N due to postheat treatment and would be expected because the local N concentration is lowered. In addition to nitride peaks, α' (BCT) martensite formed by N+ implantation can be observed, which was evident from the {110} → {110} + {101} splitting of the BCC peaks. There was some disorder in the 1e17 and 2e17 ions/cm^2 implanted samples, which the amorphous humps beneath the {110} BCC peaks indicate when compared to the unimplanted {110} BCC peaks. The disorder is due to knock-on damage. The 200°C, 1 hr anneal removed some of the disorder as the less pronounced amorphous hump indicates. As expected, the intensity of the amorphous relative to the {110} BCC peak increased with ion dose and energy. It should also be noted that cold rolling had no measurable effect on the diffraction pattern; the diffraction pattern was solely determined by implantation. The nitrogen implantations also contained detectable Fe$_3$O$_4$ in the implantations with doses greater than 1e17 ions/cm^2. XPS and TEM studies have revealed that under ambient conditions Fe$_3$O$_4$ films form on polycrystalline iron [28]. The relative peak intensities of the Fe$_3$O$_4$ lines compared to the formed nitrides were the largest at the 2e17 dose implantation indicating that implantation promoted the growth of the oxide film following implantation.

Figures 7(a) and 7(b) contain SEM images of the wear debris transferred to the unimplanted, unrolled ISF steel. The amount of transferred wear debris was quite extensive. Using Archard's wear equation (1) to calculate the wear coefficient

(a)   (b)

(c)   (d)

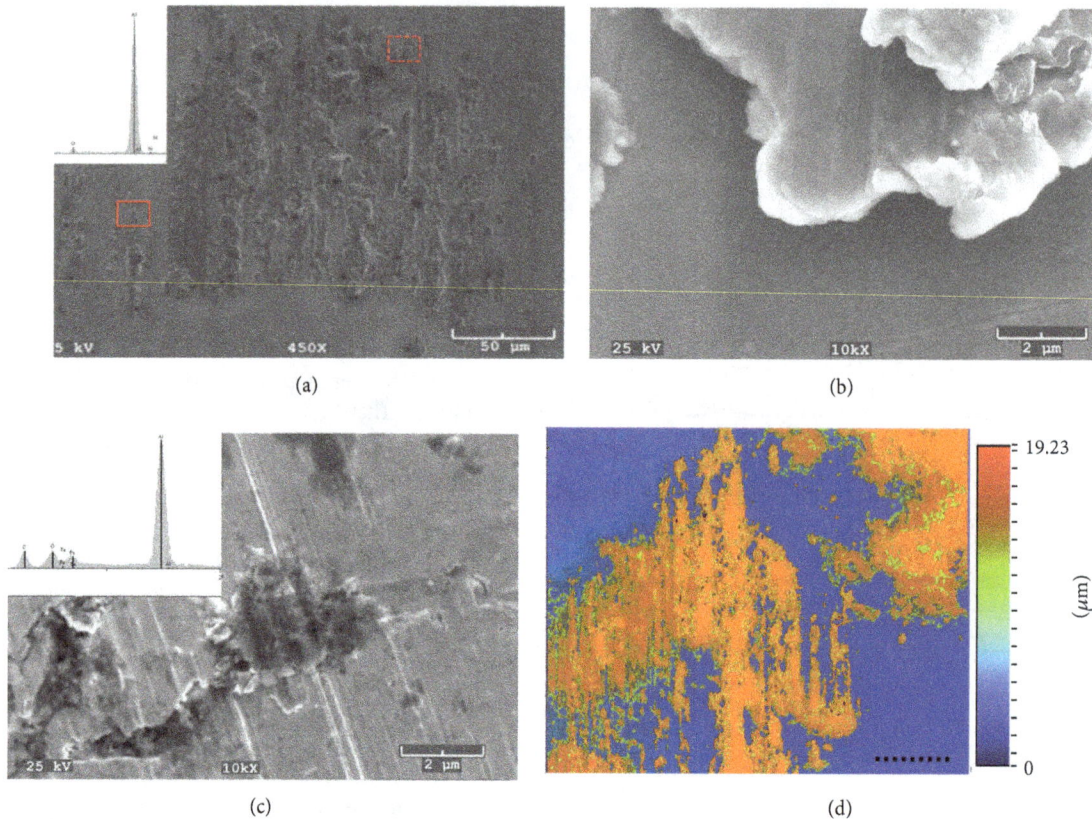

FIGURE 7: (a) An example of the metal transfer when 1100 Al is slide on a unimplanted, unrolled ISF steel surface. Spectrum is taken from the debris marked with a dotted box. (b) High magnification image of the debris marked with a dotted box in Figure 5(a). (c) High magnification image of the debris marked with a solid box with EDS spectrum in Figure 5(a). (d) Optical profilometer scan of the metal transfer when 1100 Al is slide on the unimplanted, unrolled ISF steel surface. The dotted black line in the lower right-hand corner is a 50 μm scale bar.

of the rider and using a Vickers hardness of 167 MPa for 1100 Al gave 2.9×10^{-2}. The size of the debris was also very large: Figure 7(d) gives an example of an optical profilometer scan taken from this wear track. The protrusions from the surface were almost 19 μm in some spots, but the majority of the debris height was in the 10–16 μm range. The width of the debris was 1–25 μm. It is likely that in a multipass wear test protrusions of this size will continue to build up and result in larger friction coefficients due to the added ploughing component. Certainly if the ISF steel was a metal-forming die material, protrusions of this size would damage the surface of the incoming material. In addition, the EDS analysis of the chemistry of the larger debris (Figure 7(b)) revealed no detectable iron, and the oxygen content was on par with an unworn 1100 Al ball. However, the chemistry of the smaller debris (Figure 7(c)) contained Fe and elevated amounts of oxygen. A possible reason for this is that the smaller debris rolls more than the larger debris on the surface and as a result accumulates iron as the aluminum oxides. The chemistry indicates that the smaller debris was multi-particle, while the larger debris appeared to be single bodied with evidence of plastic tearing (see Figure 7(b)) on the surface. Repeated tumbling on a surface removes the wear particle's fractographic features from the wear event. For

the larger debris, this is not the case as signs of tearing can be seen, which is in contrast with Figure 7(c) where tears were not visible on the smaller debris and the debris looked compacted. The observed tears and plate-like morphology follow the delamination mechanism proposed by Suh et al. [29] in which shear forces transmit through a welded interface. The existence of large debris means that there was sufficient contact area to transmit the required force to cause this delamination in addition to appreciable shear stresses at the interface.

Nitrogen implantation had a measureable effect on metal transfer. Figure 8(a) shows an example of the metal transfer to the 2e17 ions/cm^2, 100 kV N+ implanted ISF steel plate. The Archard wear coefficient for the rider was 3.4×10^{-4} giving two orders of magnitude reduction in Al wear compared to the unimplanted. Using optical profilometry, Figure 8(c), the debris was as high as 8 μm, which is still rather large and as a result could cause problems in tribosystems. The width of the debris was 1–8 μm. The chemistry of the debris measured by EDS also contained oxygen, iron, and nitrogen. In addition, the debris appeared to be multi-particle. The microgroove formation process likely incorporated the iron and nitrogen in the debris. Interestingly, EDS

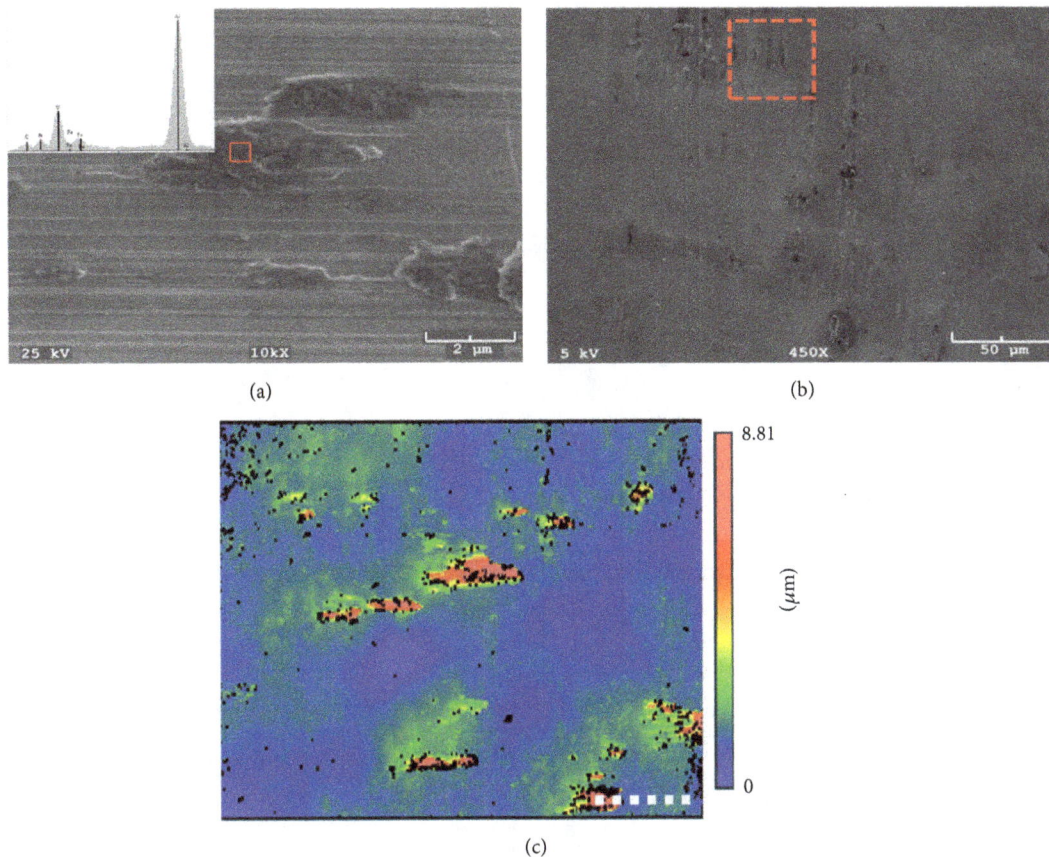

FIGURE 8: (a) SEM image of the debris transferred to the 2e17 ions/cm^2, 100 kV N+ implanted ISF steel plate. (b) High Magnification image of the debris within the dotted box. The 7 kV EDS spectra of the debris are included. In addition, Fe, O, and N were detected on the scratched ISF steel surface (red box) at 7 kV. (c) Profilometry scan of the wear debris. The protrusions on the surface exceeded 8 μm in some spots. The white dotted line is a 6 μm scale bar.

(5 kV) detected nitrogen inside the microgrooves indicating that the implanted layer was intact by some measure. Comparing the 5 kV N-Kα signal of the 100 kV implants outside of a microgroove and inside a microgroove gave $I_{\text{N-K}\alpha,\text{groove}}/I_{\text{N-K}\alpha,\text{surface}} = 0.89$. Doing the same comparison for the 50 kV implants gives $I_{\text{N-K}\alpha,\text{groove}}/I_{\text{N-K}\alpha,\text{surface}} = 0.21$. This means that the 50 kV implants lost more nitrogen than the 100 kV implants because of the microgroove formation, which is likely due to the higher concentration of the nitrogen with reference to the surface in the 50 kV implantations (2e17 ions/cm^2). Compared to the debris in Figure 7(a), the fact that all of the debris appeared to be multiparticle means that the shear stresses were reduced to the point that large debris, too large to tumble on the surface, could not form in one wear event. The debris in Figure 8(a) reveals some evidence for small particle delamination, which was not destroyed by the tumbling process.

The relationship between the transferred Al area coverage on the ISF steel and the Al volume transfer is quadratic as Figure 9 indicates. The regression curve slope is the average Al transfer height, which increases with surface coverage. Adjacent to Figure 9 is a table showing the average debris

height for each wear test as determined by the volumetric Al transfer divided by the Al covered area as described in the Methodology section. The average heights are 10.5–2.9 μm. The measured debris heights in Figure 7(d) were 10–16 μm and 4–8 μm in Figure 8(c). These measurements follow the regression curve trend of Al transfer height which increases with coverage. Note that the average debris height measured by profilometry *is not* exactly equivalent to the Al transfer-height because debris also contains wear elements of ISF steel and can be ploughed into the surface.

Microgram scale measurements of the Al rider wear gave a relationship between wear and the hardness of the ISF steel (H_{hard}) as achieved through cold rolling. The regression parameters a and b in (4) were adjusted to best fit the data. Figure 10 illustrates the regression analysis results for the unimplanted samples. The R^2 of the fit was 0.96 when a and b were 1.1×10^{-5} and 4.7×10^{-2}, respectively. The average Archard wear coefficient (b/W) was 2.6×10^{-2}. For the data points, the Archard coefficients based on (1) ranged from 2.0×10^{-2} to 2.9×10^{-2} depending on the amount of cold work—where cold rolling decreased its value. The error bars resulting from the repeatability tests (Figure 2) were included, which ends up being less than 2% of the

Test	Ave debris height (μm)	Ave debris height error (μm)
ISF, 0%, unimplanted	10.5	0.65
ISF, 7%, unimplanted	8.9	0.54
ISF, 15%, unimplanted	8.9	0.54
ISF, 30%, unimplanted	9.2	0.68
ISF, 67%, unimplanted	8.8	0.55
ISF, 100%, unimplanted	9.5	0.43
ISF, 200%, unimplanted	9.6	0.54
2e17 100 kV N	2.4	0.73
1e17 100 kV N	7.9	0.92
5e16 100 kV N	7	0.5
2e17 50 kV N	5.1	0.86
2e17 100 kV N, 100% red	3.5	1.08
2e17 50 kV N, 100% red	2.9	0.78
2e17 100 kV N, HT	3.4	0.94
2e17 100 kV N, 100% red, HT	4.5	1.49

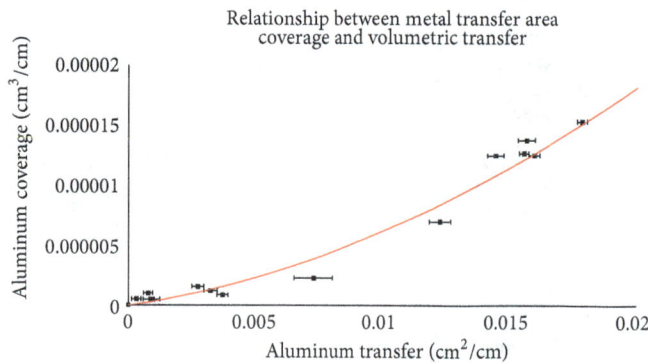

FIGURE 9: Plot of volumetric metal transfer versus area covered by transferred wear debris.

reported values. The data indicates that the hardness of the harder material had a negligible effect on rider wear rates as the wear coefficient changed less than an order of magnitude.

Figure 11 reports the mass lost and mass transfer percentage for the unimplanted and implanted tests. Depending on the dose, some of the implanted samples had lower Al rider wear rates by almost two orders of magnitude; the (1) calculated wear coefficients for all the samples ranged from 3.4×10^{-4} to 2.3×10^{-3}. Rider wear decreased with implantation dose: see A through C in Figure 11. In addition, fixing the dose and changing the accelerating voltage appeared to have a small effect on the rider wear (compare A and D in Figure 11) with the higher potential providing better wear properties. Rolling appeared to have no effect on the rider wear for the 2e17 ions/cm^2 at 100 kV dose (compare A and E in Figure 11) and the 2e17 ions/cm^2 at 50 kV dose (compare D and F in Figure 11). Unexpectedly, heat treating appeared to have no effect on rider wear (compare A and G and E and H in Figure 11). The metal transfer (red bars) decreased with implantation dose, which was the major controlling variable in reducing metal transfer for the implanted samples. Figure 11 also contains the percentage of metal transfer. Almost all of the rider wear became metal transfer for the unimplanted samples with metal transfer percentages above 90%. In general, implantation reduced the metal transfer percentage by almost 40% for some conditions. However, the metal transfer % measurements were within the same error-band for some of the samples. Based on the available data the metal transfer percentage decreased with implantation dose (compare A to C) and energy (compare A to D), but cold rolling had very little effect on this quantity.

The evaluation of friction traces helps to ascertain the nature of the interaction during sliding. The measured torque, rider offset, and load determined the coefficient of friction (COF). Figure 12(a) plots the COF for the different sets of tests. The COF for the unimplanted samples was 0.67–0.69 ± 0.005 and fell within the same margins of error. It appears that cold rolling has no detectable influence on the COF. Nitrogen implantation appeared to influence the COF; as the dose/energy increased, the COF decreased. At the lowest, the COF was 0.30 ± 0.005 for the 2e17 ions/cm^2 at 100 kV implanted samples. In addition to having a lower friction coefficient the friction noise generated against the N+ implanted plate was less. The noise levels for the unimplanted samples (20–25%) were larger than the N+ implanted samples (7–11%). As the dose increased, the percentage of noise decreased. Note that the error bars in Figure 12(a) are determined by the data recording sensitivity of the tribometer software, which was ±0.01 lbs∗inch. This gave a COF error of ±0.005 for a track diameter of 1.3″ and a load of 3 lbs.

Table 1 gives the debris chemistries along with the near surface chemistries of the different samples. For proper statistics, five EDS spectrums from different debris were used for a given sample. Note that Table 1 also reports the standard deviation next to the average chemical composition for the five-sampled debris in a given sample. Only debris that appeared to be mixed were sampled; see Figures 7(c) and 8(a) for examples. The reported near-surface chemistry in Table 1 is a 300 nm average from the surface measured by the AES depth profiles. For the averaging, 1 nm spaced data points were used. As expected, the higher the nitrogen dose, the greater the average amount of nitrogen detected in the debris. The nitrogen-implanted samples had greater amounts of oxygen detected in the debris than the unimplanted. The amount of oxygen increased with nitrogen content in the debris; the Pearson correlation coefficient between those two variables was 0.47 based on the 75 debris chemistry samples (15 test conditions × 5 debris samples). The critical Pearson correlation coefficient value based on this sample size is 0.3 for a 95% certainty, which indicates that there was a correlation between nitrogen and oxygen debris chemistries.

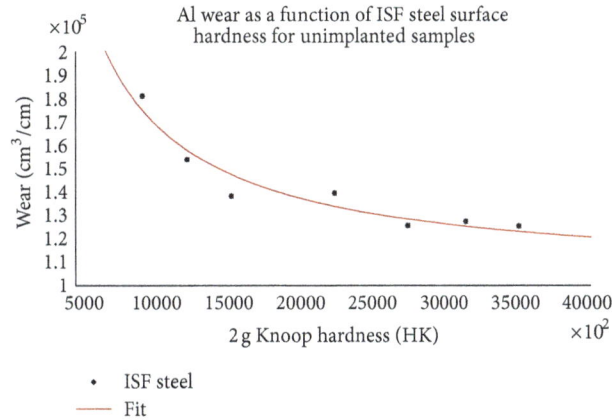

FIGURE 10: Metal transfer as a function of hardness for the unimplanted ISF Steel. The fit from (4) is included.

This indicates the nitrogen in the debris affected the tribo-oxidation that took place. Rolling had no measureable effect on the debris chemistry indicating that the fractured particles from the ISF steel going into the debris were less than the size of the implantation zone. This conclusion is based on the observation that the 2 g Knoop hardness measurements on samples **A** and **H** in Figure 12 were only 23 points apart (Sample **H** > **A**), but the 200 g Knoop hardness measurements were over 200 points apart (Sample **H** > **A**). It is expected that, if the ISF steel wear elements were larger than the implantation zone, cold rolling would affect the debris chemistry because there was a large difference in hardness values at a 200 g load. Unexpectedly, the energy of the implant affected the debris chemistry while keeping the implant dose constant. The lower energy implants produced debris with more nitrogen, but the oxygen content was within the margin of error of the 100 kV 2e17 ions/cm^2 implants. The higher concentration of nitrogen near the surface may be the cause for the larger detected amounts of nitrogen, and this gives more evidence that the particles coming from the ISF steel are smaller than the implantation zone.

4. Discussion

The real contact area was defined by W/H_{eq} in (4), but the $1/H_{hard}$ term in (3) only slightly effected the wear rates. The difference in wear rates between the unrolled and 200% reduction samples was less than 1/5th of an order of magnitude. Practically, there are no differences in wear rates between the rolled ISF steel samples. This means that the real contact area can be defined by W/H_{soft} as the hardness of the harder surface has little effect on the adhesive wear of the Al rider. This means that any reduction in wear from implantation is not related to differences in contact area generated by hardening. Kayaba et al. [30] proposed using a slip-line deformation model that the harder material asperities can yeild if an asperity tip angle is less than a critical value determined by H_{soft}/H_{hard}. These predictions have been qualitatively confimed by Fishkis [31] for steel/aluminum tests. Obviously, hard asperity yeilding

will increase the real contact area; however this effect was not observed in terms of wear rate. It is likely that the chemical polishing smoothed out the surface such that the asperity angles were well-above the critical angle. It should be mentioned that the real contact area (52816 μm^2) exceeds the initial Hertzian apparent contact area (6250 μm^2), but not the final contact area as measured by the worn flat diameter (148542 μm^2). A more detailed Hertzian contact analysis revealed that the Von-Mises stresses in the Al ball exceeded the YS of the material, meaning that the initial Hertzian relationship does not apply and that the initial apparent contact area is much higher. A reasonable estimate for the initial contact area, which is under elastoplastic contact conditions, is obtained by finite element empirical modeling [32, 33], which give apparent contact areas of 69234 and 66128 um^2, respectively. As expected, the real contact area as given by W/H_{soft} is lower than modeled apparent contact area by about 22%.

These sets of experiments are unique in that the ion implantation was not originally meant to improve the wear resistance of ISF steel, but to reduce the wear of the 1100 Al rider. In addition, volumetric wear data for single pass experiments to our knowledge have not been published. As a result, finding comparable data within the current context of experiments is difficult. The most relevent set of published, qualitative metal transfer data is provided by Antler [34] where the volumetric wear for a series of pure-metal combinations was studied under multipass wear tests. For pure Al on pure Al the rider volumetric wear rate was 1.5×10^{-6} cm^3/cm. Comparing our value for ISF steel with 1100 Al (1.7×10^{-5} cm^3/cm) the much higher value is likely due to run-in effects from the chemically polished ISF steel surface. Remember that while the Al rider is being run-in during wear testing the rider is continuously encountering a fresh, unworn ISF steel surface. The other difference is likely due to preparation methods; Anlter used mechanical grinding for sample preparation while we used chemical polishing. The real contact areas were likely larger due to the lower RMS values associated with chemical polishing. Figure 13 demonstrates that N+

TABLE 1: Debris and near-surface chemistries for the samples used in this test.

Sample	Debris chemistry			Near-surface chemistry		
	Oxygen (At%)	Carbon (At%)	Nitrogen (At%)	Oxygen (At%)	Carbon (At%)	Nitrogen (At%)
Unimplanted	7.2, 1.5	2.1, 1.3	NA	0.3	<0.1	0.0
Unimplanted, 7% reduction	8.1, 1.3	2.4, 0.5	NA	0.4	<0.1	0.0
Unimplanted, 15% reduction	6.3, 1.7	2.9, 0.9	NA	0.2	<0.1	0.0
Unimplanted, 30% reduction	6.4, 1.4	3.0, 0.8	NA	0.1	<0.1	0.0
Unimplanted, 67% reduction	6.2, 1.9	2.5, 0.5	NA	0.2	<0.1	0.0
Unimplanted, 100% reduction	7.1, 1.9	2.7, 1.1	NA	0.3	<0.1	0.0
Unimplanted, 200% reduction	6.9, 1.5	2.2, 1.5	NA	0.1	<0.1	0.0
N+ implanted, 5e16 ions/cm^2 at 100 kV	11.7, 1.4	2.4, 1.1	2.1, 0.5	1.4	<0.1	3.8
N+ implanted, 1e17 ions/cm^2 at 100 kV	12.5, 1.1	2.1, 1.7	4.0, 0.8	2.2	<0.1	7.1
N+ implanted, 2e17 ions/cm^2 at 100 kV	16.4, 1.6	2.9, 1.2	6.5, 1.1	3.3	<0.1	12.5
N+ implanted, 2e17 ions/cm^2 at 50 kV	15.1, 1.8	2.2, 0.8	9.9, 0.7	2.0	<0.1	10.1
N+ implanted, 2e17 ions/cm^2 at 100 kV, 100% reduction	16.6, 1.2	2.2, 1.4	6.3, 1.7	3.1	<0.1	12.7
N+ implanted, 2e17 ions/cm^2 at 50 kV, 100% reduction	16.2, 1.5	2.7, 1.2	9.1, 0.5	2.0	<0.1	10.3
N+ implanted, 2e17 ions/cm^2 at 100 kV heat treated	16.9, 1.3	2.1, 1.6	3.9, 1.1	3.9	<0.1	12.1
N+ implanted, 2e17 ions/cm^2 at 100 kV, 100% reduction, heat treated	16.1, 2.2	2.6, 1.7	7.0, 1.3	4.0	<0.1	12.2

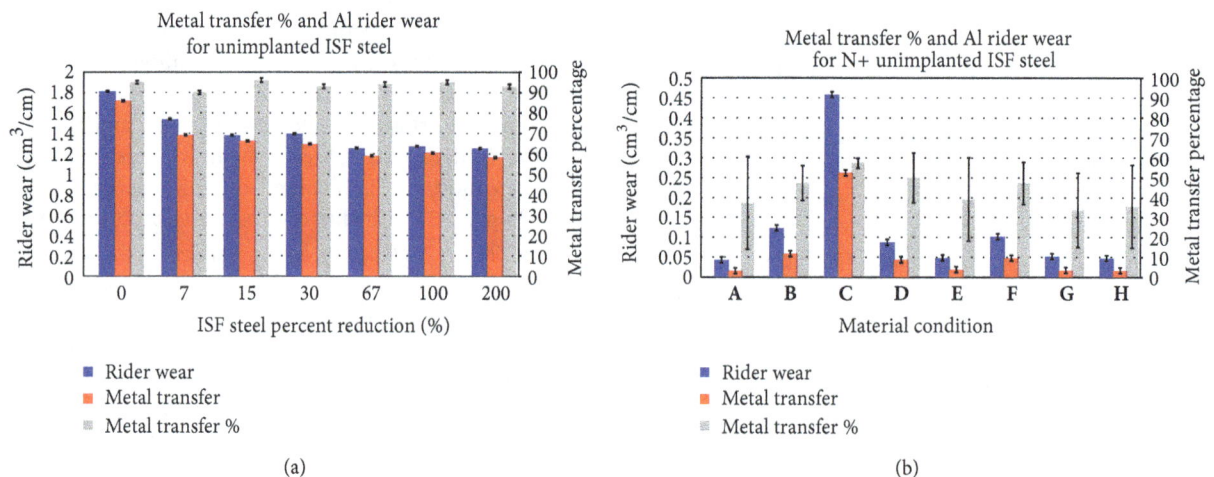

(a)

(b)

FIGURE 11: Rider wear and metal transfer percentage for the various tested conditions. The label references for the N+ implanted labels are provided at the bottom. Note that the rider wear for the N-implanted samples (b) were scaled by a factor of 10. **A** = 2e17 ions/cm^2 at 100 kV, **B** = 1e17 ions/cm^2 at 100 kV, **C** = 5e16 ions/cm^2 at 100 kV, **D** = 2e17 ions/cm^2 at 50 kV, **E** = 2e17 ions/cm^2 at 100 kV at 100% red, **F** = 2e17 ions/cm^2 at 50 kV; 100% red, **G** = 2e17 ions/cm^2; 100 kV; heat treat, and **H** = 2e17 ions/cm^2 at 100 kV; 100% red; heat treat.

ion implantation reduced the amount of rider wear and the amount of metal transfer to the plate, but the structure of the nitride as altered by heat treating the implanted zone appeared to have no effect on the rider wear. The improvement from implantation in the rider wear was at most two orders of magnitude for the 2e17 ions/cm^2 at 100 kV implantations (including rolled and heat treated). The 2e17 ions/cm^2 at 100 kV implantation wear coeficients were in the 3×10^{-4}–7×10^{-4} range, which is also in the moderate wear regime.

The rider wear mechanisms need to be discussed prior to evaluating the benefits of nitrogen implantation. The surface roughnesses of the evaluated surfaces were 0.010–0.005 μm, and by determining the number of apex points in a given area through the MetroPro software the maximum abrasive wear can be deduced. The estimated *maximum* abrasive wear rates were from 8×10^{-7} to 1×10^{-6} cm^3/cm, which is almost two orders of magnitude less than the measured wear rates for the unimplanted samples. This means that abrasive wear was initially negligible and adhesive wear via

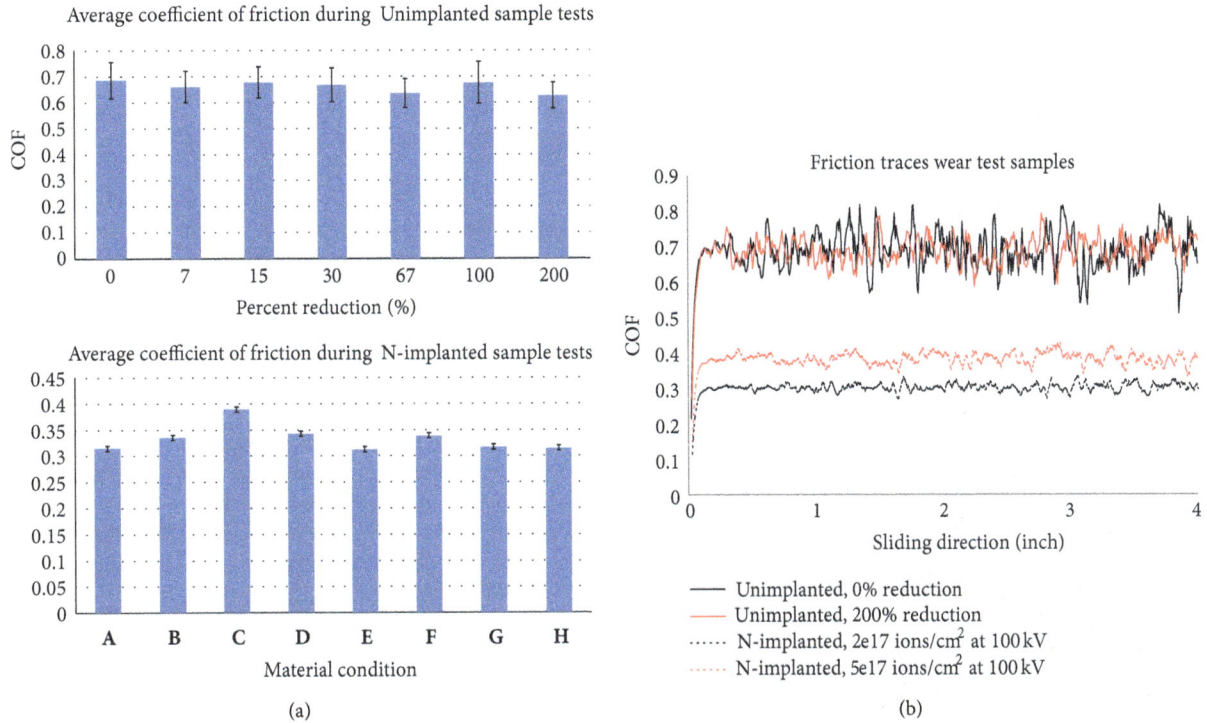

FIGURE 12: (a) Average friction coefficients of tested samples: unimplanted (Top) and N-implanted (Bottom). **A** = 2e17 ions/cm^2 at 100 kV, **B** = 1e17 ions/cm^2 at 100 kV, **C** = 5e16 ions/cm^2 at 100 kV, **D** = 2e17 ions/cm^2 at 50 kV, **E** = 2e17 ions/cm^2 at 100 kV at 100% Red, **F** = 2e17 ions/cm^2 at 50 kV; 100% red, **G** = 2e17 ions/cm^2; 100 kV; heat treat, and **H** = 2e17 ions/cm^2 at 100 kV; 100% red; heat treat. (b) Example friction traces for different samples.

delamination dominated. It is expected that if abrasive wear was a factor in these studies, cold rolling the ISF steel should have raised the wear rate of the Al rider because harder asperities will blunt less as they penetrate into the Al surface. The fact that the unimplanted samples displayed friction traces with relatively large noise levels at 20–25% indicates that a two-body stick-slip mechanism [35, 36] existed. This is also evidence of adhesive wear occuring because sufficient adhesive forces are needed to create the stick-slip condition. Stick-slip conditions arise when asperity junctions weld, and the junctions grow. The sticking condition ends, and slip begins with the formation of a wear element. The junction could also separate without wear element formation. The COF noise was reduced to 5–10% in the implanted samples. Implantation also had the effect of reducing the COF by 0.25–0.35 depending on the dose. According to Table 1, the near-surface nitrogen chemistry negatively correlated with decreases in both the friction coefficient and noise, which indicated that the nitrogen chemistry disrupted the stick-slip process by reducing the adhesive forces. Using Tabors junction growth formula, (5), given the COF and noise, the average shear strength of the interface can be determined. Equation (5) uses the shear yield strength of Al—3.26 ksi. The 2e17 ions/cm^2 at 100 kV implantation had an average interfacial shear strength of 0.45 ksi while the unimplanted sample was 0.94 ksi. Using Density Functional Theory (DFT) calculations ({100}Al | {{100}Fe interface calculations were performed using density function theory with the VASP software package. The plane augmented wave method was used for the atomic potentials. The exchange-correlation term was handled using the generalized gradient approximation. The calculations were converged to within 0.01 eV with respect to plane wave basis set size, k-point mesh density, vacuum gap, and layer size.) of the Fe | Al {100} interface along with the real contact area (W/H_{soft}) provides an upper limit in the estimation of the adhesion strength. DFT calculations indicated that the interfacial cleavage energy was rougly 50% higher than bulk aluminum. Assuming that the same relationship approximately holds for the shear strength, an estimate of the shear strength of the Fe | Al interface is 4.89 ksi. This indicates that interfacial films played a role in reducing the interfacial shear strength for *all* tests during the sticking phase. The implanted samples had a lower interfacial shear strength by almost 0.49 ksi than the unimplanted indicating that the chemistry played a role whether it be the oxide film, the implanted nitrogen, or both.

Implanting nitrogen reduced the metal transfer percentage relative to the unimplanted condition. Comparing Figure 11 with Table 1 indicates that the metal transfer percentage decreased with average debris nitrogen and oxygen content. The reasons for the reduction in metal transfer percentage is likely due to the presence of nonmetallic, abrasive particles that promote the removal of wear debris from the contact region [37]. The transfer metal debris chemisty was mixed

(a)

(b)

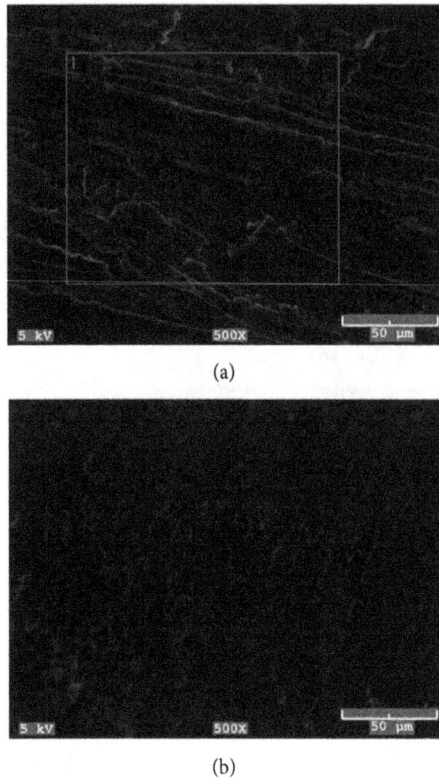

FIGURE 13: (a) Rider wear surface tested against the unimplanted ISF Steel surface. (b) Rider wear surface tested against the 2e17 ions/cm^2 at 100 kV nitrogen implantation condition.

(Fe, Al, O, and N) with the exception of the debris that included larger wear elements (see Figure 7(b)). TEM studies by Rigney et al. [38] indicated that debris transfer layers are multiparticle and multielement, which results from the mobility of the debris. Mechanically mixed and oxidized wear debris can act as a beneficial tribolayer [39–41]. According to the Mutual Material Transfer model, when a wear element forms on an opposing surface, it creates a protrusion that acts like an asperity. There is a certain probability that another wear element will form on top of previously generated wear element. In this manner, the debris will build up until it reaches a critical size in which the obstruction forces between the adhered debris and the sliding surfaces overcome the adhesion forces between the debris and the surface it is attached to resulting in the expelling of the debris from the wear track. If the debris is composed of less adherent particles such as oxides, and nitrides the adhesion force between the debris and its attached surface is reduced, which should reduce the critical-size of the debris. Figure 9 is indirect evidence to support the last statement because the reduction in wear rate was due to a reduction in adhesion from the presence of oxides and nitrides in the surface layer. Back transfer to the Al pin was also reduced by implantation. Evidence of Al back transfer to the rider is demonstated in Figure 13(a) where large Al particles can be seen sticking to the Al surface as opposed to the rider that slids on the N-implanted surface (Figure 13(b)), which had no evidence

of back transfer. EDS chemistries taken from the rider wear surfaces indicated the presence of ISF steel transfer elements.

The oxides that formed are due to tribo-oxidation [42]. There also is evidence that nitrogen implantation enhances the benefits of tribo-oxidation during high cycle wear tests [43, 44]. The amount of oxygen in the debris was positively correlated with the nitrogen debris chemistry as the 0.47 Pearson correlation coeficient indicates. This means that nitrogen implantation had an effect on tribooxidation. Due to the fact that native oxide films are self-limiting, most of the oxidation likely occurred during the mechanical mixing of the wear elements because of the higher temperatures associated with cold working the particles. There is sufficient evidence [45, 46] that N implantation effects the low-temperature oxidative properties of iron. All exposed transition metal surfaces have native oxide films on them. As mentioned above implantation has an effect on the tribo-oxidation of the debris. Figure 14 contains depth profiles of oxygen for the nitrogen implantations after 15 days of implantation with a reference unimplanted surface. The depth of the native oxide layer increased with implantation dose and energy. At the heaviest dose the film increased by as much as a factor of 3.5 with reference to the unimplanted sample. XPS (Fe 2p 3/2) surveys (unimplanted and 2e17 ions/cm^2 at 100 kV) after 400 s of sputtering to remove the surface carbon are also included. Implantation had the effect of increasing the amount of the Fe-O(III)/Fe-O(II) peak ratio relative to the unimplanted sample presumably due to the larger oxygen concentrations. The lattice defects created by implantation likely enhanced the migration rate of the changed species by the increased vaccancy count. The thicker oxidation layer present probably also helped in reducing the adhesion, which in turn reduced the size of the wear particles through delamination and the level of friction from junction growth. In addition, the larger oxide film likely helped incorporate less adherent ISF steel wear elements into the debris.

The quadratic relationship between the area covered and the volumetric transfer for a transferred material has not been established in the literature. In fact, the factors that affect debris size are usually related to the wear rate. Factors such as test load [47], roughness of the harder surface [48], and test speed [49] are related to wear rate. In some studies [50], debris size is related directly to wear rate. In this study, the load, speed, and surface roughness values are held constant. Thus, debris size changes are due to the factors that directly affect metal transfer. According to the Mutual Material Transfer mechanism debris builds up until it reaches a critical size. At that point, the obstruction forces overcome the debris surface adhesive force and cause the particle to be removed from the wear surface. Ploughing *into* the ISF steel should provide additional mechanical keying and allow larger size debris to form. Hence, according to the Mutual Material Transfer model only two factors should influence debris size: wear element size distribution and the adhesion of the debris to the surface. The Figure 10 provides a useful tool when evaluating metal transfer when the mass transferred is below the sensitivity of the microbalance.

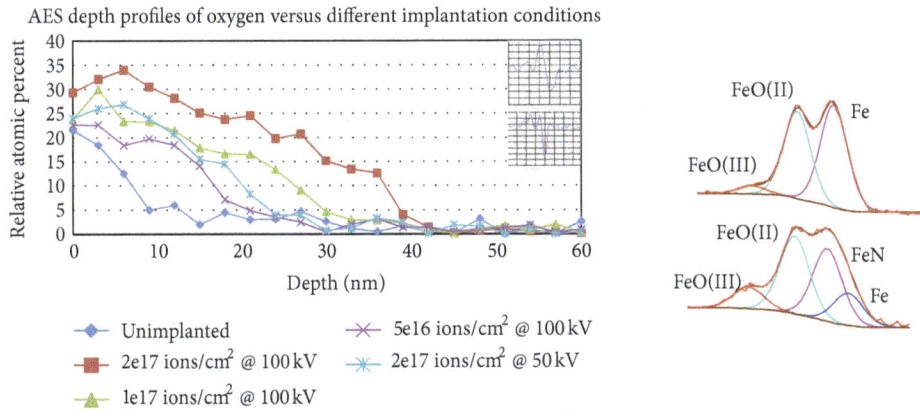

FIGURE 14: AES depth profile of oxygen for different implantation conditions. To the right are XPS (Mg Kα X-rays) surveys of the Fe 2p 3/2 photoelectron peaks after 400 s of sputtering to remove the surface carbon.

5. Conclusions

(i) Examining the metal-transfer/rider-wear data for the cold-rolled ISF steel shows that the tribological processes are not significantly impacted by hardness (less than 1/5th an order of magnitude), and the W/H_{soft} expression can be even when the hardness of the harder surface is changing.

(ii) Nitrogen implantation enhanced the native oxide film thickness by a factor of three two weeks after implantation. The enhancement is due to the knock on damage caused by implantation.

(iii) Ion implantation had the effect of reducing the Al rider wear by two orders of magnitude. The reduction in the Al rider wear was attributed to the near-surface nitrogen chemistry and the native oxide layer formed after implantation. Postimplantation heat treating caused ε-Fe_3N to decompose to γ'-Fe_4N, but this had no impact on the Al rider wear, which indicates that wear reduction was due to chemistry of the ISF steel near surface and not structure.

(iv) Implantation of nitrogen enhanced the debris tribo-oxidation process; the oxygen content in the debris showed a strong correlation with nitrogen near-surface chemistry. Due to the enhanced oxygen and nitrogen debris chemistry the fraction of Al debris ending up as metal transfer to the ISF steel surface was reduced.

(v) The noisy friction coefficient indicated that stick-slip wear mechanism was responsible for friction generation. Nitrogen implantation reduced both the friction coefficient and the friction noise. Calculations based on the Tabors junction growth formula indicate that nitrogen implantation reduced the adhesive strength of the interface with reference to the unimplanted surface.

Acknowledgments

Authors would like to thank Arcelor Mittal for supplying the ISF steel, the staff at the University of Michigan Ion Beam Laboratory and University of Albany Ion Beam Laboratory for their help in getting the implantations done, and the University of Connecticut's Institute of Materials Science and Deringer-Ney for instrumentation time. One of the authors currently has no financial ties to Arcelor Mittal or Deringer-Ney. Funding for this study was provided by the University of Connecticut's Institute of Materials Science.

References

[1] H. Wallin, *An investigation of friction graphs ranking ability regarding the galling phenomenon in dry SOFS contact [Thesis]*, Faculty of Technology and Science Department of Material Science Karlstad University.

[2] J. T. Burwell and C. D. Strang, "Metallic Wear," *Proceedings of the Royal Society of London A*, vol. 212, Mathematical and Physical Sciences, no. 1111, pp. 470–477, 1952.

[3] ASTM G98-02, 2009, Standard test method for galling resistance of materials.

[4] T. Sasada and S. Norose, "Formation and growth of wear particles through mutual material transfer," in *Proceedings of the JSLE-ASLE International Lubrication Conference*, pp. 82–91, American Elsevier Publishing, New York, NY, USA, 1976.

[5] A. Hase and H. Mishina, "Wear elements generated in the elementary process of wear," *Tribology International*, vol. 42, no. 11-12, pp. 1684–1690, 2009.

[6] A. Gåård, P. Krakhmalev, and J. Bergström, "Wear mechanisms in galling: cold work tool materials sliding against high-strength carbon steel sheets," *Tribology Letters*, vol. 33, no. 1, pp. 45–53, 2009.

[7] S. R. Hummel and B. Partlow, "Comparison of threshold galling results from two testing methods," *Tribology International*, vol. 37, no. 4, pp. 291–295, 2004.

[8] S. R. Hummel, "Development of a galling resistance test method with a uniform stress distribution," *Tribology International*, vol. 41, no. 3, pp. 175–180, 2008.

[9] U. Wiklund and I. M. Hutchings, "Investigation of surface treatments for galling protection of titanium alloys," *Wear*, vol. 251, no. 1–12, pp. 1034–1041, 2001.

[10] S. R. Hummel, "An application of frictional criteria for determining galling thresholds in line contact tests," *Tribology International*, vol. 35, no. 12, pp. 801–807, 2002.

[11] I. L. Singer, "Surface analysis, ion implantation and tribological processes affecting steels," *Applied Surface Science*, vol. 18, no. 1–2, pp. 28–62, 1984.

[12] J. K. Hirvonen, C. A. Carosella, R. A. Kant, I. Singer, R. Vardiman, and B. B. Rath, "Improvement of metal properties by ion implantation," *Thin Solid Films*, vol. 63, no. 1, pp. 5–10, 1979.

[13] T. Fujihana, Y. Okabe, and M. Iwaki, "Effects of implantation temperature on the hardness of iron nitrides formed with high nitrogen dose," *Nuclear Instruments and Methods in Physics Research Section B*, vol. 39, no. 1–4, pp. 548–551, 1989.

[14] T. Fujihana, A. Sekiguchi, Y. Okabe, K. Takahashi, and M. Iwaki, "Effects of room temperature carbon, nitrogen and oxygen implantation on the surface hardening and corrosion protection of iron," *Surface and Coatings Technology*, vol. 51, no. 1–3, pp. 19–23, 1992.

[15] H. Dimigen, K. Kobs, R. Leutenecker, H. Ryssel, and P. Eichinger, "Wear resistance of nitrogen-implanted steels," *Materials Science and Engineering*, vol. 69, no. 1, pp. 181–190, 1985.

[16] A. A. Youssef, P. Budzynski, J. Filiks, A. P. Kobzev, and J. Sielanko, "Improvement of wear and hardness of steel by nitrogen implantation," *Vacuum*, vol. 77, no. 1, pp. 37–45, 2004.

[17] W. C. Oliver, R. Hutchings, and J. B. Pethica, "The wear behavior of nitrogen-implanted metals," *Metallurgical and Materials Transactions A*, vol. 15, no. 12, pp. 2221–2229, 1984.

[18] P. Tarkowski, P. Budzynski, and W. Kasietczuk, "Adhesive character of wear processes in nitrogen-implanted iron," *Vacuum*, vol. 78, no. 2–4, pp. 679–683, 2005.

[19] R. Wei, P. J. Wilbur, W. S. Sampath, D. L. Williamson, and J. L. Wang, "Effects of Ion implantation conditions on the tribology of ferrous surfaces," *Tribology*, vol. 113, pp. 166–173, 1991.

[20] I. M. Feng, "Metal transfer and wear," *Journal of Applied Physics*, vol. 23, no. 9, pp. 1011–1019, 1952.

[21] D. A. Rigney, "The roles of hardness in the sliding behavior of materials," *Wear*, vol. 175, no. 1–2, pp. 63–69, 1994.

[22] C. C. Viáfara and A. Sinatora, "Influence of hardness of the harder body on wear regime transition in a sliding pair of steels," *Wear*, vol. 267, no. 1–4, pp. 425–432, 2009.

[23] D. W. Borland and S. Bian, "Unlubricated sliding wear of steels: towards an alternative wear equation," *Wear*, vol. 209, no. 1–2, pp. 171–178, 1997.

[24] D. Tabor, "Junction growth in metallic friction: the role of combined stresses and surface contamination," *Proceedings of the Royal Society of London A*, vol. 251, no. 1266, pp. 378–393, 1959.

[25] D. Rafaja, "X-ray diffraction and X-ray reflectivity applied to the investigation of thin films," *Advances in Solid State Physics*, vol. 41, pp. 275–286, 2001.

[26] M. Kopcewicz and J. Jagielski, "Phase transformations in nitrogen-implanted α-iron," *Journal of Applied Physics*, vol. 71, no. 9, pp. 4217–4226, 1992.

[27] A. A. Youssef, P. Budzynski, J. Filiks, A. P. Kobzev, and J. Sielanko, "Improvement of wear and hardness of steel by nitrogen implantation," *Vacuum*, vol. 77, no. 1, pp. 37–45, 2004.

[28] G. Bhargava, I. Gouzman, C. M. Chun, T. A. Ramanarayanan, and S. L. Bernasek, "Characterization of the "native" surface thin film on pure polycrystalline iron: a high resolution XPS and TEM study," *Applied Surface Science*, vol. 253, no. 9, pp. 4322–4329, 2007.

[29] N. P. Suh, N. Saka, and S. Jahanmir, "Implications of the delamination theory on wear minimization," *Wear*, vol. 44, no. 1, pp. 127–134, 1977.

[30] T. Kayaba, K. Kato, and K. Hokkirigawa, "Theoretical analysis of the plastic yielding of a hard asperity sliding on a soft flat surface," *Wear*, vol. 87, no. 2, pp. 151–161, 1983.

[31] M. Fishkis, "Metal transfer in the sliding process," *Wear*, vol. 127, no. 1, pp. 101–110, 1988.

[32] R. L. Jackson and I. Green, "A finite element study of elasto-plastic hemispherical contact against a rigid flat," *Journal of Tribology*, vol. 127, no. 2, pp. 343–354, 2005.

[33] L. Kogut and I. Etsion, "A finite element based elastic-plastic model for the contact of rough surfaces," *Tribology Transactions*, vol. 46, no. 3, pp. 383–390, 2003.

[34] M. Antler, "Processes of metal transfer and wear," *Wear*, vol. 7, no. 2, pp. 181–203, 1964.

[35] M. Antler, "Wear, friction, and electrical noise phenomena in severe sliding systems," *ASLE Transactions*, vol. 5, no. 2, pp. 297–307, 1962.

[36] A. D. Berman, W. A. Ducker, and J. N. Israelachvili, "Origin and characterization of different stick-slip friction mechanisms," *Langmuir*, vol. 12, no. 19, pp. 4559–4562, 1996.

[37] T. Sasada, M. Oike, and N. Emori, "The effect of abrasive grain size on the transition between abrasive and adhesive wear," *Wear*, vol. 97, no. 3, pp. 291–302, 1984.

[38] D. A. Rigney, L. H. Chen, M. G. S. Naylor, and A. R. Rosenfield, "Wear processes in sliding systems," *Wear*, vol. 100, no. 1–3, pp. 195–219, 1984.

[39] D. A. Rigney, "Sliding wear of metals," *Annual Review of Materials Science*, vol. 18, pp. 141–163, 1988.

[40] J. Jiang, F. H. Stott, and M. M. Stack, "A mathematical model for sliding wear of metals at elevated temperatures," *Wear*, vol. 181–183, no. 1, pp. 20–31, 1995.

[41] R. L. Deuis, C. Subramanian, and J. M. Yellup, "Dry sliding wear of aluminium composites—a review," *Composites Science and Technology*, vol. 57, no. 4, pp. 415–435, 1997.

[42] I. M. Hutchings, *Tribology: Friction and Wear of Engineering Materials*, Arnold, London, UK, 1992.

[43] H. J. Kim, A. Emge, S. Karthikeyan, and D. A. Rigney, "Effects of tribooxidation on sliding behavior of aluminum," *Wear*, vol. 259, no. 1–6, pp. 501–505, 2005.

[44] P. Budzynski, A. A. Youssef, Z. Surowiec, and R. Paluch, "Nitrogen ion implantation for improvement of the mechanical surface properties of aluminum," *Vacuum*, vol. 81, no. 10, pp. 1154–1158, 2007.

[45] A. Galerie, M. Caillet, and M. Pons, "Oxidation of ion-implanted metals," *Materials Science and Engineering*, vol. 69, no. 2, pp. 329–340, 1985.

[46] G. Dearnaley, "The alteration of oxidation and related properties of metals by ion implantation," *Nuclear Instruments and Methods*, vol. 182–183, no. 2, pp. 899–914, 1981.

[47] K. M. Jasim and E. S. Dwarakadasa, "SEM studies of wear debris in Al-Si alloys," *Journal of Materials Science Letters*, vol. 8, no. 11, pp. 1285–1287, 1989.

[48] H. J. Cho, W. J. Wei, H. C. Kao, and C. K. Cheng, "Wear behavior of UHMWPE sliding on artificial hip arthroplasty materials," *Materials Chemistry and Physics*, vol. 88, no. 1, pp. 9–16, 2004.

[49] K. M. Jasim and E. S. Dwarakadasa, "Effect of sliding speed on adhesive wear of binary Al-Si alloys," *Journal of Materials Science Letters*, vol. 12, no. 9, pp. 650–653, 1993.

[50] U. Beerschwinger, T. Albrecht, D. Mathieson, R. L. Reuben, S. J. Yang, and M. Taghizadeh, "Wear at microscopic scales and light loads for MEMS applications," *Wear*, vol. 181–183, no. 1, pp. 426–435, 1995.

12

Structure and Abrasive Wear of Composite HSS M2/WC Coating

S. F. Gnyusov,[1] V. G. Durakov,[2] and S. Yu. Tarasov[2]

[1] *Tomsk Polytechnic University, 634050 Tomsk, Russia*
[2] *Institute of Strength Physics and Materials Science SB RAS, 634055 Tomsk, Russia*

Correspondence should be addressed to S. Yu. Tarasov, tsy@ispms.ru

Academic Editor: Huseyin Çimenoğlu

Features of phase-structure formation and abrasive wear resistance of composite coatings "WC-M2 steel" worn against tungsten monocarbide have been investigated. It was established that adding 20 wt.% WC to the deposited powder mixture leads to the increase in M_6C carbide content. These carbides show a multimodal size distribution consisting of ~5.9 μm eutectic carbides along the grain boundaries, ~0.25 μm carbides dispersed inside the grains. Also a greater amount of metastable austenite (~88 vol.%) is found. The high abrasive wear resistance of these coatings is provided by $\gamma \rightarrow \alpha'$-martensitic transformation and multimodal size distribution of reinforcing particles.

1. Introduction

A tendency is to develop and build new equipment at lower costs using sparingly alloyed low carbon steels, which, however, demand surface hardening either by nitridation or carbonization. A primary task in this situation is to improve both endurance and reliability of working components of machines and technology equipment. In this connection, a solution is related not only to improving the wear resistance only but also to a totality of problems including but not limited to corrosion resistance, contact endurance, small plastic strain, and heat resistances.

Effective approach to improving the wear resistance of materials is cladding and modification metals by concentrated energy fluxes. These methods are used in practice and based on fast quenching of a melted pool at cooling rates 10^4 to 10^9 K/s. Electron beam surfacing in vacuum is a good candidate to obtain a hard coating on the surface of low carbon substrate. Such a processing provides some advantages [1] which include the ability to feed composite surfacing powders directly to the melted pool, vacuum refinement of the melted metal, gradual and accurate adjustment of the electron beam power to provide both minimal fusion penetration to the substrate, constant chemical composition, and small pool's size at electron beam power density up to 10^5 W/cm². All these parameters may be optimized to

achieve the pool's overheating required to obtain over saturated solution of the alloying elements and fine-grain structures in cooling. Taking it into account we believe that electron beam surfacing in vacuum may be applied for depositing a composite coating after final heat treatment and main mechanical grinding.

In modern practice, both hard and superhard composite coatings made of stellite, sormite, or cast tungsten carbide are used to improve wear resistance of working surfaces of machine components. A disadvantage of these materials is that they contain 30 to 90% wt. of costly tungsten carbide. Also coatings made by depositing these materials are brittle because of high hard phase content, nonuniform distribution of these particles throughout the bulk of the coating, especially for their content in the range 30 to 50% wt., and network cracking on the coating's surface. All of it prevents using them in a row of applications when a totality of these properties is demanded.

In particular, there is a problem of quick failure of pinion shafts of heavy loaded reduction gearboxes due to intense wear of nitrided journals in needle bearings. A design specificity of the reduction gearboxes is that their pinion shafts are positioned in close proximity to each other and traditional design of bearings consisting of external and internal rings with rollers between them is not usable. Therefore, one has to make a journal bearing directly on the pinion-shaft surface.

These journal surfaces must possess high wear resistance, contact endurance, low plastic strains, fracture toughness, which dictates the necessity for application of homogeneous multifunctional coatings.

The specific features of coatings obtained by multipass electron beam cladding with M2 steel powder are described elsewhere [2]. It is established that a multimodal size distribution of reinforcing particles is generated in a carbide subsystem of the coating. The volume contents of both secondary carbide M_6C and retained austenite can be regulated within the ranges 4.5 to 7.5% wt. and и 5 to 30% wt., respectively, as depending on the thermal cycling conditions created during the surfacing. Wear resistance of the coatings improves with the volume content of the retained austenite because of $\gamma \rightarrow \alpha'$ strain-induced martensitic transformation and fine carbides precipitated in the matrix's grains. We can say safely that the higher is the content of the retained metastable austenite in a coating, the higher is the wear resistance. A route to go is admixing the tungsten monocarbide to the HSS powder. Since WC is of high solubility in a steel matrix, it may provide 80 to 90% vol. content of austenite as well as higher volume content of precipitated fine carbides.

Contribution of strain-induced phase transformation in abrasive wear resistance of trip steels with metastable austenite has been assessed [3] on a basis of both developed model and experimental data. Within the framework of this model the energy balance equations were derived and then served to determine the fracture work values for samples with either stable or metastable austenite. It was found out that the fracture work value for samples with metastable austenite matrix is by a factor of 7 higher as compared to that of without ability to experience strain-induced phase transformation under the same wear test conditions.

The objective of this work is to study effect of tungsten carbide content in the source mixture on the structure, phase contents, and abrasive wear resistance of HSS M2/WC composite coatings.

2. Materials, Equipment, and Experimental Methods

Source materials for electron beam surfacing were HSS M2 powder mixtures added with 5 to 350 μm WC powder of contents 10, 15, 20, 25, 30, 40, and 50% wt. Chemical composition of M2 steel was as follows: C—1%, Cr—4%, W—6.5%, Mo—5%, V—1.5%, Si < 0.5%, Mn < 0.55%, Ni < 0.4%, S < 0.03%, O_2 < 0.03%, Fe—balance. Composite mixtures were prepared by mixing the above components, compacting and sintering the mixtures in vacuum at partial pressure not higher 10^{-2} Pa, followed by milling and sieving the cakes by fractions. Surfacing was carried out on flat 20 × 30 × 200 mm samples of 0.3% C steel by applying four passes per each plate. The scanning-line length was 20 mm. The thickness of clad metal layer per pass was 1 mm. Some samples were subjected to either single or double tempering carried out at 570±°C for 1 hour.

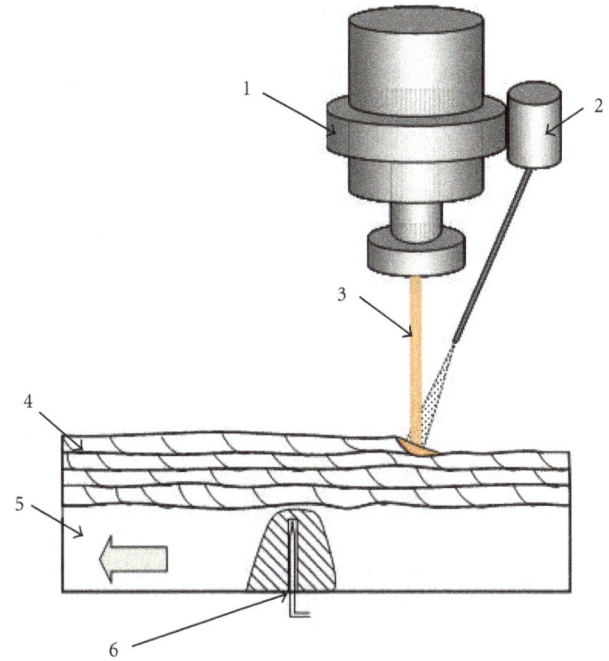

FIGURE 1: Schematic of the electron beam surfacing. 1: electron gun; 2: powder feeder; 3: electron beam; 4: coating; 5: substrate; 6: thermocouple.

A machine for electron beam cladding in vacuum was operated in automatic mode so that samples were loaded into its chamber and secured there in manipulators. These manipulators are driven in rotation and displacement modes by external electrodrive system. The chamber was evacuated to reach the residual pressure 10^{-1} Pa. Electron beam generated by the electron gun was scanned over the surface of the sample thus creating a melted pool (Figure 1). Simultaneously, the powder mixture was fed to the pool by means of a measure feeder.

Accelerating voltage, diameter, electron beam sweep length, and sample feed velocity were 28 kV, 1 mm, and 20 mm, 2 mm/s, respectively, and did not change during the experiment.

The microstructure of the deposited coatings in the longitudinal and transverse microsections was examined using optical microscope (OM) *Olympus GX 51* equipped with 700 SIAMS analysis device, SEM instrument *Philips SEM 515* equipped with microanalysis device *EDAX ECON IV*, and TEM Tecnai G2 FEI instrument equipped with a microanalysis device.

Traditional method of preparing the microsections such as mechanical grinding and polishing with diamond pastes of different grades was used in this work. Chemical etching was performed in 4% solution of HNO_3 in alcohol. Determining the quantitative characteristics of the microstructure including number, size, shape, and distribution of various phases was performed using the *SIAMS* software package. TEM foils of thickness 150 μm were cut from coatings using an electrospark machine. After mechanical and electrolytic finishing, to achieve the foils' thicknesses 70 to 90 μm, the final ion thinning was carried out to achieve 200 nm thickness.

Phase composition of the samples both after surfacing and abrasion wear tests was investigated by XRD using *Shimadzu XRD 6000* diffractometer operated in the 2Θ range 30 to 120 degrees at 0.02° step and with filtered CoKα-radiation. Integrated intensities of the diffraction peaks were used for the quantitative phase analysis. The volume content of a phase in a multicomponent system was determined according to an expression as follows:

$$V_\alpha = \frac{K_\alpha\left(I_\alpha/I_\beta\right)}{1 + K_\alpha\left(I_\alpha/I_\beta\right) + K_\gamma\left(I_\gamma/I_\beta\right) + K_\varepsilon\left(I_\varepsilon/I_\beta\right) + \cdots},$$

$$K_\alpha = \frac{Q_\alpha^2(P\Phi F^2)_\beta}{Q_\beta^2(P\Phi F^2)_\alpha},$$

(1)

where α, β, γ, and ε stand for phases, Φ is the angle factor, P is the multiplicity factor, and F and Q are the phase's structural factor and cell volume, respectively.

2.1. Microhardess and Wear. Microhardness numbers (H_μ) of both the coating and the substrate were measured using PMT-3 microhardness tester operating at 100 μm step between the indents and load of 0.981 N. The measurements were carried out on two parallel indentation paths shifted by 50 μm in depth with respect to each other. The distance between the paths was 200 μm. Such a procedure allowed obtaining the microhardness in-depth profiles of 50 μm step in the coating.

To determine the wear rate (mg/hour) we used method of loose abrasive particles (Figure 2) and measured the mass losses hourly. The mean wear rate was then calculated from five experimental results. Faceted quartz sand particles with sharp corners and edges and round electrocorundum particles of mean size $d_m = 100$ to 300 μm were used as abrasive material for wear testing at 60 RPM. Normal load was 44.1 $\pm$ 0.25 N. The analysis of microstructure and microhardness of the specimens after wear tests was performed on taper sections made at 2° angle between the top surface and the section plane.

The analysis of microstructure and microhardness of the specimens after wear tests was performed on taper sections made at 2° angle between the top surface and the section plane.

3. Results and Discussion

HSS M2 + WC mixtures of WC contents 10, 15, 20, 25, 30, 40, and 50% wt. were prepared by ball milling for 24 hours. Sintering the mixtures was carried out in a vacuum oven at 1200°C for 1 hour. If using the lower temperatures, the particles do not sinter one to each other and no quality composite cakes are obtained. At temperatures above 1200°C the cakes are very hard to mill and the powder yield is too low. The resulting cakes were milled and sieved to particle size fractions. The yield of usable milled composite 30 to 350 μm

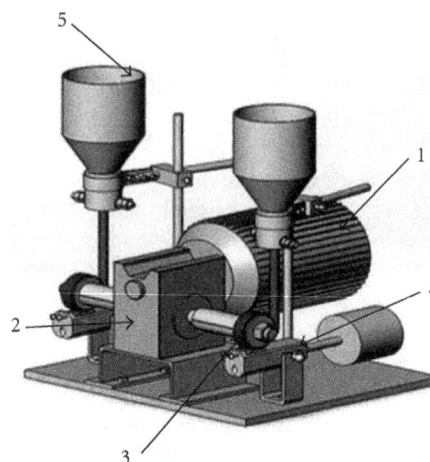

FIGURE 2: Schematic of abrasive wear test setup. 1: electromotor; 2: worm gear; 3: rubber counterbody; 4: tested sample.

fraction was 85 to 90%. The smaller particle fractions were discarded.

The microstructure of a composite cake intended for surfacing is shown in Figure 3. As seen, the isolated particles of M2 steel are surrounded either by islet shape phase (a) only or angular 1 to 2 μm particles in combination with the islets (b) as depending on the WC content in the mixture. Increasing the source WC content from 10 to 40% wt. results in increasing the amount of angular particles from 0 to 20–25% vol. in the cake.

According to XRD (Figure 4), two carbides such as WC and M_6C were found in the cakes in addition to the matrix's phases α and γ. It means that the source powder particles interact with each other in the course of sintering to produce a composite. That such is true is supported by the results reported elsewhere [4]. The authors [4] show that M_6C carbide's formation starts after heating the composite WC-(Ni-Al) and WC-NiTi mixtures to 700°C. Taking into account the morphology, we can state that α-phase found in cakes is in the form of martensite needles.

Examining the microstructures as well as the XRD data obtained from the surfaced samples (Figure 4), we can see that coatings are composed of α-martensite, retained austenite (γ-phase), 0.9 vol.% of $d_1 = 0.65$ μm VC carbides, and M_6C carbide (Figure 5). This M_6C carbide is of two morphological types. The first type I is presented by dendrite-like eutectic 3.8 μm carbides (Figures 5(a) and 5(b)) found at the grain boundaries of solid solution. Second type II carbides look like fine elongated particles inside the grains (Figure 5(d)). Table 1 shows the data on chemical composition of structures detected in the points shown in Figures 5(c) and 5(d). The data in Table 1 allow conclusion on essential reduction in the vanadium content of secondary M_6C carbides as compared to that of the eutectic carbides.

On admixing more tungsten carbide to the source mixture, the amount of M_6C carbides grows (Figure 6(a)) and starting from WC 30% wt. and higher some amount of the source WC retained in addition to the above shown.

TABLE 1: Chemical composition of carbides found in composite coatings.

No. point	Phase	Chemical composition, % wt.					
		C	V	Cr	Fe	W	Mo
1	M_6C (eutectic)	5.0	27.0	2.8	2.3	44.6	18.3
2	M_6C (secondary)	1.6	7.0	6.1	5.1	54.9	25.3
3	M_6C (secondary)	1.8	6.6	5.4	5.3	55.1	25.8
4	Matrix	1.8	1.1	3.0	77.4	10.8	5.9
5	Matrix	2.2	0.9	2.9	77.2	10.2	6.5

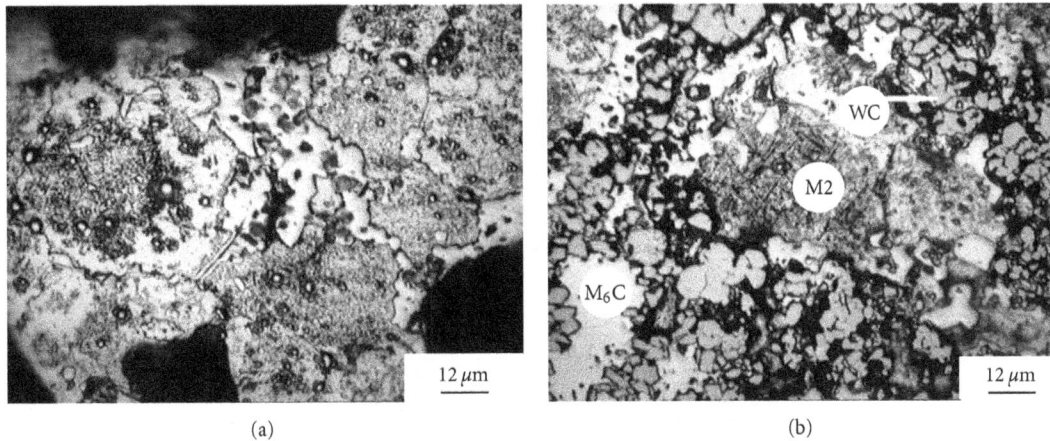

(a)   (b)

FIGURE 3: The postsintering microstructure of composite HSS M2 + WC intended for electron beam surfacing. (a) WC 20% wt.; (b) WC 40% wt.

(Figures 6(b) and 6(c)). M_6C carbide in coatings obtained using mixtures of WC content below 30% wt. was found in the form of dendrite-like precipitations at the austenite grain boundaries having a mean size $d_2 = 5.9 \mu m$ and content ~15 vol.%. In addition to that, it exists in the form of fine $d_3 = 0.24 \mu m$ precipitations inside the austenitic grains of total content ~8% vol. It is made clear that electron beam cladding composite coatings with up to 30% wt. WC may result in generation of a multimodal size distribution of hard phase carbide particles in the coating.

Coarse faceted particles of M_6C carbide were found in coatings obtained from the source powder mixtures of WC content 40% wt. (Figure 6(c)). These particles grow even larger when the WC content grows up to 50% wt. (Figure 6(d)). Furthermore, it exists in two morphological types such as globular and elongated (platelets) particles. These coarse mixed carbide particles form a developed framework. The retained WC particles are surrounded by M_6C precipitations and built in the framework. Binder phase was found in the form of isolated particles. It is inconceivable to suggest that such a framework structure of the reinforcing phase will hardly serve to improve the wear resistance. The coarse carbides will plausibly have positive effect on abrasive wear resistance. The wear resistance of a framework built of fine M_6C carbides will be determined by the phase composition of the matrix.

The dependence of the austenite's amount on the tungsten carbide's content in the source powder is a curve with a maximum in the range 20 to 25% wt. of WC (Figure 7(a)).

Both single and double tempering the deposited coating result in partial $\gamma \rightarrow \alpha$ martensitic transformation. It is established that electron beam surfacing with deposition of both pure M2 steel powder and M2 + WC 40 to 50% wt. mixtures gives only 3 to 4% vol. of retained austenite. However, if we used M2 + WC 20 to 30% wt. mixtures, the composite coatings contained 30 to 40% vol. of retained austenite even after conducting the double tempering.

It is notable that the coatings made using M2 + WC 0 to 30% wt. showed no network cracking on their surfaces.

The ascending part of the curve in Figure 7(a) shows the amount of the retained austenite increasing up to 82% vol. with adding 20 to 25% WC wt. to the source powder mixture. Such an effect is provided by better solubility of WC and, therefore, the greater effect of alloying the γ-solid solution both by carbon and carbide-forming elements.

According to the EDAX data, the solubility of tungsten in the coating's matrix grows from 3.5 to 11% wt. with the WC content in the source mixture. This involves reducing the onset temperature of martensitic transformation and increasing the amount of metastable retained austenite after cooling (quenching). It follows from [5, 6] that, increasing the temperature of quenching for HSS M2, we simultaneously increase the content of both carbon and alloying elements, which reduce the martensitic transformation point. In our situation, we actually deal with quenching from a liquid micropool that serves to additionally reduce the martensitic transformation temperature interval. The higher content of retained austenite provides for minimal hardness

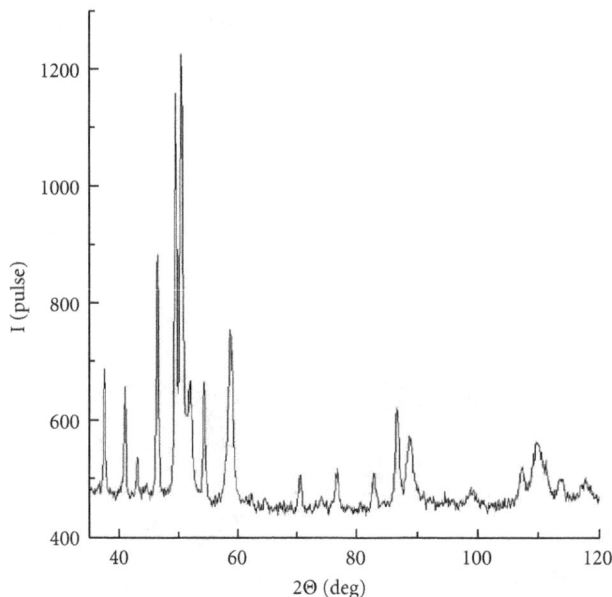

No.	2Θ (deg)	d (nm)	Phase
1	37.734	0.27660	M_6C (400)
2	41.265	0.25384	M_6C (331)
3	43.343	0.24222	VC (111)
4	46.682	0.22576	M_6C (422)
5	49.684	0.21291	M_6C (511)
6	50.652	0.20910	γ-Fe (111), VC (200)
7	52.008	0.20401	α-Fe (110)
8	54.435	0.19557	M_6C (440)
9	58.870	0.18201	γ-Fe (200)
10	70.556	0.15487	M_6C (551)
11	76.777	0.14404	α-Fe (200), M_6C (553)
12	82.821	0.13523	M_6C (733)
13	86.655	0.13036	M_6C (822)
14	88.747	0.12790	γ-Fe (220), M_6C (555)
15	98.808	0.11780	α-Fe (211)
16	107.139	0.11117	M_6C (933)
17	109.682	0.10941	γ-Fe (311)
18	111.162	0.10843	M_6C (862)
19	113.484	0.10697	M_6C (733)
20	116.984	0.10491	α-Fe (220)
21	117.652	0.10454	γ-Fe (222)

FIGURE 4: XRD data for composite coating M2 + 20% wt. WC.

of M2 + 20% wt. WC composite coatings being at the level of 6 GPa (Figure 7(b), curve 1).

As seen from the descending part of curve in Figure 7(a), the amount of retained austenite is reduced when the WC content in the source mixture grows from 30 to 50% wt. The rationale behind that is incomplete dissolution of both M_6C and WC particles, which then serve either as crystallization nuclei or substrates for epitaxial growth of mixed carbide from austenite. Simultaneously, the same effect reduces the overheating of the melted pool as well as, consequently, the solubility of carbides in it. Moreover, the reduced content of a binder phase in the coating facilitates relaxation of stress generated by the difference in phase volumes of α and γ phases. Also this serves for more full $\gamma \rightarrow \alpha$ transformation. The discussed here abnormal behavior of austenite in coatings must have effect on their wear resistance.

3.1. Abrasive Wear. The wear rate by quartz particles is reduced quickly for hard coatings containing up to 20% wt.

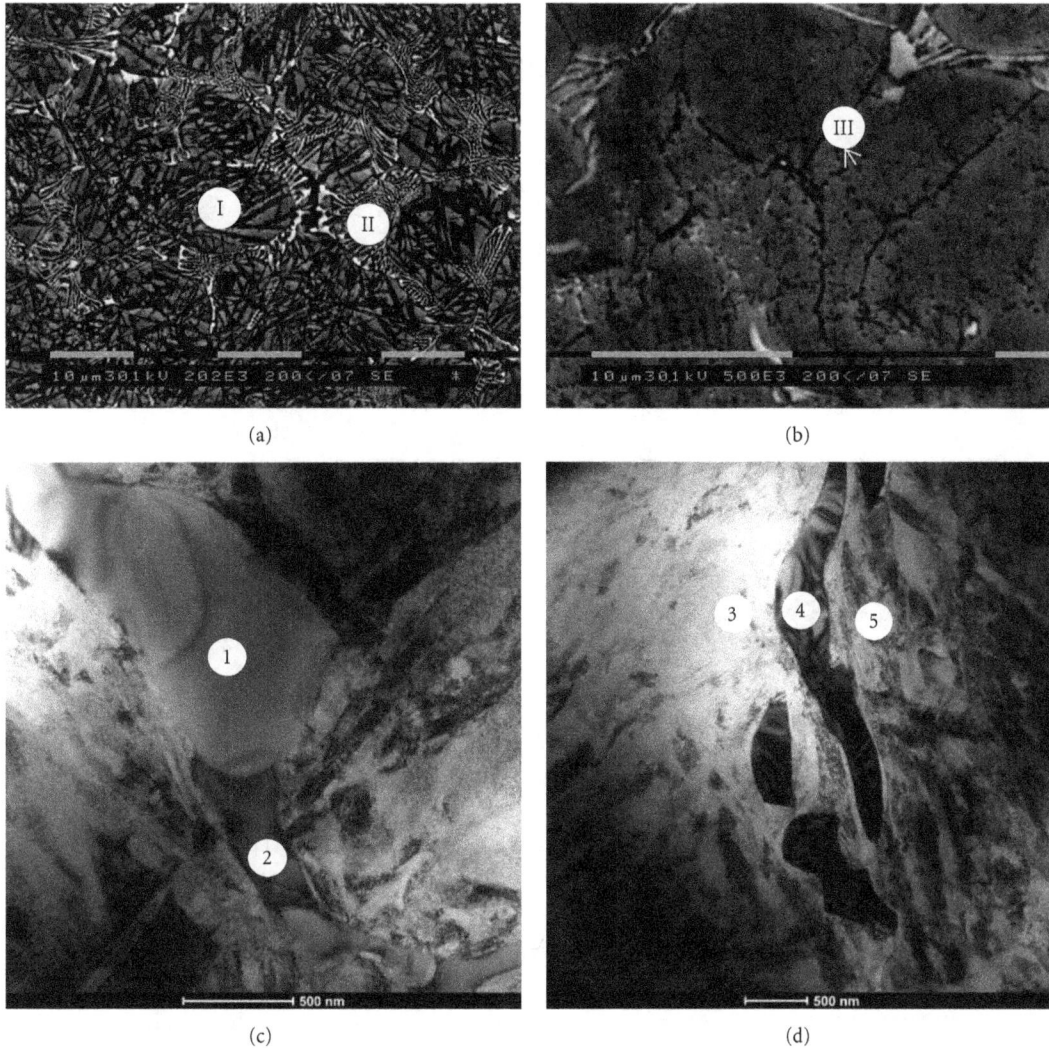

FIGURE 5: The microstructure of M2 coating clad on the substrates. ((a), (b) SEM, (c), (d) TEM).

WC whereas some growth is observed for 25 to 30% wt. WC coatings. However, the wear rate continues falling for higher WC content (Figure 8(a)). The local wear rate maximum (Figure 8(a)) may find its explanation in structural changes occurring in the coating when depositing 20% wt. WC powder mixtures. In this case, a carbide framework structure starts forming involving source WC particles. Also the use of faceted shape of abrasive quartz particles may add to this effect. When using electrocorundum, the wear rate is reduced with the WC content growth. This is especially the case for coatings deposited from 20% wt. WC powder mixtures (Figure 8(b)). Another aspect is that using more hard abrasive particles provides higher wear rate by a factor of ~20 as compared to quartz sand.

Taking into account the microstructures in Figures 5 and 6, volume content of austenite in Figure 7(a), hardness (Figure 7(b)), and wear rate dependencies in Figure 8 as functions of WC content in the source powder mixture, we can say that the wear resistance is improved due to higher amount of austenite in samples surfaced to obtain the multimodal (d_1, d_2, and d_3) size distribution of hardening

carbide phases. Such a distribution serves to provide thinner intercarbide binder layers and, thus, to reduce their selective wear and prevent carbides from spalling.

Also since the metastable austenite is capable of partial strain-induced martensitic $\gamma \rightarrow \alpha'$ transformation, this provides for extra stress relieve effect during wear. In accordance to XRD data, the amount of α'-phase in M2 + 20% WC coating has grown by 40% vol. after abrasive wear test. This martensitic transformation in combination with substructural work hardening results in growth of hardness from 6 to 10 GPa in the subsurface of coatings at 40 μm depth below the worn surface (Figure 7(b)).

Another structural factor important for improving the wear resistance of composite coating with metastable austenite matrix is its higher ability to hold brittle carbide phases as compared to martensite and, thus, prevent them against spalling both in microcutting and fatigue wear. This is true both for primary dendrite carbides found at the grain boundaries and secondary equiaxed intragrain fine carbides [4]. Neither single nor double tempering the coatings is able to change the dependence of wear rate on the tungsten carbide

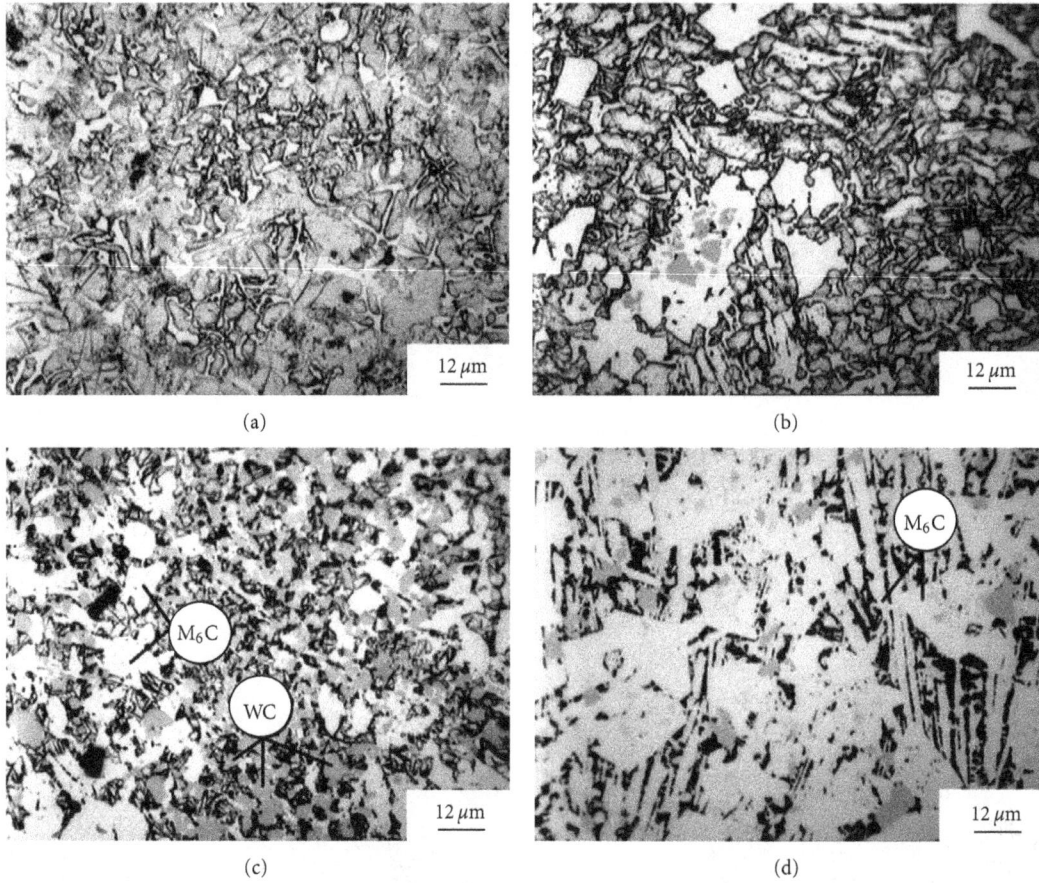

FIGURE 6: The microstructure of composite coating clad on the substrates; (a) "M2 + 20% WC"; (b) "M2 + 30% wt. WC"; (c) "M2 + 40% wt. WC"; (d) "M2 + 50% wt. WC".

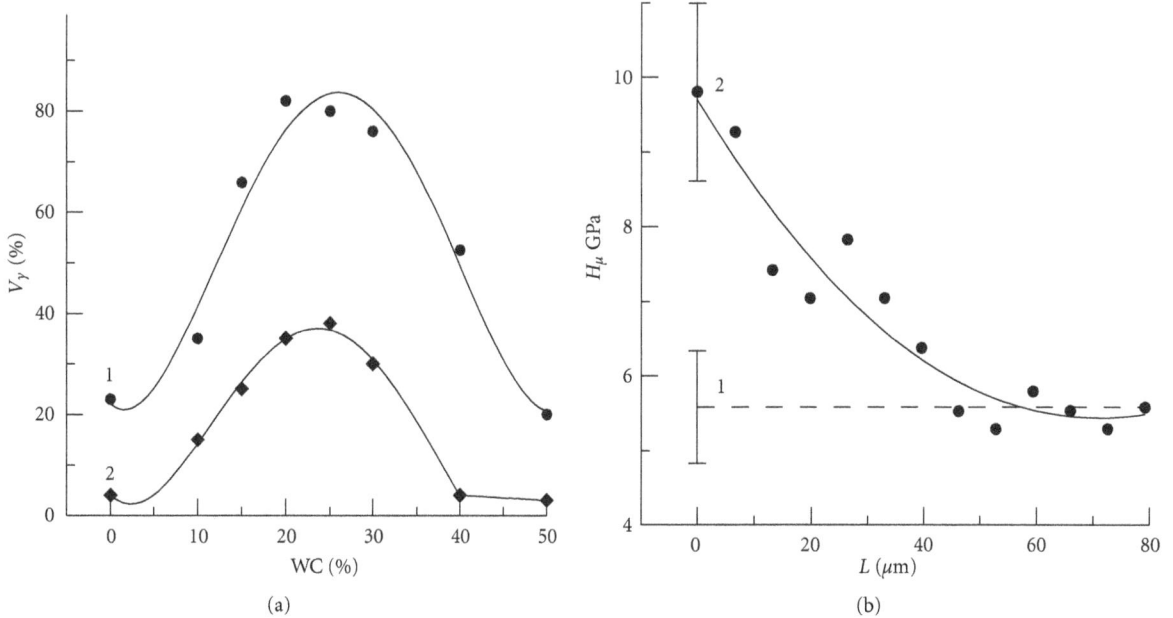

FIGURE 7: Volume content of austenite in the coating's matrix versus WC content in the source powder mixture (a), curve 1 is for deposited coating, curve 2 is for double tempering, and microhardess numbers as a function of depth below the worn surface for the coating made using M2—20 wt.% WC mixture. ((b), curve 2) 1: initial microhardness.

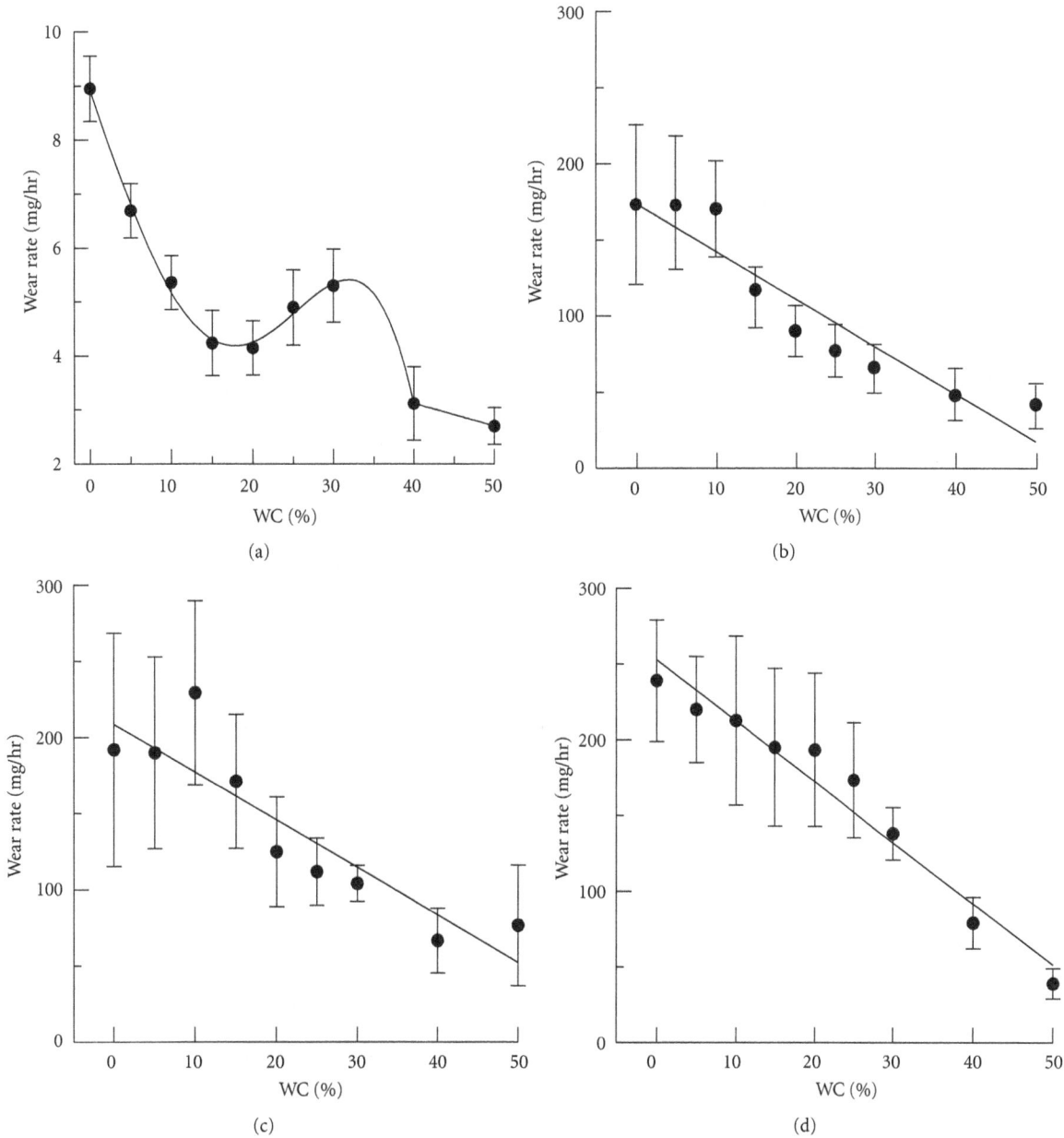

FIGURE 8: Abrasive wear rate versus WC content in the source powder mixture. (a) Quartz sand, ((b), (c), (d)) electrocorundum; (a), (b) just after cladding; (c) single tempering; (d) double tempering.

content albeit the absolute values are somewhat increased as compared to those after surfacing (Figures 8(b), 8(c), and 8(d)). Such a behavior may be related to the fact that tempering partially transforms austenite to martensite and, therefore, reduces the effect of austenite for relieving stresses in $\gamma \rightarrow \alpha'$ transformation. From this standpoint, tempering is not a heat treatment desired after surfacing.

The wear rate of M2 + 40% wt. WC coating is greatly influenced by the heat treatment. The microstructure of this coating reveals a carbide framework composed of M_6C and WC carbides (Figure 6(c)). The binder of a just sur-faced coating contains about 50% vol. of austenite (Figure 7(a)), which has positive effect both on stress relieving and preventing spalling of the carbide particles.

Since tempering results in partial $\gamma \rightarrow \alpha'$ transformation, it serves to impair both resistance to spalling and carbide retention ability. Finally, the wear rate of M2 + 40% wt. WC coatings is increased by 40% for wear by electrocorundum particles.

The carbide framework formation is observed in M2 + 50% WC wt. coating (Figure 6(d)). However, these M_6C carbide particles are coarser as compared to those found in M2 + 40% wt. WC coating (Figure 6(c)). Binder phase is found in the form of isolated particulates and, therefore, has no great effect on wear resistance of this coating. When comparing the wear resistances of deposited and heat-treated coatings, we note the wear rate increased by ~20% in the latter case. We suggest that spalling of coarse carbides particles is not easy and the carbide framework is stable with respect to pure abrasive wear. However, the impact loading from abrasive particles will increase the wear rate of both carbide framework and more brittle matrix. Also the carbide crack network is formed in the deposited M2 + 50% wt. WC coatings that will add to the process.

It follows from here that high wear resistance of the previously shown composite coatings is due to high content of the retained austenite. Such a conclusion is supported by numerous data found in the literature sources [2, 4, 7, 8]. Another factor contributing to wear resistance is the multimodal size distribution of carbide particles. Apart from the retained austenite, there are another important factors that determine the wear resistance of HSS such as origin, quantity, character of distribution, and size of carbides [8–10]. It is known [11] that higher wear resistance of cast HSS as compared to wrought HSS is explained by the presence of hard eutectic carbide framework formed along γ-solid solution boundaries. The carbide framework was found to be more efficient for better wear resistance as compared to isolated coarse carbide particles under conditions of both abrasive and oxidative wear.

To prevent wear of relatively soft matrix grains, one of the possible solutions is to provide their reinforcement by fine carbides. The literature search devoted to WC-Co hard metal [12–14] shows that when successively reducing the carbide phase size from micro- to submicro- and then to nanosize for Co contents kept constant, we reduce the intercarbide binder layer and simultaneously increase the hardness. The wear resistance grows both in abrasive and sliding wear tests due to achieving smaller carbide grains and thinner intercarbide binder layers which serve to limit the selective binder wear and further spalling of carbide particles [12, 13].

In such a manner, M2 + 20% WC coating is most preferable for surfacing pinion-shaft journals since it shows good levels of wear resistance, hardness, no surface cracking, structural homogeneity of coating, and high amount of retained austenite which is one of the main factors to reduce the cyclic external stresses.

4. Conclusions

(1) It is established that WC particles interact with HSS M2 particles in the powder mixture during sintering the powder cakes and thus form M_6C carbide. The WC content in a source powder mixture increased from 10 to 40% wt. corresponding to WC content 20 to 25% vol. in the cake.

(2) As shown, admixing 20 to 25% wt. WC into the surfacing mixture results in higher content of M_6C carbide found in the forms of eutectic 5.9 μm and fine 0.25 μm carbides inside the grains (multimodal size distribution) in addition to VC 0.65 μm carbides and higher amount of austenite—88% vol.

(3) Increasing the WC content in the source powder mixture serves to reduce the abrasive wear rate irrespective of the abrasive particle type used in test (Figure 8). The most intensive wear resistance growth

is observed for WC content 20% wt. coatings. This is related both to mechanical stress relief by strain-induced $\gamma \rightarrow \alpha'$ martensitic transformation during the wear test and effect of multimodal size distribution of carbides.

Acknowledgment

This work is carried out with financial support of State Task of NIR TPU no. 8.3664.2011.

References

[1] V. E. Panin, S. I. Belyuk, V. G. Durakov, G. A. Pribytkov, and N. G. Rempe, "Electron-beam deposition in a vacuum: the equipment, technology, properties of the coatings," *Svarochnoe Proizvodstvo*, no. 2, pp. 34–38, 2000.

[2] S. F. Gnyusov, A. A. Ignatov, and V. G. Durakov, "Structure and wear resistance of R6M5 steel based coatings," *Technical Physics Letters*, vol. 36, no. 8, pp. 745–748, 2010.

[3] T. V. Smyshlyaeva, "Assessment of the fracture energy in trip-steel during abrasive wear," *Journal of Friction and Wear*, vol. 22, no. 3, pp. 295–298, 2001.

[4] S. N. Kulkov and S. F. Gnyusov, *Carbide Steels Based on Carbides of Titanium and Tungsten*, Nauchno-Tekhnicheskoy Literatury, Tomsk, Russia, 2006.

[5] A. P. Gulyaev, *Superplasticity of Steel*, Metallurgiya, Moscow, Russia, 1982.

[6] S. G. Koop, *Heat Treatment of HSS*, Metallurgiya, Moscow, Russia, 1956.

[7] A. V. Makarov, L. G. Korshunov, I. Y. Malygina, and A. L. Osintseva, "Effect of laser quenching and subsequent heat treatment on the structure and wear resistance of a cemented steel 20KhN3A," *Physics of Metals and Metallography*, vol. 103, no. 5, pp. 507–518, 2007.

[8] E. Badisch and C. Mitterer, "Abrasive wear of high speed steels: influence of abrasive particles and primary carbides on wear resistance," *Tribology International*, vol. 36, no. 10, pp. 765–770, 2003.

[9] E. Pippel, J. Woltersdorf, G. Pöckl, and G. Lichtenegger, "Microstructure and nanochemistry of carbide precipitates in high-speed steel S 6-5-2-5," *Materials Characterization*, vol. 43, no. 1, pp. 41–55, 1999.

[10] S. F. Gnyusov, I. O. Khazanov, B. F. Sovetchenko et al., *The Application of the Superplasticity Effect of Steels in the Tool Manufacturing*, STB, Tomsk, Russia, 2008.

[11] A. S. Chaus and M. Hudáková, "Wear resistance of high-speed steels and cutting performance of tool related to structural factors," *Wear*, vol. 267, no. 5–8, pp. 1051–1055, 2009.

[12] K. Jia and T. E. Fischer, "Sliding wear of conventional and nanostructured cemented carbides," *Wear*, vol. 203-204, pp. 310–318, 1997.

[13] K. Jia and T. E. Fischer, "Abrasion resistance of nanostructured and conventional cemented carbides," *Wear*, vol. 200, no. 1-2, pp. 206–214, 1996.

[14] B. H. Kear and L. E. McCandlish, "Chemical processing and properties of nanostructured WC–Co materials," *Nanostructured Materials*, vol. 3, no. 1–6, pp. 19–30, 1993.

13

Surface Layer States of Worn Uncoated and TiN-Coated WC/Co-Cemented Carbide Cutting Tools after Dry Plain Turning of Carbon Steel

Johannes Kümmel,[1] Katja Poser,[1] Frederik Zanger,[2] Jürgen Michna,[2] and Volker Schulze[1,2]

[1] *Institute of Applied Materials (IAM-WK), Karlsruhe Institute of Technology (KIT), Kaiserstraße 12, 76131 Karlsruhe, Germany*
[2] *Institute of Production Science (wbk), Karlsruhe Institute of Technology (KIT), Kaiserstraße 12, 76131 Karlsruhe, Germany*

Correspondence should be addressed to Johannes Kümmel; johannes.kuemmel@kit.edu

Academic Editor: Meng Hua

Analyzing wear mechanisms and developments of surface layers in WC/Co-cemented carbide cutting inserts is of great importance for metal-cutting manufacturing. By knowing relevant processes within the surface layers of cutting tools during machining the choice of machining parameters can be influenced to get less wear and high tool life of the cutting tool. Tool wear obviously influences tool life and surface integrity of the workpiece (residual stresses, surface quality, work hardening, etc.), so the choice of optimised process parameters is of great relevance. Vapour-deposited coatings on WC/Co-cemented carbide cutting inserts are known to improve machining performance and tool life, but the mechanisms behind these improvements are not fully understood. The interaction between commercial TiN-coated and uncoated WC/Co-cemented carbide cutting inserts and a normalised SAE 1045 steel workpiece was investigated during a dry plain turning operation with constant material removal under varied machining parameters. Tool wear was assessed by light-optical microscopy, scanning electron microscopy (SEM), and EDX analysis. The state of surface layer was investigated by metallographic sectioning. Microstructural changes and material transfer due to tribological processes in the cutting zone were examined by SEM and EDX analyses.

1. Introduction

At machining metals it is important to know about the wear behaviour of the cutting tool. This importance arises due to the fact that the surface integrity of the machined workpiece is influenced by tool wear [1, 2]. In this case surface integrity is described by three main parameters: the surface roughness, the residual stress state, and the work hardening in the surface zone [1]. For further improvement in the knowledge of wear behaviour of cutting tools the surface layer states of the worn tools are important to distinguish between different wear mechanisms acting in the cutting zone. One possibility at investigating wear with respect to the applied cutting parameters (e.g., cutting speed, feed rate, and depth of cut) is the idea proposed by Lim and Ashby with the wear mechanism maps [3]. Here the most important aspects of wear (seizure, delamination wear, mild wear, severe wear, etc.) are displayed with respect to the parameters varied in the wear tests (sliding velocity, pressure, etc.). The wear map approach is also applied to metal machining, and therefore the wear is displayed as a function of cutting speed v_c and feed rate f [4, 5]. In this work the aim is to develop a better understanding of the wear characteristics for different cutting parameters and different tool materials and to get therefore a deeper knowledge of the wear processes acting in the cutting zone. The surface layer states in the uncoated and in the TiN-coated cutting tool generated during the metal machining operation are mainly addressed. These surface layer states are important for the wear mechanisms that will lead to the degradation of the cutting tool. By knowing these surface

TABLE 1: Chemical composition of workpiece material (SAE 1045) in weight %.

C	Si	Mn	P	S	Cr	Ni	Mo
0.420	0.285	0.663	0.021	0.035	0.153	0.107	0.021

TABLE 2: Cutting parameter combinations used in cutting experiments for the uncoated tool (all values) and for the TiN coated tool (values marked with symbol *).

Cutting speed (m/min)	50*	100*	125	150*	100	100	100	100	100*
Feed rate (mm/rev)	0.2*	0.2*	0.2	0.2*	0.1	0.15	0.25	0.3	0.5*

layer states a knowledge based-metal cutting operation can be achieved. Nowadays some more complex coatings than TiN coatings are also used in the metal cutting of plain carbon steels (e.g., TiAlCN, AlCrN + TiAlCN coatings, etc.). For this examination the TiN-coating was used to have a better possibility of surface layer state characterisation by metallographic methods and the SEM examination [6].

2. Experimental Setup

For the experiments the chosen workpiece material was SAE 1045 plain carbon steel in a normalised state. The workpiece material had a ferritic-pearlitic microstructure with a mean grain size of ferrite and pearlite of 16 μm. The hardness of the steel was 194 HV1 measured by the Vickers hardness testing method. The chemical composition of the workpiece material is given in Table 1.

The workpieces were cylinders with length 100 mm and diameter 58 mm. The machining was done by dry plain turning down to a diameter of 24 mm. The dry plain turning was performed with a machining centre Heller MC 16. The cutting tool used for the experiments was industrial fine-grained (grain size of WC is 0.5 μm) cemented carbide (K10) with a composition of 94 volume-% WC and 6 volume-% Co in an uncoated state and in a TiN coated state. The TiN-coating (thickness 3 μm) was deposited directly on the cemented carbide substrate. The designation of the tool is SNMA 120408 according to the standard DIN ISO 1832 without any chip breakers to provide better wear measurement. The cutting tool has a square geometry and a wedge angle β of 90°. The corner radius r_ε is 0.8 mm and the cutting edge radius r_β of the cutting tools is 30 μm. The entering angle between the main cutting edge and feed direction κ_r is 45°, and the used clearance angle α was 7° with a rake angle γ of −7°. The varied cutting parameters (cutting speed and feed rate) used in the experiments are shown in Table 2.

The wear measurements were conducted after different cutting lengths by light optical microscopy and according to the standard ISO Norm 3685 [7]. The examination of surface layer states was done by scanning electron microscopy (SEM) and chemical analysis by EDX. For further examination the cutting tools were sectioned in the worn zone by using a diamond wire saw for acquiring a metallographic section. These specimens were also examined by SEM and EDX analyses.

3. Results

3.1. Wear Examination of Uncoated Cutting Tools for Different Cutting Parameters.
The wear was documented after different cutting lengths for the different cutting parameter combinations used, which were mentioned in Table 2. In Figure 1 there is a display of the wear evolution for the parameters v_c = 100 m/min, f = 0.25 mm/rev, and a_p = 0.1 mm. The reason for the detailed examination of wear measurement is the proper determination of tool wear curves [8, 9] and to get the possibility to relate the wear of the cutting tool to the surface layer state in the resulting workpiece surface.

With an increasing cutting length the tool has an increasing crater wear depth and an increase in flank and notch wear. Another wear characteristic is the good adhesion between the workpiece material and the cutting tool. In Figure 1 the formation of built-up edges is visible and some built-up layer zones are shown. Those are present due to good adhesion tendency between steel and cemented carbide. Fragments of these built-up edges (or dead zones) which are described fundamentally in [10, 11] are visible also on the chip surface and on the workpiece surface. These layers cause relatively rough surfaces on the workpiece with a severely deformed microstructure. For the constant cutting speed of 100 m/min the feed rate was varied according to Table 2. The results of the wear measurement are shown in Figure 2. An increase in the wear intensity with increasing feed rate can be seen.

The same measurements were done for constant feed rate of 0.2 mm/rev and varying cutting speed. The results are shown in Figure 3.

For varying cutting speed there is also a strong tendency to increase wear intensity with increasing cutting speed. For the cutting speed of 100 m/min there is a slight minimum, considering the final state of the worn cutting tool after the same amount of material removal.

3.2. Wear Examination of TiN-Coated Cutting Tools.
For the comparison of the uncoated and TiN-coated cutting tools four sets of parameters (feed rate and cutting speed) were chosen. These sets of parameters are also shown in Table 2 marked with the symbol *.

In Figure 4 a comparison of the wear of uncoated cutting tools and TiN-coated cutting tools is shown for one of the four parameter sets. It can be seen that the TiN coating is highly improving the wear behaviour of the cemented carbide cutting tools. During the cutting process there is no built-up

Surface Layer States of Worn Uncoated and TiN-Coated WC/Co-Cemented Carbide Cutting Tools after Dry Plain
Turning of Carbon Steel

115

Cutting length (m)	719	3312	5894	8340
Crater wear				Minor cutting edge / Main cutting edge
Flank wear and notch wear				Minor cutting edge / Main cutting edge

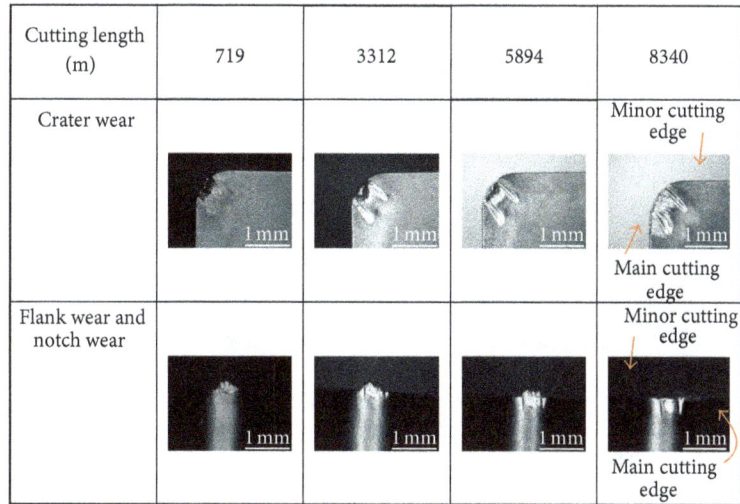

FIGURE 1: Overview of wear evolution for the cutting parameters v_c = 100 m/min, f = 0.25 mm/rev, and a_p = 0.1 mm.

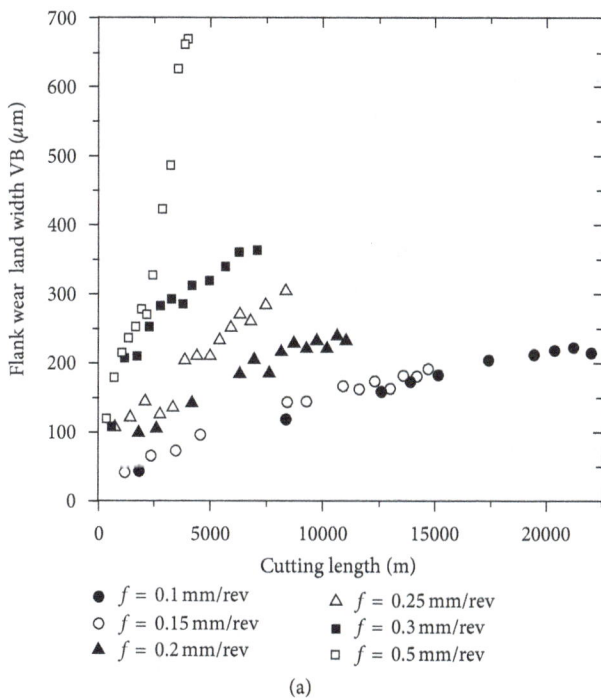

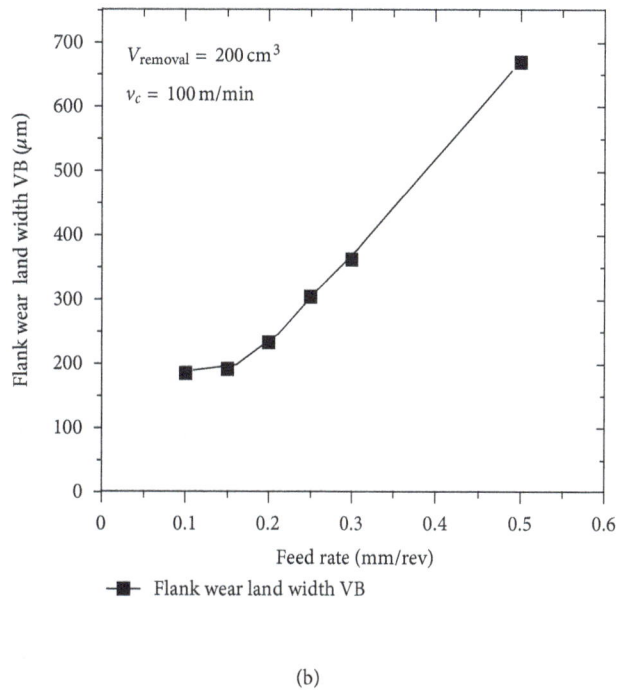

FIGURE 2: Flank wear land width VB measurement with respect to the cutting length for different feed rates (0.1 mm/rev–0.5 mm/rev) and comparison of the final state of tool wear after material removal of 200 cm^3.

edge formation on the TiN-coated cutting tool, and the wear intensity is much smaller than for the uncoated cutting tool. For the TiN-coated tool there is only little wear visible in Figure 4.

The wear was also measured in the case of the TiN-coated cutting tool and is displayed in Figure 5 with respect to the cutting length, and the final wear states are also matched in a 3D-view of flank wear land width VB with regard to the cutting parameters: cutting speed and feed rate. From

Figure 5 one can clearly see that the TiN coating is highly improving the wear resistance for the cutting tool in dry plain turning application. Especially in the higher cutting speed regime (150 m/min) for the own chosen cutting parameters the TiN coating improves wear behaviour to a great extent.

The dashed line "linear fit TiN" in Figure 5 is a linear regression line for the TiN-coated cutting tool used with the parameters v_c = 150 m/min and f = 0.2 mm/rev. The solid line "linear fit WC/Co" in Figure 5 is a linear regression

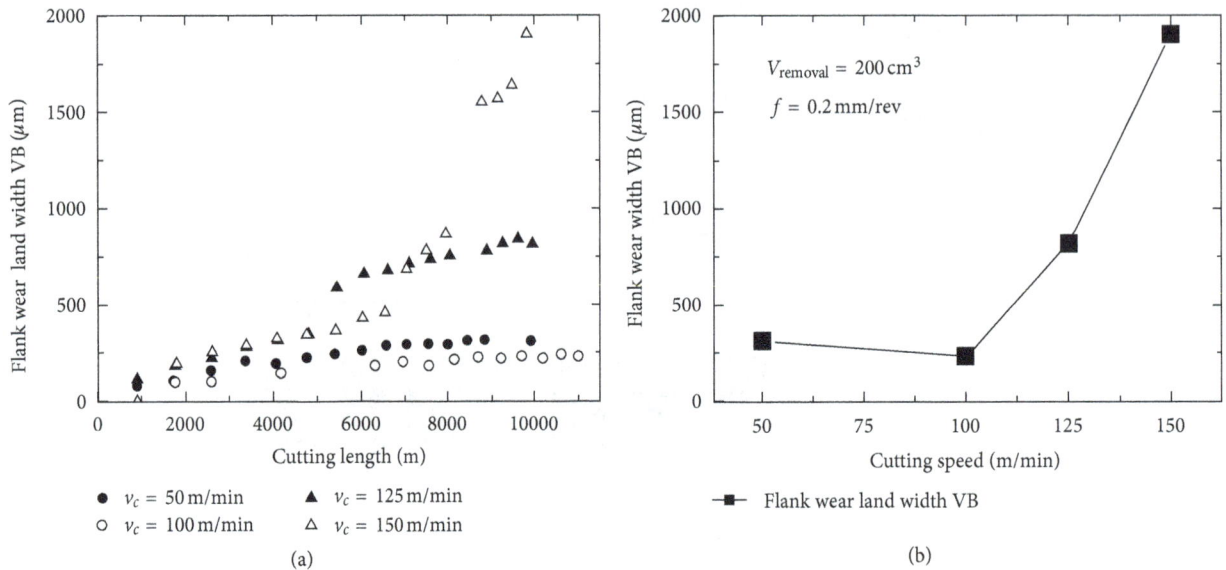

FIGURE 3: Flank wear land width VB with respect to cutting length (for varying cutting speed 50 m/min–150 m/min) and constant feed rate and comparison of the final state of tool wear after material removal of 200 cm^3.

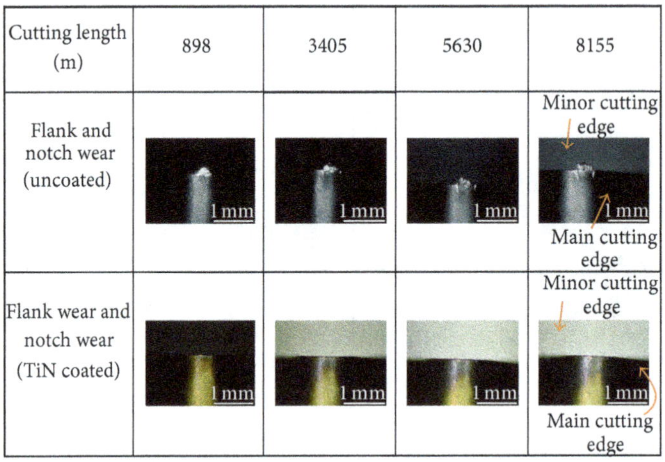

Cutting length (m)	898	3405	5630	8155
Flank and notch wear (uncoated)				Minor cutting edge / Main cutting edge
Flank wear and notch wear (TiN coated)				Minor cutting edge / Main cutting edge

FIGURE 4: Comparison of wear of uncoated and TiN-coated cutting tools for set of parameter: v_c = 100 m/min, f = 0.20 mm/rev, and a_p = 0.1 mm.

line for the linear wear regime up to a cutting length of 8000 m for the uncoated cutting tool used with the same parameters.

The examination of wear in the final state of the cutting process was also done by SEM. Two different states, one for a lower (100 m/min) and one for a higher (150 m/min) cutting speed, were examined.

In Figure 6 the surface structure of the worn rake face of the uncoated cutting tool is shown. In the crater wear region the most important wear mechanisms are the adhesion of workpiece material adhering to the rake face and tungsten carbide grain pullouts. This is due to the strong adhesion tendency of steel to the cemented carbide. This strong adhesive effect is also responsible for the formation of built-up edges. In comparison with Figure 6(a) the higher cutting speed applied during the metal cutting process leads to a smoother worn surface in the crater wear region which is seen in a higher magnification in Figure 6(b).

The rake face of the TiN-coated cutting tool that is shown in Figure 7 is less worn than for the uncoated cutting tool. Only in the region next to the cutting edge, where the highest intensity of chip flow is assumed, there are some areas without the TiN coating. In these parts of the rake face the cemented carbide substrate is visible. Further examination by energy dispersive X-ray spectroscopy (EDX) on the rake face gives some chemical information of the worn surface. In Figure 8 there is a picture shown from the rake face with a delaminated TiN-coating structure and an EDX line scan showing the chemical analysis across the worn surface. In the brighter region of the worn tool (WC/Co substrate) there·is

Surface Layer States of Worn Uncoated and TiN-Coated WC/Co-Cemented Carbide Cutting Tools after Dry Plain
Turning of Carbon Steel

117

- ■ WC/Co $v_c = 50$ m/min; $f = 0.2$ mm/rev
- ● WC/Co $v_c = 100$ m/min; $f = 0.2$ mm/rev
- ▲ WC/Co $v_c = 150$ m/min; $f = 0.2$ mm/rev
- ◆ WC/Co $v_c = 100$ m/min; $f = 0.5$ mm/rev
- □ TiN $v_c = 50$ m/min; $f = 0.2$ mm/rev
- ○ TiN $v_c = 100$ m/min; $f = 0.2$ mm/rev
- △ TiN $v_c = 150$ m/min; $f = 0.2$ mm/rev
- ◇ TiN $v_c = 100$ m/min; $f = 0.5$ mm/rev
- —— Linear fit WC/Co
- ·—·— Linear fit TiN

(a)

(b)

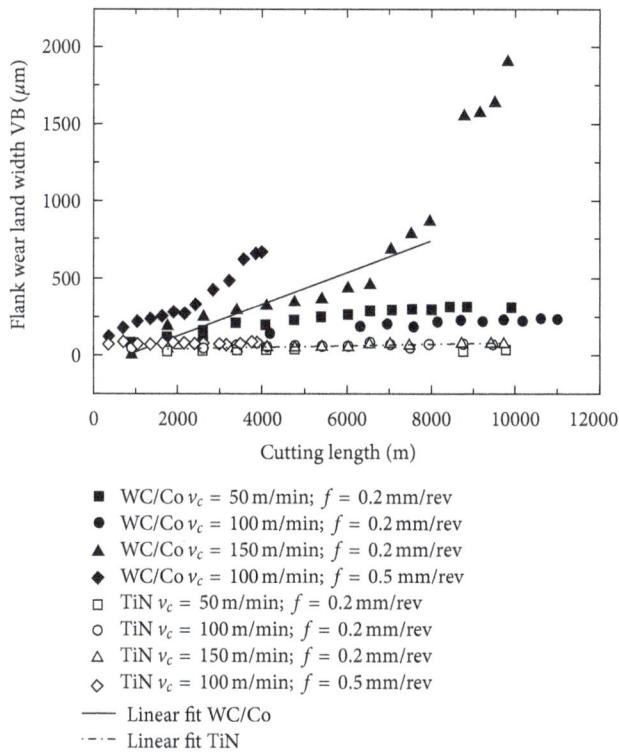

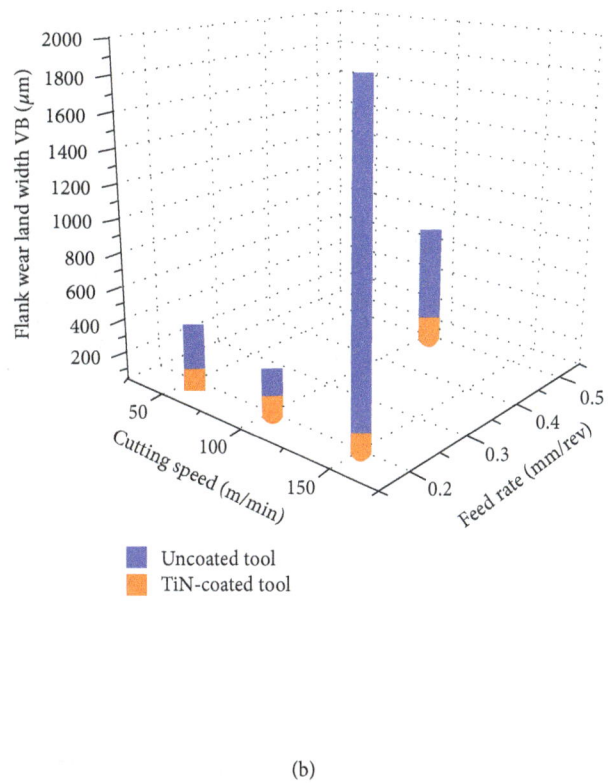

FIGURE 5: Comparison of flank wear land width VB with respect to the cutting length for the uncoated (black closed symbols) and TiN-coated (open symbols) cutting tool. The diagram on the right side shows a comparison (uncoated versus coated cutting tool) of the flank wear land width VB in the final state of the cutting tool with respect to cutting speed and feed rate.

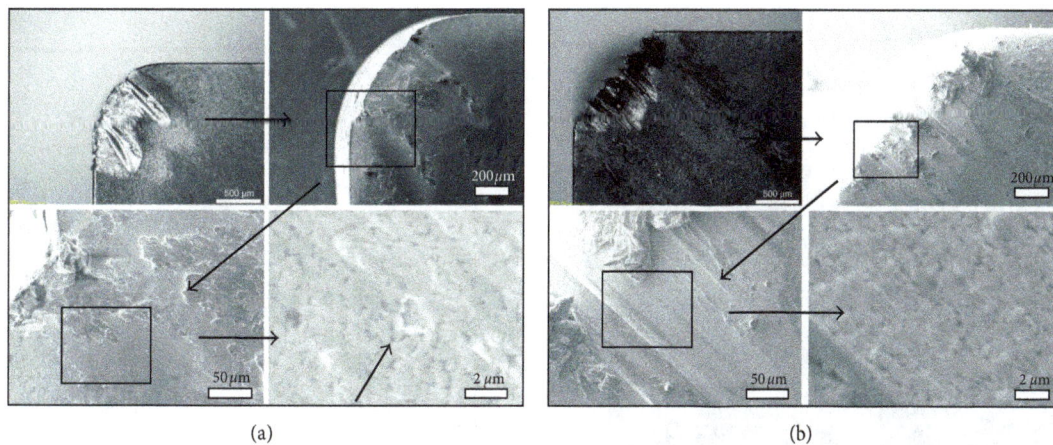

(a)

(b)

FIGURE 6: SEM examination of worn rake face of uncoated cutting tools. In (a) the crater wear is visible for the cutting parameters $v_c = 100$ m/min, $f = 0.25$ mm/rev, and $a_p = 0.1$ mm. In the highest magnification there is a WC-grain pullout visible on the rake face. On the right (b) there is a SEM examination of worn rake face of uncoated cutting tool with $v_c = 150$ m/min, $f = 0.20$ mm/rev, and $a_p = 0.1$ mm.

a strong tendency of Fe adhesion from the workpiece (peaks of high intensity of Fe). The TiN-coated areas (darker parts in Figure 8(a)) show no such intensive Fe peaks in the EDX spectrum. Therefore, the adhesion of iron to the TiN coating is less strong.

In Figure 8(b) two important intensity curves obtained from an EDX line scan are shown. The black closed symbols denote the titanium peak intensity and the open symbols denote the intensity of Fe. In the region without TiN coating there are high intensities from Fe, showing strong adhesion

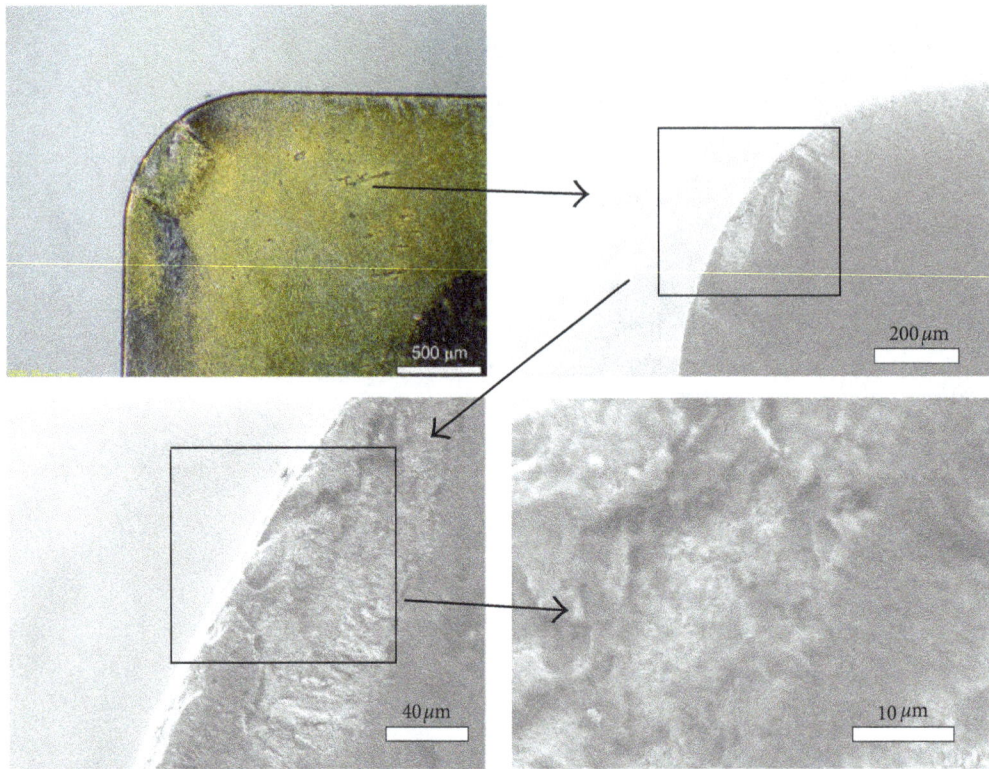

FIGURE 7: SEM examination of worn rake face of TiN-coated cutting tool: $v_c = 150$ m/min, $f = 0.20$ mm/rev, and $a_p = 0.1$ mm.

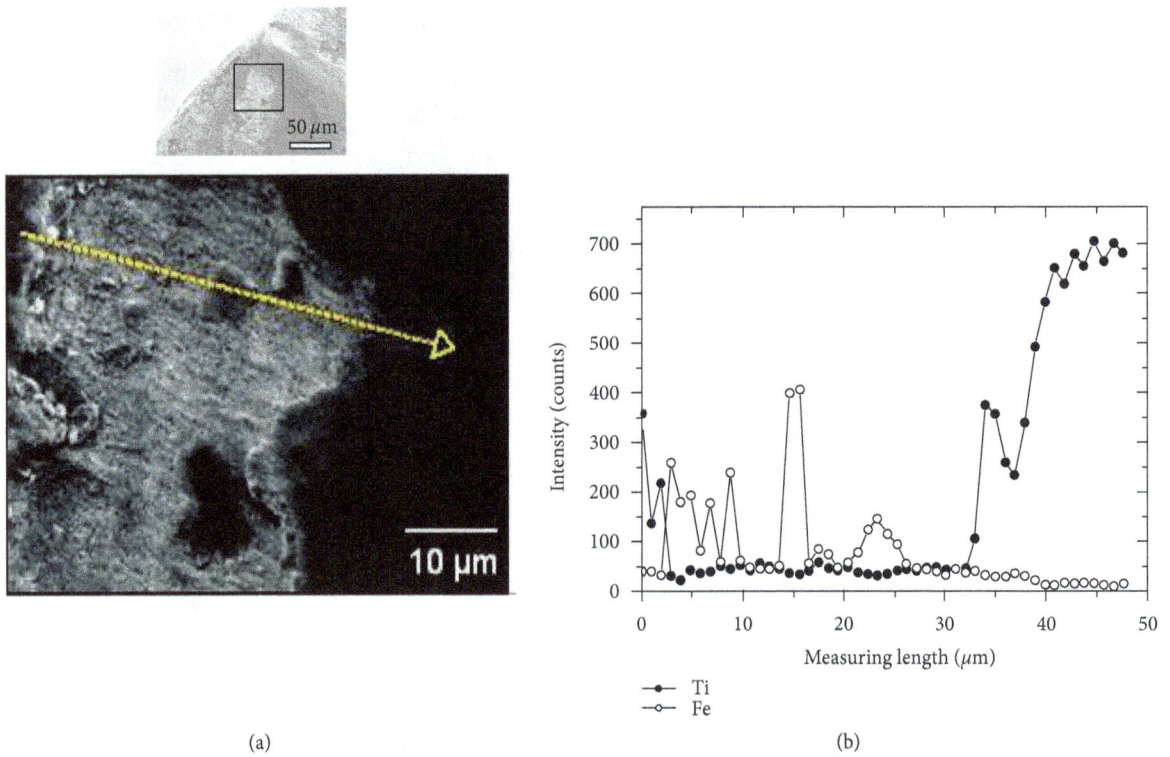

(a)   (b)

FIGURE 8: (a) SEM image worn rake face of TiN-coated cutting tool: $v_c = 150$ m/min, $f = 0.20$ mm/rev, and $a_p = 0.1$ mm. The arrow depicts the path of the EDX line scan shown in (b).

Surface Layer States of Worn Uncoated and TiN-Coated WC/Co-Cemented Carbide Cutting Tools after Dry Plain
Turning of Carbon Steel

119

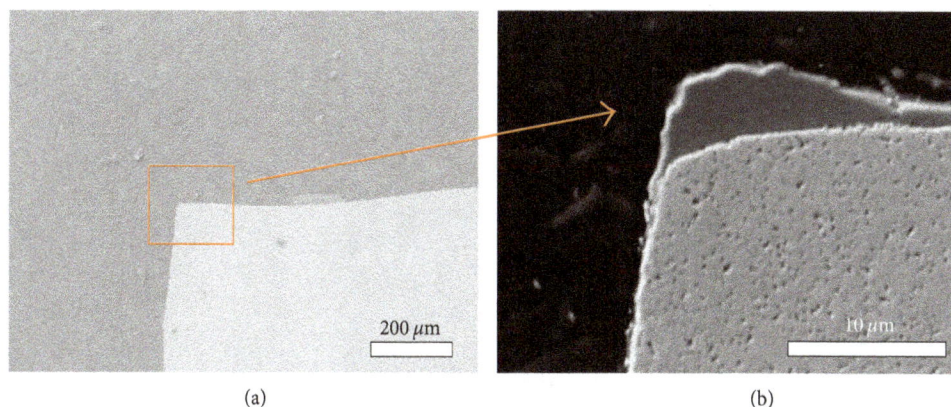

(a)   (b)

FIGURE 9: SEM picture of worn uncoated cutting tool in final state (after cutting length of 7084 m) with cutting parameters as follows: v_c = 100 m/min, f = 0.30 mm/rev, and a_p = 0.1 mm. Due to the high temperatures acting in the cutting zone some evidence is seen for plastic lowering of the cutting edge [9].

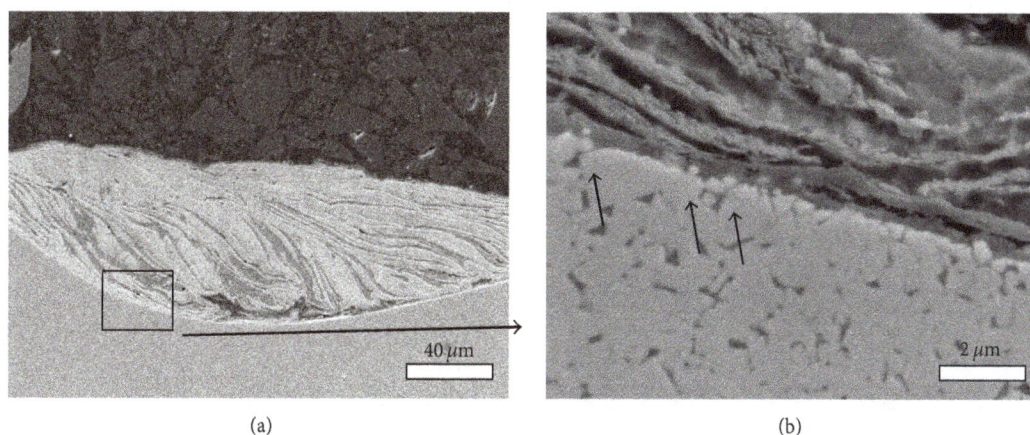

(a)   (b)

FIGURE 10: SEM picture of worn uncoated cutting tool in final state (after cutting length of 14669 m) with cutting parameters as follows: v_c = 100 m/min, f = 0.15 mm/rev, and a_p = 0.1 mm.

of workpiece material on the rake face, because of the worn TiN coating.

3.3. Examination of Surface Layer States in Uncoated and TiN-Coated Cutting Tool.

For the examination of the surface layer states in the cutting tool metallographic sectioning was done [6], where the worn tool is carefully cut by the use of a diamond wire saw and is prepared by metallographic methods. The sectioned worn cutting tool was embedded in a thermoset resin for metallographic grinding and polishing. Some examples of metallographic sections prepared this way are shown in Figures 9 and 10.

In Figure 9 an overview of a worn WC-Co cemented carbide cutting tool in uncoated state is shown with a built-up edge on the cutting edge. In the metallographic sections the crater wear, flank wear and material transfer from the work piece can be detected.

In Figure 10 the worn surface state is shown, and with the aid of the back scatter electron detector (BSE) some little WC-grain fragments are visible in the interface between the built-up layer and the uncoated cutting tool. Concerning

the wear mechanisms acting for the uncoated cutting tool, microcracking of WC grains is found. These small particles cause abrasive wear due to their hardness. This abrasive wear is caused by the small wear debris in the interfacial area.

The same procedure for the examination of cutting tool surface layer states of TiN-coated cutting tools was applied for the cutting parameter set v_c = 100 m/min, f = 0.5 mm/rev, and a_p = 0.1 mm.

In Figure 9(a) the flank wear and crater wear are both visible. On the right side in Figure 9(b) there is a detailed picture showing the built-up edge in a higher magnification.

In Figure 10(a) the crater wear is visible with a large built-up layer of work material. In Figure 10(b) the interface between the cemented carbide substrate and the etched SAE 1045 built-up layer is shown (etching agent: Nital).

In Figure 11 it is shown that the TiN coating has a protective effect on the wear behaviour of the cutting tool. Only locally there are some workpiece fragments adhering on the cutting tool substrate, where the TiN coating is worn, and the cemented carbide substrate is exposed to the workpiece material.

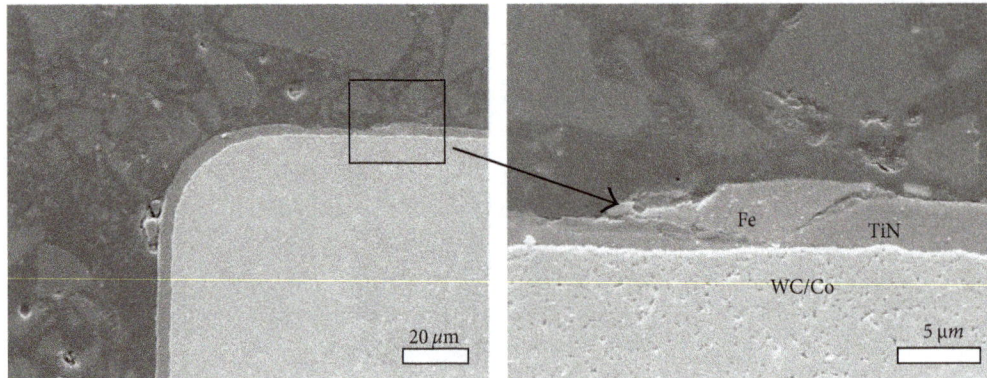

FIGURE 11: SEM picture of slightly worn TiN-coated cutting tool in final state with cutting parameters as follows: $v_c = 100 \, \text{m/min}$, $f = 0.5 \, \text{mm/rev}$, and $a_p = 0.1 \, \text{mm}$. The cutting edge is shown, and on the right hand side the detail is shown, with a piece of workpiece material adhering to the WC/Co substrate.

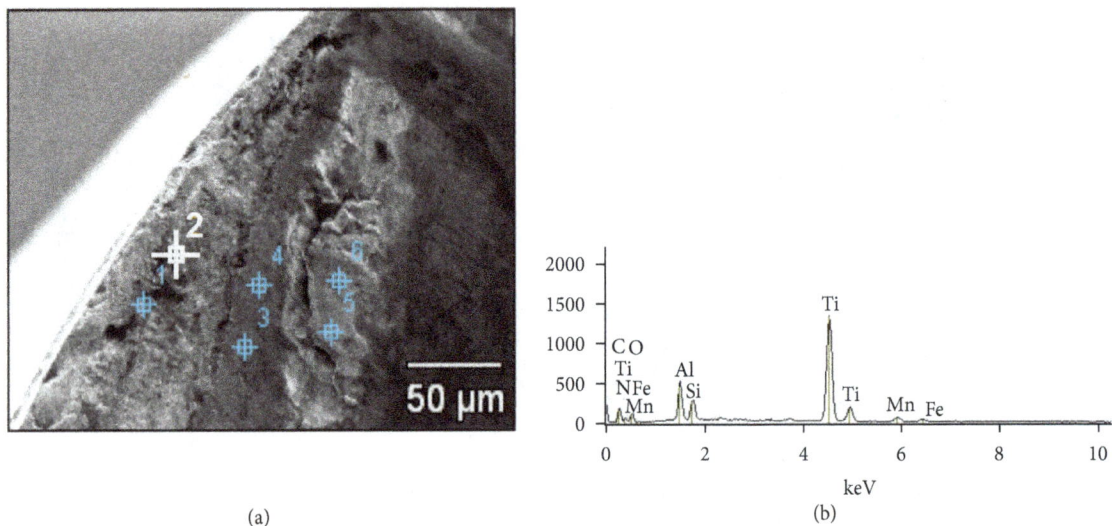

(a)

(b)

FIGURE 12: EDX spectrum on rake face of a worn TiN-coated cutting tool for point 2 near the cutting edge. Some elements from the workpiece material were found (cutting conditions: $v_c = 150 \, \text{m/min}$, $f = 0.2 \, \text{mm/rev}$, and $a_p = 0.1 \, \text{mm}$).

4. Discussion

The examination of surface layer states of uncoated WC/Co-cutting tools is one important aim of this paper. The main wear mechanisms for all chosen parameters in Table 2 are adhesive wear and a three-body abrasive wear for the uncoated cutting tool. The adhesive and three-body abrasive wear was also shown in further studies [12]. Other wear mechanisms like diffusion are also possible [13, 14]. The morphology of crater wear on the rake face for lower cutting speeds and feed rates, as seen in Figure 6, can be explained by the chip side flow because of small cutting depth chosen in the experiments. The chips show a curled structure, so that they can produce notch-like crater wear.

The examination of surface layer states in the TiN-coated cutting tool is another important aim of this paper. From Figure 5 it can be seen that the TiN coating is highly improving the wear behaviour of cemented carbide cutting tools. The wear intensity is reduced in the highest cutting speed regime by a factor of 20 due to the TiN coating for the comparison of the highest cutting speed of 150 m/min in the linear wear regime in Figure 5 (the wear intensities as flank wear land width VB per cutting length were calculated by linear regression of the flank wear distribution in Figure 5). This can be explained by different reasons. TiN has got a higher hardness (2300 HV0.05 [15]) than cemented carbide (fine-grained cemented carbide 94 volume-% WC, 6 volume-% Co (1850 HV30) [16]). The thermal conductivity of TiN is lower (29 $\text{Wm}^{-1} \text{K}^{-1}$ [16]) than that of cemented carbide (68.9 $\text{Wm}^{-1} \text{K}^{-1}$ [16]). Therefore, more heat will be conducted through the chip and the work piece material, and the TiN-coated cutting tool will see lower temperatures.

The most important difference between the wear behaviour of uncoated and TiN-coated tools is, however, the adhesion between workpiece and tool. In the case of steel workpieces, this adhesion to cemented carbide substrate is much stronger than to TiN coating. The adhesive wear is correlated to the atomic bonding of the different materials

Surface Layer States of Worn Uncoated and TiN-Coated WC/Co-Cemented Carbide Cutting Tools after Dry Plain
Turning of Carbon Steel

121

FIGURE 13: Small region of flank wear (resp. notch wear) of TiN-coated cutting tool, which was used under the cutting conditions of $v_c = 100$ m/min, $f = 0.5$ mm/rev, and $a_p = 0.1$ mm.

between the different atoms of the workpiece material, and the cutting tool material and a smooth surface is necessary [17]. The different wear mechanisms acting in the cutting zone are abrasion and adhesion.

As shown, there is less-adhesive effect of the work material on the TiN coating. The wear mechanisms acting, when turning with a TiN-coated tool, were not examined in detail. There are some further possible reasons, why TiN has got a better wear behaviour. Some protective adhesion layers were found or assumed on TiN and AlCrN coatings [6, 18]. On the worn TiN-layers some elements (Al, Si, etc., see Table 1) from the workpiece material were found (see Figure 12). These could also form a protective tribolayer during the cutting process.

The TiN coating changes its surface structure from a rougher surface to a smoother surface during the metal cutting process which can be attributed to the wear of chip flow respectively, the tool-workpiece interaction (see Figure 13). In Figure 13 the worn region of the TiN coating has got a smoother surface structure than the unworn TiN-coating.

In Figure 13, a small part of the flank face is shown where some notch wear is detected via SEM. There are three main parts of the cutting tool region: the surrounding part is consisting of unworn TiN coating. This as-deposited coating has got a rougher surface structure than the worn part of the tool, where sliding between chip/workpiece and the tool occurs. Here the sliding process leads to a smoothening of the TiN coating. In the middle of the worn region, the coating is fully worn. Cemented carbide substrate can be seen (brighter area in Figure 13), and adhering workpiece material (Fe) can be detected. This notch wear can be attributed to the high wear intensity that is acting at the beginning of the cutting process, when the cutting edge enters the workpiece.

5. Conclusions/Summary

An increasing load (higher feed rate or higher cutting speed) subjected to the cutting tool results in an increasing wear rate with respect to flank wear.

The formation of built-up edges, built-up layers, and dead zones is detected for the uncoated cemented carbide cutting tool. The reason for that is a strong adhesion of steel (workpiece material) to the cemented carbide.

The wear was examined by metallographic sectioning, and it was revealed that the wear mechanisms are adhesion and abrasion.

Improving wear behaviour of the TiN-coated cemented carbide cutting tools is due to the low adhesion of the selected workpiece material (SAE 1045) to the TiN coating. Hence there is no built-up edge or built-up layer formation on the TiN coating. On delaminated, respectively, fully worn TiN coating there is again a strong adhesion detected to the cemented carbide substrate, and a local increase in wear rate is observed.

Acknowledgment

The authors gratefully acknowledge the company OC Oerlikon Balzers for the deposition of the TiN coating.

References

[1] I. S. Jawahir, E. Brinksmeier, R. M'Saoubi et al., "Surface integrity in material removal processes: recent advances," *CIRP Annals—Manufacturing Technology*, vol. 60, no. 2, pp. 603–626, 2011.

[2] Q. Xie, A. E. Bayoumi, and L. A. Kendall, "On tool wear and its effect on machined surface integrity," *Journal of Materials Shaping Technology*, vol. 8, no. 4, pp. 255–265, 1990.

[3] S. C. Lim and M. F. Ashby, "Wear-mechanism maps," *Acta Metallurgica*, vol. 35, no. 1, pp. 1–24, 1987.

[4] S. C. Lim, Y. B. Liu, S. H. Lee, and K. H. W. Seah, "Mapping the wear of some cutting-tool materials," *Wear*, vol. 162-164, pp. 971–974, 1993.

[5] S. C. Lim, "Recent developments in wear-mechanism maps," *Tribology International*, vol. 31, no. 1–3, pp. 87–97, 1998.

[6] S. Karagöz and H. F. Fischmeister, "Metallographic observations on the wear process of TiN-coated cutting tools," *Surface and Coatings Technology*, vol. 81, no. 2-3, pp. 190–200, 1996.

[7]   "ISO 3685: tool-life testing with single-point turning tools,"
      1993.

[8]   V. P. Astakhov, *Tribology of Metal Cutting*, vol. 52 of *Tribology
      and Interface Engineering Series*, Elsevier, 2006.

[9]   V. P. Astakhov, "The assessment of cutting tool wear," *Interna-
      tional Journal of Machine Tools and Manufacture*, vol. 44, no. 6,
      pp. 637–647, 2004.

[10] S. Jacobson and P. Wallén, "A new classification system for dead
      zones in metal cutting," *International Journal of Machine Tool
      Design and Research*, vol. 28, no. 4, pp. 529–538, 1988.

[11] P. K. Philip, "Built-up edge phenomenon in machining steel
      with carbide," *International Journal of Machine Tool Design and
      Research*, vol. 11, no. 2, pp. 121–132, 1971.

[12] H. Opitz and M. Gappisch, "Some recent research on the wear
      behaviour of carbide cutting tools," *International Journal of
      Machine Tool Design and Research*, vol. 2, no. 1, pp. 43–73, 1962.

[13] J. A. Arsecularatne, L. C. Zhang, and C. Montross, "Wear and
      tool life of tungsten carbide, PCBN and PCD cutting tools,"
      International Journal of Machine Tools and Manufacture, vol. 46,
      no. 5, pp. 482–491, 2006.

[14] H. O. Gekonde and S. V. Subramanian, "Tribology of tool-
      chip interface and tool wear mechanisms," *Surface and Coatings
      Technology*, vol. 149, no. 2-3, pp. 151–160, 2002.

[15] "Oerlikon Balzers product information BALINIT A," 2011.

[16] "Springer Materials: the Landolt Börnstein database, the worlds
      largest resource for physical and chemical data," 2009.

[17] Valentin L. Popov, *Kontaktmechanik und Reibung: Von der Nan-
      otribologie bis zur Erdbebendynamik*, Springer, Berlin, Germany,
      2010.

[18] J. Gerth, M. Larsson, U. Wiklund, F. Riddar, and S. Hogmark,
      "On the wear of PVD-coated HSS hobs in dry gear cutting,"
      Wear, vol. 266, no. 3-4, pp. 444–452, 2009.

14

The Head-Disk Interface Roadmap to an Areal Density of 4 Tbit/in^2

Bruno Marchon,[1] Thomas Pitchford,[2] Yiao-Tee Hsia,[3] and Sunita Gangopadhyay[2]

[1] *HGST, San Jose, CA 95135, USA*
[2] *Seagate Technology, Minneapolis, MN 55435, USA*
[3] *Western Digital, San Jose, CA 95138, USA*

Correspondence should be addressed to Bruno Marchon; bruno.marchon@hgst.com

Academic Editor: Tom Karis

This paper reviews the state of the head-disk interface (HDI) technology, and more particularly the head-medium spacing (HMS), for today's and future hard-disk drives. Current storage areal density on a disk surface is fast approaching the one terabit per square inch mark, although the compound annual growth rate has reduced considerably from ~100%/annum in the late 1990s to 20–30% today. This rate is now lower than the historical, Moore's law equivalent of ~40%/annum. A necessary enabler to a high areal density is the HMS, or the distance from the bottom of the read sensor on the flying head to the top of the magnetic medium on the rotating disk. This paper describes the various components of the HMS and various scenarios and challenges on how to achieve a goal of 4.0–4.5 nm for the 4 Tbit/in^2 density point. Special considerations will also be given to the implication of disruptive technologies such as sealing the drive in an inert atmosphere and novel recording schemes such as bit patterned media and heat assisted magnetic recording.

1. Introduction

As the areal density of commercial hard disk drives is quickly approaching the terabit per square inch milestone [1–5] (Figure 1), the need to improve the reliability of the head-disk interface (HDI) and to further decrease the head-medium spacing (HMS) is becoming evermore critical [3, 6, 7]. Low HMS is a necessary enabler to good writability as well as strong read-back signal integrity [8, 9]. It is estimated that the HMS will soon need to cross the 7 nm mark in order to reach this terabit per square inch density point [2, 6]. It is remarkable to realize that the error rate of the stored digital signal that is being read back improves approximately by about 2x for every 0.3–0.5 nanometer of decreased HMS. In addition to relentless demand for novel, ultrathin protecting films of overcoat and lubricant, and subnanometer air gap between the disk and the head, alternative recording technologies presently being contemplated involve heating the disk to over 500°C (heat-assisted magnetic recording or HAMR) [10–12] and/or physically isolating magnetic bits on small islands of sub-30 nm in physical dimensions (bit-patterned recording or BPR) [13–16].

In this paper, the roadmap to an areal density of 4 terabits per square inches will be discussed. Particular emphasis will be given to the various spacing components that comprise the HMS budget and their physical limits. The various implications of recording technologies incorporating HAMR and/or BPR will also be addressed.

2. Historical Perspective

Hard drive technology has constantly evolved to achieve consistent areal density growth. As one technology such as longitudinal recording has reached its limit, another such as perpendicular recording has taken over [18]. Along with advances in heads, media, signal processing, and servo technology, areal density growth is sustained through improvements in the head-disk interface (HDI). Head-media spacing (HMS) is the most important HDI parameter related to areal density growth [6]. Continued developments in the tribological design of disk drives have maintained the reliability of the head/disk interface despite decreased spacing.

Analysis of long-term trends shows that HMS has steadily declined over time (Figure 2). The historical trends indicate

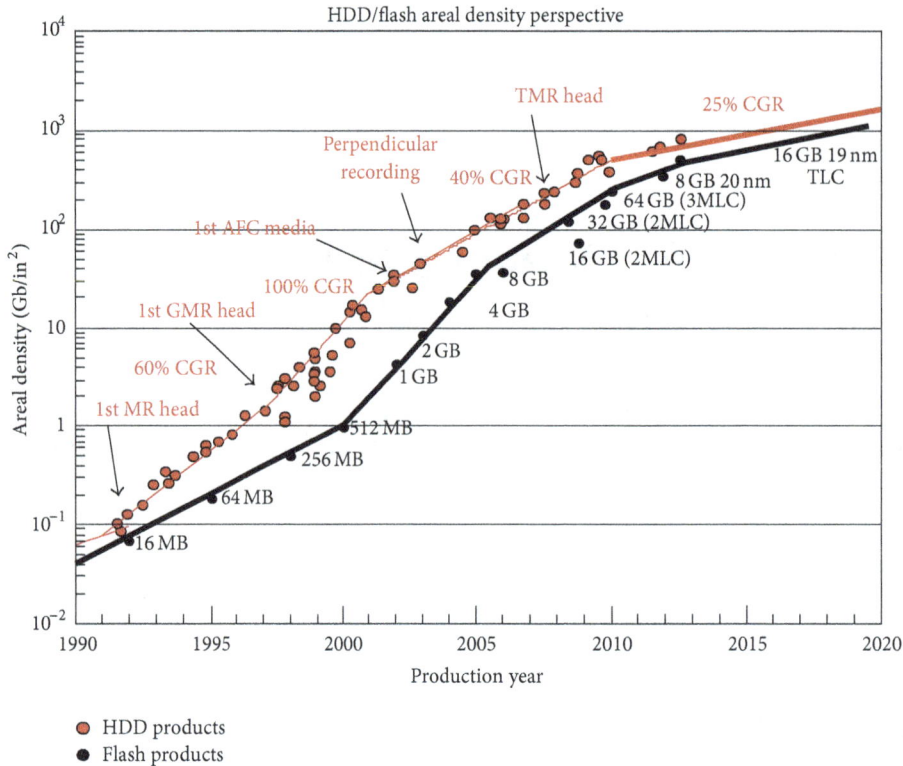

FIGURE 1: Areal density evolution of HDD and flash memory. After Grochowski [17], with permission.

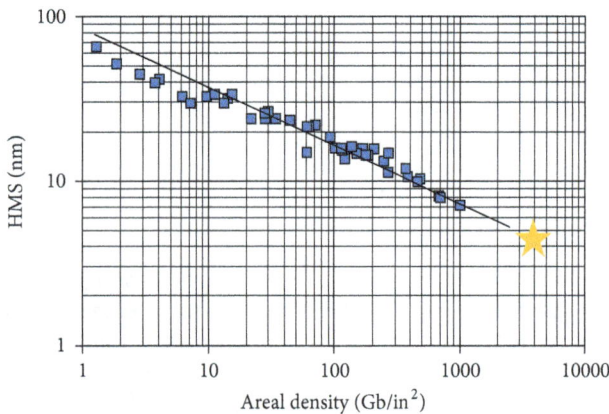

FIGURE 2: Historical variations of HMS versus areal density from [6]. The yellow star is an extrapolation to 4 Tb/in^2.

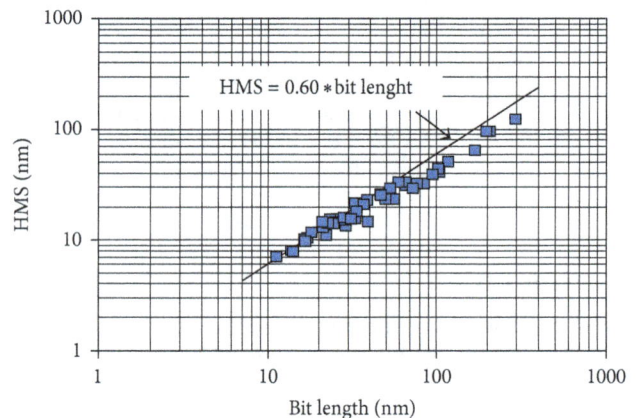

FIGURE 3: Historical variations of HMS versus bit length from [6].

that the HMS of recent products is ~60% of bit length (Figure 3) [6]. The HMS for recording demonstrations has been typically more aggressive, at ~50% of bit length.

Research consortia such as the Information Storage Industry Consortium (INSIC—https://www.insic.com/), Storage Research Consortium (SRC—http://www.srcjp.gr.jp/), and the Advanced Storage Technology Consortium (ASTC—http://www.idema.org/?page_id=3193) have included collaborations on HDI technology development. For each areal density, the HMS target has typically been set by the Recording Subsystem (RSS) groups. The past and current HMS trends for the different consortia are shown in Figure 4. The HMS trends have typically followed the overall trends of Figures 2 and 3 fairly well. For the HMS roadmaps released in 2003–2010, the INSIC trend tended to be lower than for SRC. For the current update (2012), the ASTC trend is higher than SRC.

For ASTC the main areal density targets are 2 Tb/in^2 and 4 Tb/in^2, with scheduled product introductions of 2016 and 2020, respectively. The current HMS targets for these two density points is ~5-6 and 4-5 nm, respectively, that is, less aggressive than those for INSIC. The 4 Tb/in^2 HMS goal will be discussed in more details in Section 4.

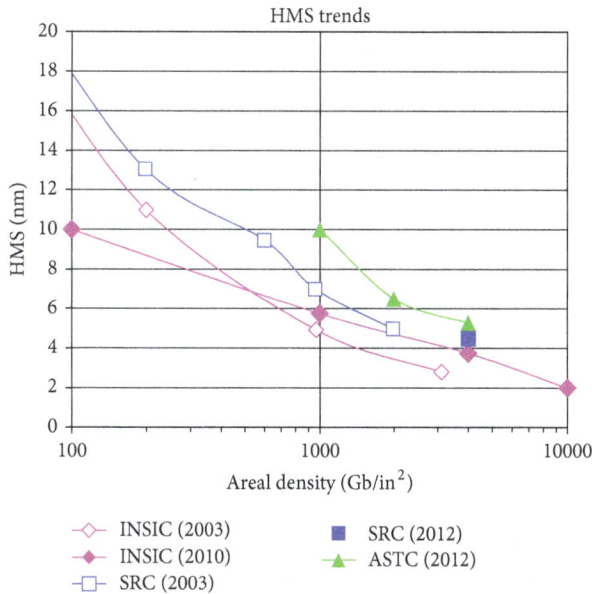

FIGURE 4: HMS trends of industry research consortia.

3. Definition of the Head-Media Spacing Components

As the capacity and performance of disk drives has improved, the mechanical interface has evolved. The spacing between the head and the media has steadily decreased to achieve the rapid improvements in areal density. Figure 5 is a diagram of head-media spacing (HMS). The HMS is comprised of the fly height and coatings (head and disk overcoats, disk lubricant).

In the diagram of Figure 5(b), the surfaces are assumed to be perfectly smooth—they do not include factors such as surface topography and variability of the fly height. The fly height denotes the spacing between the centerline surfaces of the head and disk. The clearance denotes the spacing between the close points of the surfaces. The distinction between fly height and clearance is depicted in Figure 6.

Below are some definitions of terms commonly used by the HDI/tribology community.

(1) Head-media spacing (HMS): the spacing between the top of the magnetic layer and the surface of the transducer.

(2) Flying clearance: difference or margin in fly height between nominal operation and contact between the head and disk.

(3) Glide avalanche/touchdown height (TDH): the lowest slider flying height above the mean roughness level without significant slider-disk contact

(4) Flying height = flying clearance + touchdown height

(5) Media lubricant thickness: the average thickness of the lubricant, assumed to be on top of the media overcoat

(6) Media overcoat thickness: the average thickness of the media overcoat, assumed to follow the topography of the underlying surface

(7) Head overcoat: head adhesion layer (e.g., silicon) + diamond-like carbon (DLC)

$$HMS = \text{head overcoat} + TDH + \text{clearance} + \text{media lubricant} + \text{media overcoat} \quad (1)$$

For current hard drives, HMS reductions have been achieved mostly through reductions in the clearance and disk and head overcoats. While this trend will continue for the immediate future, additional HMS reductions will soon require notraditional HDI designs in order to meet the demanding recording requirements for areal densities beyond 1 Tb/in².

In older drive designs, the air bearing design and passive topography of the slider determined the fly height and clearance of the transducer. In newer drive designs, the fly height is actively controlled by a signal that changes the shape of the slider. The most common method used for this control is to embed an electrical resistive heater in the slider that will cause the transducer area to protrude closer to the disk, as is shown in Figure 6 [19–21]. At some point the interface may need to be designed to withstand intermittent or continuous contact [22–27].

Current products have HMS on the order of 10 nm or slightly below [6]. For the 1 Tb/in² products under development, a Hi-Lo range bracket can be defined, since this areal density point is no longer precompetitive. For beyond 1 Tb/in², achievable and stretch values were estimated. The technological development required to achieve the targets will be discussed in Section 4. The assessment indicated that there is promise for achieving 4–6 nm HMS in the long term. While advancements in HDI design could enable significant reduction in HMS, new recording schemes could add to the challenge due to new sources of HMS loss: heat assisted magnetic recording (HAMR) will experience effects of high temperature at the interface [28–30], and bit-patterned magnetic recording (BPR) could have issues with added disk topography that would add to the touchdown height [31–34].

4. ASTC HDI Roadmap to 4 Tb/in² and Major Research Challenges

As shown in Figure 2, a good estimation of the HMS is ~60% of the length of the bit, and some rationale for this was recently proposed [6]. This offers a convenient way to project future HMS values, as the following simple equations can be derived. If one defines areal density (AD) in bits per square inch as the product of the linear density in bit per inch (BPI) and track density (TPI), $AD = BPI \cdot TPI$. Furthermore, the bit aspect ratio BAR is usually defined as $BAR = BPI/TPI$; hence, the bit length (BL), in inches, can be expressed as

$$BL = (AD \cdot BAR)^{-1/2}. \quad (2)$$

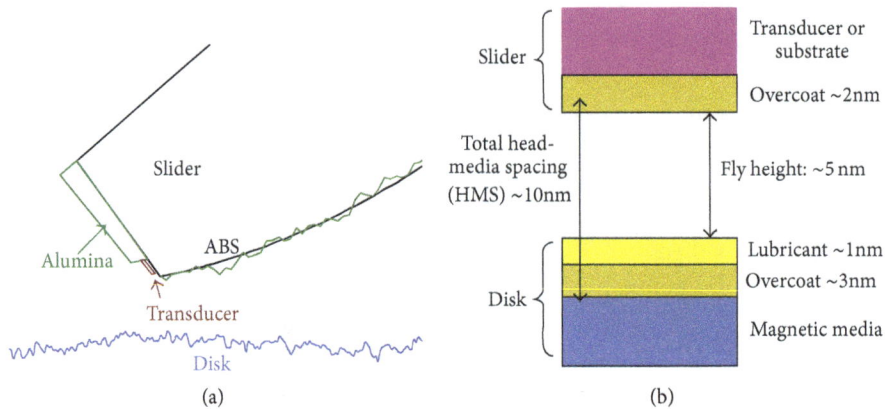

FIGURE 5: Components of head-disk spacing: (a) schematic of trailing end of head as it flies over disk. (b) Idealized HMS stacks up.

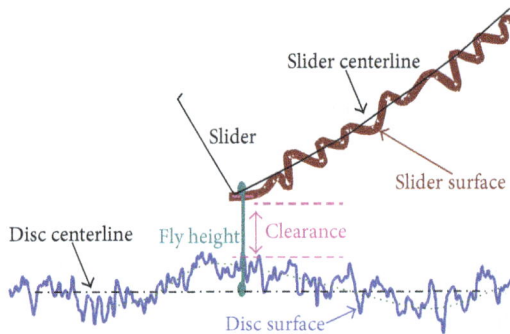

FIGURE 6: Parameters related to head-disk clearance and fly height.

TABLE 1: Estimated HMS (nm) values as a function of AD and BAR.

AD (Tb/in^2)	BAR			
	3.0	3.5	4.0	4.5
2	6.2	5.8	5.4	5.1
4	4.4	4.1	3.8	3.6

With the HMS approximation of 60% of BL, we now have a very convenient expression linking HMS (in nanometer) to AD (in Tb/in^2), for a given BAR:

$$HMS\,(nm) \approx 15 \cdot (AD \cdot BAR)^{-1/2}. \qquad (3)$$

Numerical HMS values based on (3) are reproduced in Table 1 below.

As shown in Table 1, achieving a 4 Tb/in^2 areal density will require an HMS in the vicinity of ~4.4 nm, assuming a bit aspect ratio (BAR) of 3. Table 2 offers two scenarios for the HMS breakdown into its various components. In the table, the 2010 projection from INSIC is also presented, as well as a realistic range for the 1 Tb/in^2 density point that sponsor companies are working towards. For the 4 Tb/in^2 point, both "Achievable" and "Stretch" values are offered, corresponding to roughly 60–100% and <60% chance of success, respectively. It is clear from the table that achieving the ~4.4 nm goal is a high risk, and it will require evolutionary as well as revolutionary changes in materials, processes, and clearance

control schemes. HDD architecture (e.g., sealing in inert atmosphere) might also be needed [35]. This is discussed in the following sections.

4.1. Materials: Disk and Head Overcoat. It is clear from Table 2 that the biggest contributor to today's HMS is the carbon overcoat, both on the disk and on the head. Combined, they amount to about half of today's (aka 1 Tb/in^2) HMS budget. Historically, overcoat thickness reduction on both components has been enabled by denser carbon, as well as smoother underlying surfaces for the magnetic medium (disk) and RW elements (head). Head carbon has evolved from sputtered to plasma-enhanced chemical vapor deposition (PE-CVD) to filtered cathodic arc (f-CAC) carbon [36–39], which increased the density, sp^3/sp^2 ratio, and hardness [40]. On the disk side, f-CAC technology has not yet been made manufacturable [41, 42], mostly because of particle and deposition rate challenges, and most if not all disks shipped today are coated with some sort of PE-CVD deposited amorphous carbon overcoat. It is believed that disk deposition tools will soon need to offer new carbon technologies, able to closely emulate a high sp^3 bonded, f-CAC-type carbon film. On the head side, evolutionary optimization of the overcoat technology might allow to reach the 4 Tb/in^2 goal of ~1 nm, believed to be the ultimate coverage limit of f-CAC. Finally, although many attempts have been made by disk manufacturers to develop and ship noncarbon overcoated disks, it remains an open question whether any thin film materials other than carbon can be made as hard, dense, and chemically inert as carbon films [43–47]. The latter issue could be alleviated if HDD's can be sealed in an inert environment [35]. It is also believed that such a drastic change in the drive mechanical platform could help reduce overcoat thickness on both heads and disks, as magnetic medium and read/write alloy corrosion could then be suppressed, or at least drastically reduced. Our estimate of such benefit is in the range of 0.2–0.3 nm or greater, for an overall reduction of ~0.5 nm, that is, a substantial amount of about 10% of the total HMS. Another potential alternative is to develop a disk surface modification process such that the highly corrosion-susceptible media material surface is

TABLE 2: HMS breakdown scenarios for 4 Tb/in^2.

HMS Budget Components Component	1 Tb/in^2			4 Tb/in^2			HMS adder (effect of HAMR, BPM)	
	INSIC 2010	Target HMS by ~2014		INSIC 2010	Target HMS by ~2020-21		BPM	HAMR
		Hi	Lo		Medium risk	High risk		
TDH	1.8	2.0	1.0	1.4	1.1	0.6	0.3	0.5
Disc overcoat	0.9	2.5	2.0	0.6	1.8	1.5	0.3	0.6
Disk lubricant	0.9	1.2	1.0	0.8	1.0	0.8		
Clearance	1.3	1.2	1.0	0.6	0.6	0.5	0.2	0.3
Head overcoat	1.0	2.0	1.5	0.7	1.1	0.9		
Total	5.8	8.9	6.5	4.0	5.6	4.3	0.8	1.4

The HMS adder columns corresponding to HAMR and BPM will be discussed in Section 4.

treated or modified without contributing to spacing loss with a creation of a nonfunctional or weakened layer of magnetic material on the media surface [48, 49].

4.2. Materials: Disk Lubricant. Table 2 shows that a 0.2 nm reduction of lubricant thickness from 1.0–1.2 to 0.8–1.0 is needed. This does not seem like much, but with today's extremely tight reliability/HMS margins, this change is actually significant, and it is believed that inventions will be needed to reach those goals [50, 51]. The lubricant industry is now more diversified than before, and new lubricant structures are now routinely offered by at least three different companies. It remains to be seen whether conventional lubricant chemistries (functionalized perfluoropolyether) will be able to achieve this thickness goal. Perhaps unconventional approaches, such as direct surface treatment/functionalization of the disk overcoat will be needed [48].

4.3. TDH: Topography. Touchdown height (TDH) globally defines all residual disk and head topographies that prevent the head from coming into close proximity to the magnetic medium (Figure 6). It is affected by waviness (~1–1000 μm wavelength range) and roughness (<1 μm) of the substrate [52–54], as well as the nanoroughness of the magnetic film-overcoat structure of the disk [55]. Unlike materials thicknesses (lubricant, overcoat), TDH is not a requirement for proper HDD reliability and could, in theory, be brought to zero. It assumed that engineering evolutionary optimization of polishing (disk substrate, slider) and deposition (media/overcoat) processes will be possible to achieve the HMS goal. Finally, it has been proposed that disk lubricant "roughness", both at the nanoscale (conformation) [56] and the microscale (thickness modulation or "moguls" [57] and "ripples" [58, 59]), also contributes to TDH, and lubricant optimization, as discussed previously, will be needed to also lower its contribution.

4.4. Head-Disk Clearance. Of all the HMS contributors, clearance has probably exhibited the largest decrease in the last 10 years or so, thanks to the advent of thermal flying height control (TFC-Figure 7) [19, 21]. This revolutionary approach has allowed the HDD industry to achieve a 10-fold reduction in clearance from ca. 10 nm ten years ago to ~1.5 nm

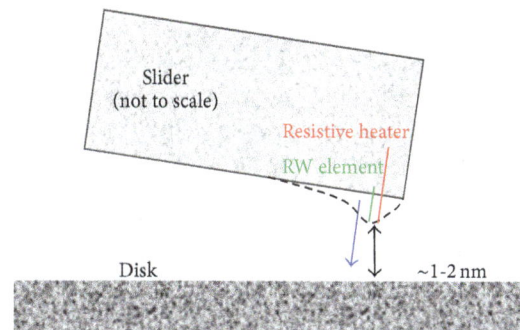

FIGURE 7: Schematics of the active fly height control scheme using an embedded resistive heater.

today (Table 2). To achieve the 0.6 nm ASTC clearance goal and to easily compensate for clearance changes induced by temperature, altitude, or humidity, further enhancement of clearance control will be needed such as closed-loop TFC control. "Surfing" of light contact recording has also been suggested [26], but it remains to be seen whether this approach can be made reliable.

5. Tribology and HDI Challenges for Alternate Technologies

As discussed earlier, to advance HDD magnetic recording beyond the ~1 Tb/in^2 believed achievable with PMR/SMR, HAMR and BPM technologies are being developed. It is believed that HAMR will first be introduced. To go beyond ~6 Tb/in^2, the ASTC roadmap shows that HAMR will be augmented with BPM recording technology [11]. While these will enable significant gains in areal density, they will present challenges for the HDI.

In terms of HMS, the effect of these technologies is shown in the "HMS Adder" columns of Table 2. The touchdown height is increased due to added media roughness for HAMR or residual topography for BPM [32, 60]. The increased roughness also drives the need for added media overcoat thickness to maintain adequate coverage. With regards to clearance, HAMR would require added margin as a guardband for possible thermal protrusion of the near-field transducer (NFT) [10], whereas BPM will require clearance margin as a guardband against residual topography that could induce head wear. It is believed that BPM will not require head overcoat thickness adder.

As the HDD industry moves towards the introduction of HAMR recording technologies, there is a high demand for the HDI to demonstrate robustness at higher temperatures [30, 61–63]. If the current disk overcoat and lubricant films are not capable of withstanding the high recording temperature (>500°C), new materials will be needed. On the head side, it is currently believed that the temperature should be limited to ~ 150°C in order to provide long-term interface reliability to the current overcoat material. In any event, further development in tribological materials for both heads and media seem likely needed in order to insure proper reliability of the HAMR interface.

The later introduction of BPM recording technology will present further challenges for HDI design. Residual topography left behind by the BPM manufacturing process challenges air bearing designers to design a slider that can better follow residual topography [34, 64, 65] as well as media manufacturers to have a uniform overcoat and lubricant coverage of the media surface. Moreover, the residual media topography could lead to excessive induced lubricant roughness on the microscale (lubricant moguls and ripples) unless a new lubricant can be designed.

6. Conclusion

In summary, head-media spacing continues to be an enabler for continued march to increase the areal density to fend off the assault and encroachment by nonvolatile solid-state memory technologies. To maintain HDD leadership and competitive edge in the data storage arena, the industry must continue to develop new component technologies to support this areal density advance. However, to be successful, the design of the head-disk interface must be an integrated effort where the component technologies are developed in concert to solve the system problem. The success of the areal density growth will depend on how successful we, as an industry, are in integrating this development.

References

[1] R. W. Wood, J. Miles, and T. Olson, "Recording technologies for terabit per square inch systems," *IEEE Transactions on Magnetics*, vol. 38, no. 4, pp. 1711–1718, 2002.

[2] R. Wood, "The feasibility of magnetic recording at 1 Terabit per square inch," *IEEE Transactions on Magnetics*, vol. 36, no. 1, pp. 36–42, 2000.

[3] C. M. Mate, Q. Dai, R. N. Payne, B. E. Knigge, and P. Baumgart, "Will the numbers add up for sub-7-nm magnetic spacings? Future metrology issues for disk drive lubricants, overcoats, and topographies," *IEEE Transactions on Magnetics*, vol. 41, no. 2, pp. 626–631, 2005.

[4] M. E. Schabes, "Micromagnetic simulations for terabit/in² head/media systems," *Journal of Magnetism and Magnetic Materials*, vol. 320, no. 22, pp. 2880–2884, 2008.

[5] M. Mallary, A. Torabi, and M. Benakli, "One terabit per square inch perpendicular recording conceptual design," *IEEE Transactions on Magnetics*, vol. 38, no. 4, pp. 1719–1724, 2002.

[6] B. Marchon and T. Olson, "Magnetic spacing trends: from LMR to PMR and beyond," *IEEE Transactions on Magnetics*, vol. 45, no. 10, pp. 3608–3611, 2009.

[7] J. Gui, "Tribology challenges for head-disk interface toward 1 Tb/in²," *IEEE Transactions on Magnetics*, vol. 39, no. 2, pp. 716–721, 2003.

[8] R. L. Wallace, "The reproduction of magnetically recorded signals," *Bell System Technical Journal*, vol. 30, pp. 1145–1173, 1951.

[9] B. Marchon, K. Saito, B. Wilson, and R. Wood, "The limits of the Wallace approximation for PMR recording at high areal density," *IEEE Transactions on Magnetics*, vol. 47, pp. 3422–3425, 2012.

[10] M. H. Kryder, E. C. Gage, T. W. Mcdaniel et al., "Heat assisted magnetic recording," *Proceedings of the IEEE*, vol. 96, no. 11, pp. 1810–1835, 2008.

[11] B. C. Stipe, T. C. Strand, C. C. Poon et al., "Magnetic recording at 1.5 Pbm-2 using an integrated plasmonic antenna," *Nature Photonics*, vol. 4, no. 7, pp. 484–488, 2010.

[12] W. A. Challener, C. Peng, A. V. Itagi et al., "Heat-assisted magnetic recording by a near-field transducer with efficient optical energy transfer," *Nature Photonics*, vol. 3, pp. 220–224, 2009.

[13] E. A. Dobisz, Z. Z. Bandić, T. W. Wu, and T. Albrecht, "Patterned media: nanofabrication challenges of future disk drives," *Proceedings of the IEEE*, vol. 96, no. 11, pp. 1836–1846, 2008.

[14] B. D. Terris, T. Thomson, and G. Hu, "Patterned media for future magnetic data storage," *Microsystem Technologies*, vol. 13, no. 2, pp. 189–196, 2007.

[15] X. Yang, S. Xiao, W. Wu et al., "Challenges in 1 Teradotin. 2 dot patterning using electron beam lithography for bit-patterned media," *Journal of Vacuum Science & Technology B*, vol. 25, no. 6, pp. 2202–2209, 2007.

[16] A. Kikitsu, "Prospects for bit patterned media for high-density magnetic recording," *Journal of Magnetism and Magnetic Materials*, vol. 321, no. 6, pp. 526–530, 2009.

[17] E. Grochowski, Future Technology Challenges for NAND Flash and HDD Products, Flash Memory Summit, 2012.

[18] S. I. Iwasaki, "Principal complementarity between perpendicular and longitudinal magnetic recording," *Journal of Magnetism and Magnetic Materials*, vol. 287, pp. 9–15, 2005.

[19] M. Suk, K. Miyake, M. Kurita, H. Tanaka, S. Saegusa, and N. Robertson, "Verification of thermally induced nanometer actuation of magnetic recording transducer to overcome mechanical and magnetic spacing challenges," *IEEE Transactions on Magnetics*, vol. 41, no. 11, pp. 4350–4352, 2005.

[20] K. Miyake, T. Shiramatsu, M. Kurita, H. Tanaka, M. Suk, and S. Saegusa, "Optimized design of heaters for flying height adjustment to preserve performance and reliability," *IEEE Transactions on Magnetics*, vol. 43, no. 6, pp. 2235–2237, 2007.

[21] T. Shiramatsu, M. Kurita, K. Miyake et al., "Drive-integration of active flying-height control slider with micro thermal actuator," *IEEE Transactions on Magnetics*, vol. 42, no. 10, pp. 2513–2515, 2006.

[22] J. Itoh, Y. Sasaki, K. Higashi, H. Takami, and T. Shikanai, "An experimental investigation for continuous contact recording technology," *IEEE Transactions on Magnetics*, vol. 37, no. 4, pp. 1806–1808, 2001.

[23] C. M. Mate, P. C. Arnett, P. Baumgart et al., "Dynamics of contacting head-disk interfaces," *IEEE Transactions on Magnetics*, vol. 40, no. 4, pp. 3156–3158, 2004.

[24] S. C. Lee and A. A. Polycarpou, "Microtribodynamics of pseudo-contacting head-disk interfaces intended for 1 Tbit/in²," *IEEE Transactions on Magnetics*, vol. 41, no. 2, pp. 812–818, 2005.

[25] B. Liu, M. S. Zhang, S. K. Yu et al., "Towards fly- and lubricant-contact recording," *Journal of Magnetism and Magnetic Materials*, vol. 320, no. 22, pp. 3183–3188, 2008.

[26] B. Liu, M. S. Zhang, S. K. Yu et al., "Lube-surfing recording and its feasibility exploration," *IEEE Transactions on Magnetics*, vol. 45, no. 2, pp. 899–904, 2009.

[27] W. Hua, B. Liu, S. K. Yu, and W. D. Zhou, "Contact recording review," *Microsystem Technologies-Micro-and Nanosystems-Information Storage and Processing Systems*, vol. 16, pp. 493–503, 2010.

[28] L. Wu and F. E. Talke, "Modeling laser induced lubricant depletion in heat-assisted-magnetic recording systems using a multiple-layered disk structure," *Microsystem Technologies*, vol. 17, no. 5-7, pp. 1109–1114, 2011.

[29] Y. S. Ma, L. Gonzaga, C. W. An, and B. Liu, "Effect of laser heating duration on lubricant depletion in heat assisted magnetic recording," *IEEE Transactions on Magnetics*, vol. 47, no. 10, pp. 3445–3448, 2011.

[30] W. D. Zhou, Y. Zeng, B. Liu, S. K. Yu, W. Hua, and X. Y. Huang, "Evaporation of polydisperse perfluoropolyether lubricants in heat-assisted magnetic recording," *Applied Physics Express*, vol. 4, no. 9, Article ID 095201, 3 pages, 2011.

[31] L. Wu, "Lubricant distribution and its effect on slider air bearing performance over bit patterned media disk of disk drives," *Journal of Applied Physics*, vol. 109, no. 7, Article ID 074511, 2011.

[32] B. E. Knigge, Z. Z. Bandic, and D. Kercher, "Flying characteristics on discrete track and bit-patterned media with a thermal protrusion slider," *IEEE Transactions on Magnetics*, vol. 44, no. 11, pp. 3656–3662, 2008.

[33] S. Shen, B. Liu, S. Yu, and H. Du, "Mechanical performance study of pattern media-based head-disk systems," *IEEE Transactions on Magnetics*, vol. 45, no. 11, pp. 5002–5005, 2009.

[34] L. Li and D. B. Bogy, "Dynamics of air bearing sliders flying on partially planarized bit patterned media in hard disk drives," *Microsystem Technologies*, vol. 17, no. 5–7, pp. 805–812, 2011.

[35] W. D. Zhou, B. Liu, S. K. Yu, and W. Hua, "Inert gas filled head-disk interface for future extremely high density magnetic recording," *Tribology Letters*, vol. 33, no. 3, pp. 179–186, 2009.

[36] J. Robertson, "Requirements of ultrathin carbon coatings for magnetic storage technology," *Tribology International*, vol. 36, no. 4–6, pp. 405–415, 2003.

[37] X. Shi, Y. H. Hu, and L. Hu, "Tetrahedral amorphous carbon (ta-C) ultra thin films for slider overcoat application," *International Journal of Modern Physics B*, vol. 16, no. 6-7, pp. 963–967, 2002.

[38] G. G. Wang, X. P. Kuang, H. Y. Zhang et al., "Silicon nitride gradient film as the underlayer of ultra-thin tetrahedral amorphous carbon overcoat for magnetic recording slider," *Materials Chemistry and Physics*, vol. 131, pp. 127–131, 2011.

[39] N. Yasui, H. Inaba, K. Furusawa, M. Saito, and N. Ohtake, "Characterization of head overcoat for 1 Tb/in² magnetic recording," *IEEE Transactions on Magnetics*, vol. 45, no. 2, pp. 805–809, 2009.

[40] J. Robertson, "Ultrathin carbon coatings for magnetic storage technology," *Thin Solid Films*, vol. 383, no. 1-2, pp. 81–88, 2001.

[41] T. Yamamoto and H. Hyodo, "Amorphous carbon overcoat for thin-film disk," *Tribology International*, vol. 36, no. 4–6, pp. 483–487, 2003.

[42] C. Y. Chan, K. H. Lai, M. K. Fung et al., "Deposition and properties of tetrahedral amorphous carbon films prepared on magnetic hard disks," *Journal of Vacuum Science & Technology A*, vol. 19, pp. 1606–1610, 2001.

[43] B. K. Yen, R. L. White, R. J. Waltman, C. Mathew Mate, Y. Sonobe, and B. Marchon, "Coverage and properties of a-SiNx hard disk overcoat," *Journal of Applied Physics*, vol. 93, no. 10, pp. 8704–8706, 2003.

[44] Y. Hijazi, E. B. Svedberg, T. Heinrich, and S. Khizroev, "Comparative corrosion study of binary oxide and nitride overcoats using in-situ fluid-cell AFM," *Materials Characterization*, vol. 62, no. 1, pp. 76–80, 2011.

[45] E. B. Svedberg and N. Shukla, "Adsorption of water on lubricated and non lubricated TiC surfaces for data storage applications," *Tribology Letters*, vol. 17, no. 4, pp. 947–951, 2004.

[46] F. Rose, B. Marchon, V. Rawat, D. Pocker, Q. F. Xiao, and T. Iwasaki, "Ultrathin TiSiN overcoat protection layer for magnetic media," *Journal of Vacuum Science & Technology A*, vol. 29, Article ID 051502, 11 pages, 2011.

[47] M. L. Wu, J. D. Kiely, T. Klemmer, Y. T. Hsia, and K. Howard, "Process-property relationship of boron carbide thin films by magnetron sputtering," *Thin Solid Films*, vol. 449, no. 1-2, pp. 120–124, 2004.

[48] M. A. Samad, E. Rismani, H. Yang, S. K. Sinha, and C. S. Bhatia, "Overcoat free magnetic media for lower magnetic spacing and improved tribological properties for higher areal densities," *Tribology Letters*, vol. 43, pp. 247–256, 2011.

[49] E. Rismani, S. K. Sinha, H. Yang, and C. S. Bhatia, "Effect of pretreatment of Si interlayer by energetic C+ ions on the improved nanotribological properties of magnetic head overcoat," *Journal of Applied Physics*, vol. 111, Article ID 084902, 10 pages, 2012.

[50] X. C. Guo, B. Knigge, B. Marchon, R. J. Waltman, M. Carter, and J. Burns, "Multidentate functionalized lubricant for ultralow head/disk spacing in a disk drive," *Journal of Applied Physics*, vol. 100, no. 4, Article ID 044306, 2006.

[51] X. C. Guo, B. Marchon, R. H. Wang et al., "A multidentate lubricant for use in hard disk drives at sub-nanometer thickness," *Journal of Applied Physics*, vol. 111, Article ID 024503, 7 pages, 2012.

[52] D. Gonzalez, V. Nayak, B. Marchon, R. Payne, D. Crump, and P. Dennig, "The dynamic coupling of the slider to the disk surface and its relevance to take-off height," *IEEE Transactions on Magnetics*, vol. 37, no. 4, pp. 1839–1841, 2001.

[53] Z. Jiang, M. M. Yang, M. Sullivan, J. L. Chao, and M. Russak, "Effect of micro-waviness and design of landing zones with a glide avalanche below 0.5μ" for conventional pico sliders," *IEEE Transactions on Magnetics*, vol. 35, no. 5, pp. 2370–2372, 1999.

[54] B. Marchon, D. Kuo, S. Lee, J. Gui, and G. C. Rauch, "Glide avalanche prediction from surface topography," *Transactions of the ASME Journal of Tribology*, vol. 118, no. 3, pp. 644–650, 1996.

[55] Q. Dai, U. Nayak, D. Margulies et al., "Tribological issues in perpendicular recording media," *Tribology Letters*, vol. 26, no. 1, pp. 1–9, 2007.

[56] M. F. Toney, C. M. Mate, and K. A. Leach, "Roughness of molecularly thin perfluoropolyether polymer films," *Applied Physics Letters*, vol. 77, no. 20, pp. 3296–3298, 2000.

[57] R. Pit, B. Marchon, S. Meeks, and V. Velidandla, "Formation of lubricant "moguls" at the head/disk interface," *Tribology Letters*, vol. 10, no. 3, pp. 133–142, 2001.

[58] X. Ma, H. Tang, M. Stirniman, and J. Gui, "Lubricant thickness modulation induced by head-disk dynamic interactions," *IEEE Transactions on Magnetics*, vol. 38, no. 1, pp. 112–117, 2002.

[59] Q. Dai, F. Hendriks, and B. Marchon, "Modeling the washboard effect at the head/disk interface," *Journal of Applied Physics*, vol. 96, no. 1, pp. 696–703, 2004.

[60] I. Takekuma, H. Nemoto, H. Matsumoto et al., "Capped L1(0)-ordered FePt granular media with reduced surface roughness," *Journal of Applied Physics*, vol. 111, Article ID 07B708, 3 pages, 2012.

[61] N. Wang and K. Komvopoulos, "Thermal stability of ultrathin amorphous carbon films for energy-assisted magnetic recording," *IEEE Transactions on Magnetics*, vol. 47, pp. 2277–2282, 2011.

[62] N. Tagawa, H. Tani, and K. Ueda, "Experimental investigation of local temperature increase in disk surfaces of hard disk drives due to laser heating during thermally assisted magnetic recording," *Tribology Letters*, vol. 44, pp. 81–87, 2011.

[63] N. Tagawa, H. Andoh, and H. Tani, "Study on lubricant depletion induced by laser heating in thermally assisted magnetic recording systems: effect of lubricant thickness and bonding ratio," *Tribology Letters*, vol. 37, no. 2, pp. 411–418, 2010.

[64] C. Choi, Y. Yoon, D. Hong, Y. Oh, F. E. Talke, and S. Jin, "Planarization of patterned magnetic recording media to enable head flyability," *Microsystem Technologies*, vol. 17, pp. 395–402, 2011.

[65] H. Li and F. E. Talke, "Numerical simulation of the head/disk interface for bit patterned media," *IEEE Transactions on Magnetics*, vol. 45, no. 11, pp. 4984–4989, 2009.

15

Tribochemistry of Ionic Liquid Lubricant on Magnetic Media

Hirofumi Kondo

R&D Division, Sony Chemical & Information Device Corporation, 1078 Kamiishikawa, Kanuma 3228503, Japan

Correspondence should be addressed to Hirofumi Kondo, hirofumi.kondo@jp.sony.com

Academic Editor: Arvind Agarwal

The newly synthesized perfluoropolyether (PFPE) ionic liquid whose terminal group is an ammonium salt with a carboxylic acid has better frictional properties when compared to the conventional PFPEs. Stick-slip motion was not observed even for the smooth surface for the modified PFPE tape. The friction is almost independent of the PFPE structure, but depends on the amine structures. The ammonium salt being tightly anchored to the rubbing surface covers uniformly, which leads to better lubricity. The higher dispersive interaction of the hydrophobic group of the amine is endowed with a compensating friction reduction. Steric hindrance of the hydrophilic group causes a high friction. Based on these findings, a saturated long chain ammonium salt is the best selection. Moreover, the modified PFPEs are dissolved in alcohol and hexane, which makes practical use convenient without any environmental problems. These ionic lubricants invented around 1987 have been used for magnetic tapes for about a quarter century because of their good lubricity and are reviewed in this paper.

1. Introduction

Magnetic recording systems have been responsible for the widespread and inexpensive recording of sound, video, and information processing. Despite the availability of other means of storing data, such as optical recordings and semiconductor devices, magnetic recording media have the advantages of low cost, stable storage, a relatively higher data transfer rate, a relatively short seek time, and high volumetric storage capacity [1].

2. Hard Disk Drive Systems (HDDs)

In current hard disk drive (HDD) systems, a rigid disk is rotated by a spindle motor at a speed of 10000 revolutions per minute (rpm). Information is written and read by a magnetic head with a tiny electric current attached to the end of the slider. The physical spacing between the magnetic sensors and the disk is down to almost 10 nm in recent systems, and it will be necessary to be within 5–7 nm for areal densities in the Tbin^{-2} range [2]. The read/write magnetic heads are mounted in the slider and travel across the data zone during the reading and writing operations. However, when the drive stops, this head assembled device rests in the landing zone which is typically textured in order to reduce wear during contact-start-stop (CSS) operations. Most drives require that the static and kinetic friction forces at the head-media-interface (HMI) remain low under extreme environmental conditions and after the required number of CSS that is usually 10,000 or greater [3].

The rigid disk consists of an Al-Mg alloy or glass substrate, undercoat layer, a magnetic multilayer, a carbon overcoat, and a very thin lubricant layer as illustrated in Figure 1. Nowadays, the magnetic media are perpendicular media, which consist of a Co-Cr-based film [4]. A carbon overcoat is used to enhance the wear and corrosion resistance. Finally a molecular thin lubricant, which is the topic of this paper, is coated to further reduce both the wear and stiction between the HMI.

3. Tape Drive Systems

Magnetic tape media are divided into two categories [5]; that is, particulate media in which magnetic particles are dispersed in a polymer matrix with some additives and coated onto the polyethylene terephthalate (PET) substrate, and thin film media in which monolithic magnetic thin films are deposited onto the substrate in a vacuum, which is discussed

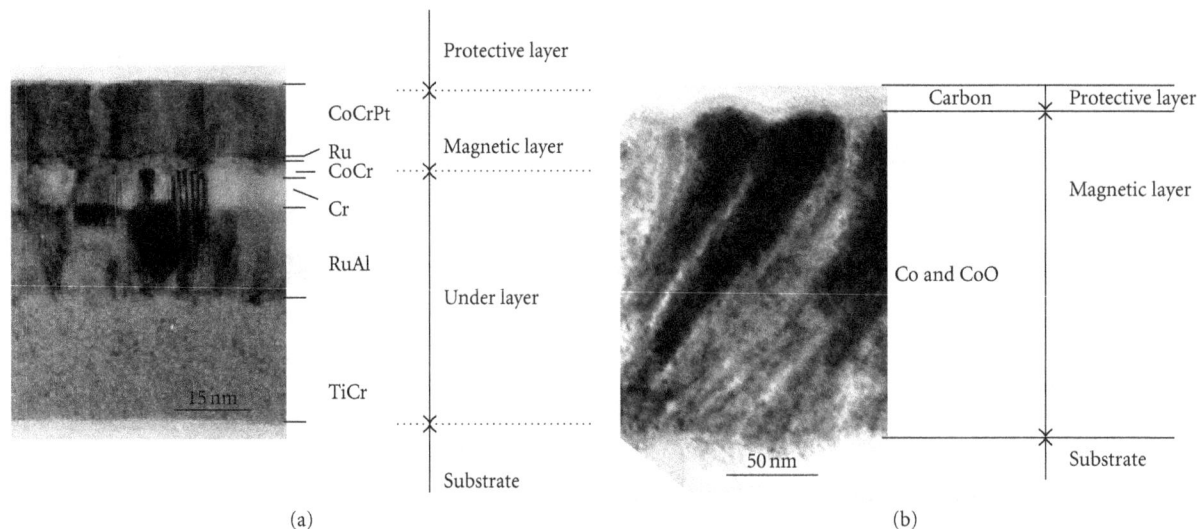

FIGURE 1: Cross-sectional TEM image of magnetic layer for rigid disk (a) and magnetic tape (b).

in this chapter. For magnetic tape helical scan systems, the tape is driven by a pinch roller and a tension (0.2–0.5 N) is applied [6, 7]. Much higher data transfer rates can be achieved using this rotating head drive system [8]. The tape used is a PET substrate with an evaporated film of Co which is typically 100 nm thick. The magnetic layer fabricated with oxygen gas has obliquely aligned fine particles (Figure 1) [9, 10], which leads to higher electromagnetic characteristics [11, 12]. The carbon overcoat and the lubricant layer are deposited onto the magnetic layer in a similar way to the rigid disks [13].

The continuous demand for increasingly high recording densities has led to the development of monolithic magnetic thin films. In the most recently established system, the recording density has increased about ten times in this decade using the highly sensitive magnetoresistive head systems [14–16]. From the viewpoint of recording density, a smoother media surface exhibits a higher carrier to noise ratio, which makes the higher recording density possible. However, the smoother surface results in a higher real area of contact and higher friction coefficient [17, 18].

4. Demand for Lubricant of Thin Film Magnetic Media

In conventional magnetic recording, thin film media typically have their surfaces lubricated to reduce friction and wear resulting from contacts between the read/write magnetic head and media surface. In practice, to avoid adhesion related problems, lubrication has to be achieved with a molecularly thin lubricant film [19]. However, the main challenge in selecting the best lubricant for a magnetic media surface is finding a material which provides wear protection while the media surface is exposed to various environmental situations. It is important that the lubricants remain on the media surface over the life of the file without being subject to desorption, spin-off, or chemical degradation. This problem has become more difficult with the advent of

a very smooth thin film surface, because thin film media do not have a mechanism for lubricant replenishment [20, 21]. Furthermore, lubricant adhesion to the overcoat surface is often insufficient to prevent lubricant depletion that eventually results in accelerated wear.

The presence of an excess lubricant is often deemed necessary to replenish itself after sliding events. Increasing the amount of a lubricant enhances the durability, but exceeding the surface roughness of the tape generally leads to adhesion-related problems, such as deleterious stiction. In order to reduce this trade-off, novel lubricants must be designed and synthesized for the smoother surface magnetic thin film media. The very large body of patents relating to the lubricity of perfluoropolyethers (PFPEs) on thin film magnetic media shows the importance of this problem to manufacturers [22–27].

New types of PFPE lubricants whose chemical structure are summarized in Table 1 have been reported to enhance the performance and reliability. Z-DOL has hydroxyl groups at both chain ends, which has been widely used for the rigid disk application. With the additional functional hydroxyl groups in the middle of the PFPE backbone chain, Z-tetraol multidentate (ZTMD) can achieve a reduced clearance, while still achieving an overall drive reliability [28, 29].

The solid lubricants are used in high temperature and extreme high pressure environments, whereas liquid lubricants typically will not survive. Topical lubrication of solid lubricants such as graphite and molybdenum disulfide (MoS_2) has not been successful because the solid lubricating layer is often found to interfere with the sensitive magnetic transducing process, and because most solid lubricants have a poor wear resistance, they tend to wear away in the tracks under the head and generate debris [21]. Liquid lubricants have the advantage that they will creep across the surface to replenish a portion of the layer which has been removed by abrasion or head wear. However, because of their mobility, liquid lubricants may suffer the disadvantage of spinning off from the disk surfaces during operation, especially

Table 1: Functionalized PFPE lubricants [30].

$X–CF_2(OCF_2CF_2)_n(OCF_2)_m\,OCF_2–X\ (0.5 < n/m < 1)$	
Z	$X = –OCF_3$
Z-DOl	$X = –CH_2OH$
Z-DIAC	$X = –COOH$
Z-tetraol	$X = –CH_2\,OCH_2CHCH_2\,OH$ OH
AM2001	

at higher operating temperatures. These lubricants may also slowly evaporate with time at the high temperatures, thereby reducing their protection. The use of higher viscosity, low-volatility liquid lubricants may help to decrease the evaporation rate and prolong their life.

Over the past decade, ionic liquids have received a great deal of attention as a class of green solvents with a wide range of potential applications including organic and inorganic synthesis [31], energy storage devices [32], separations [33, 34], and catalysis [35–37]. The term ionic liquid is broadly used to describe a large class of low melting fused salts that are liquids below 100°C. The most notable characteristics of many ionic liquids are their low vapor pressure, nonflammability, thermal stability, wide liquid range, and solvating properties for diverse substances. Limited results from very recent studies have shown the potential for using ionic liquids as a new class of lubricants. Friction and wear reductions have been reported on metallic and ceramic surfaces lubricated by selected ionic liquids compared to the conventional hydrocarbon lubricants [38, 39]. Ammonium-based ionic liquids provide friction reduction from elastohydrodynamic to boundary lubrication regimes compared to the fully-formulated base oil [40]. Ionic liquids have also been studied to determine their effectiveness as additives for base oil and water, and the chemical and tribochemical reactions have been evaluated to understand the lubrication mechanisms [41–44].

Ionic liquids, which possess an octadecyl ammonium salt with pentadecafluoro octanate, significantly reduce the friction compared to the corresponding amide and Z-DOL [45–50]. The modified PFPEs having the same hydrophilic group have also been synthesized and also show better frictional properties, which have been used as a lubricant for magnetic thin film media for a long time [51–53]. This type of ionic liquids are named protic ionic liquids which are a subset of ionic liquids formed by the stoichiometric (equimolar) combination of a Bronsted acid with a Bronsted base [54–57]. Relevant investigations into the molecular interactions of carboxylic acids and amines were conducted by Kohler et al., and the complexes of acid and amine with the molecular ratio of 1 : 1 can be found [58, 59]. In this paper, a series of ionic lubricants having the same hydrophilic group stated above are deposited on the magnetic thin film media and the effect of their molecular structures on the frictional properties is systematically investigated.

The lubricant is required to be very thin on the order of a monomolecular layer. Therefore, the frictional properties depend not only on the molecular structure, but also on the microscopic structure of the lubricant film [60, 61]. Microscopic coverage of this alkylammonium-based protic ion liquid film on the medium surface is also examined using FTIR and X-ray photoelectron spectroscopy (XPS) and related to the spectra to the frictional properties.

5. Materials

Three types of lubricants which possess both the perfluoroalkyl group and long chain hydrocarbon, that is an ester, amide, and carboxylic acid ammonium salt, were synthesized by the following Scheme 1 in Figure 2. The ester and the amide were prepared by the addition of carboxylic acid chloride to the hexane solution of the corresponding alcohol and amine in the presence of a base agent. The perfluorocarboxylic acid ammonium salts are prepared by warming the mixture of the perfluorocarboxylic acid and the amine to 80°C until the complete dissolution was obtained (Scheme 2) [45, 47]. The ammonium salts of long chain fatty acid were synthesized in the same manner (Scheme 2). They are then recrystallized from n-hexane.

The ammonium salts with PFPE carboxylate lubricants were synthesized according to Scheme 2 and Scheme 3 in Figure 2 by merely warming the mixture of the above carboxylic acid and a 5% excess of the long chain alkyl amine to 80°C with stirring until complete dissolution is obtained. Three different types of PFPEs, which possess a carboxylic acid group as the end group, are used as the raw materials. K-lubricant is a homopolymer of perfluoro-isopropylene oxide, and D-lubricant is a homopolymer of perfluoro-n-propylene oxide. K-lubricant and D-lubricant have one end group. Z-lubricant is a random copolymer of the perfluoro oxymethylene and oxyethylene oxide monomers, which have two identical end groups. The average molecular weight of the PFPEs is about 2000. Since most commercial PFPEs have a fairly broad and asymmetrical molecular weight distribution, a small excess of the alkyl amine is used, which is removed by washing with n-hexane after the reaction [52]. The chemical structure is determined by its infrared spectra: 3200–2800 cm^{-1} (N$^+$H$_3$ stretching), 2918 cm^{-1} and 2958 cm^{-1} (CH$_2$ stretching), 1674 cm^{-1} (CO stretching), 1280–1110 cm^{-1} (CF stretching). The CO stretching moved from 1800 cm^{-1} to 1674 cm^{-1}, and the N$^+$H$_3$ stretching at 3200–2800 cm^{-1} appears, thus identifying the ammonium salt with a carboxylate structure. The synthesized lubricants are summarized in Table 3. Each lubricant was deposited on a magnetic layer by a dip-coating method.

6. Friction Properties

6.1. Friction Measurement Apparatus. The apparatus shown in Figure 3 was used to measure the CSS friction characteristics of the rigid disks. Friction at the head slider was measured by a strain gauge for each CSS operation during the starting of the spindle motor with a 10 g load at 25°C, 50% relative humidity.

Scheme 1

$$R_f\text{–}COCl + R_1\text{–}XH \longrightarrow R_f\text{–}COXR_1$$

$$X = O, NH$$
$$R_f = F\text{–}(CF_2)_m\text{–} \quad m = 7, 9$$
$$R_1 = C_nH_{2n+1,2n-1}$$

Scheme 2

$$R_f\text{–}COOH + R_1\text{–}N(R_2)_2 \longrightarrow R_f\text{–}COO^-N^+H(R_2)_2R_1$$

$$R_f = F\text{–}(CF_2)_m\text{–} \quad m = 7, 9$$
$$R_f = C_nH_{2n+1,2n-1,2n-5}$$
$$R_f = F\text{–}(CF_2CF_2CF_2O)_n\text{–}CF_2CF_2\text{–} \quad \text{for D–lube}$$
$$R_f = F\text{–}(CF_2CFO)_n\text{–}CF_2\text{–} \quad \text{for K–lube}$$
with CF_3 branch
$$R_1, R_2 = CH_3, C_nH_{2n+1,2n-1,2n-5}$$

Scheme 3

$$HOCO\text{–}R_f\text{–}COOH + R_1\text{–}N(R_2)_2 \longrightarrow (R_2)_2R_1N^+HO^-COR_f\text{–}COO^-NH^+(R_2)_2R_1$$

$$R_f = \text{–}CF_2O\text{–}(CF_2O)_m\text{–}(CF_2CF_2O)_n\text{–}CF_2O\text{–}CF_2\text{–} \quad \text{for Z–lube}$$
$$R_1, R_2 = CH_3, C_nH_{2n+1,2n-1,2n-5}, C_6H_5$$

Figure 2: Synthetic scheme for the new ionic liquid lubricant and the reference compound.

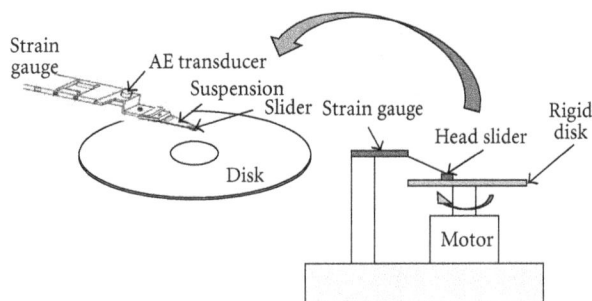

Figure 3: Friction measuring apparatus for rigid disk.

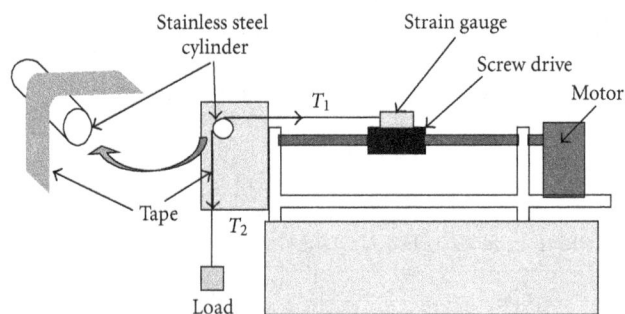

Figure 4: Schematic diagram of the friction measurement apparatus for ME tapes.

A schematic diagram of the friction measurement apparatus for the magnetic tapes is shown in Figure 4. The coefficients of kinetic friction are measured for 8-mm wide tapes sliding around a quadrant of a 4-mm diameter polished stainless steel (SUS 304) cylinder. The friction coefficient was calculated from the change in the sliding of the tension (T_1) exerted by a 20-g weight (T_2) hanging from the tape sliding on the cylinder. A 50-mm section of the tape is made to slide against the cylinder at a speed of 5 mms^{-1} in a reciprocating motion at 25°C, 60% relative humidity.

6.2. Frictional Performance of the Newly Synthesized Ionic Lubricant with Ammonium Salt for Magnetic Media

6.2.1. Rigid Disks.
The new lubricants exhibited a high performance compared to the conventional PFPE (Z-DOL) as shown in Figure 5. Frictional coefficient of disk coated with

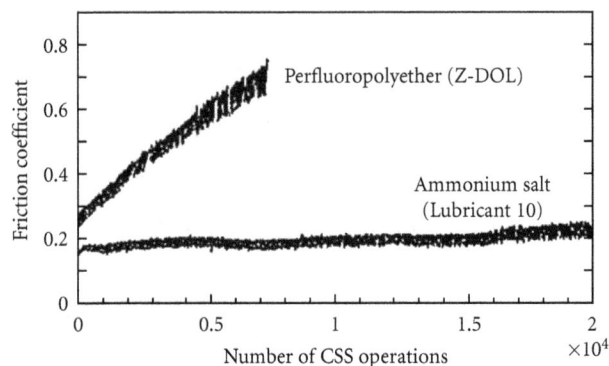

Figure 5: Frictional coefficient of disk coated with lubricant (10) versus number of CSS operations. The functional PFPE (Z-DOL) shown for comparison.

TABLE 2: Molecular structures and melting points of ionic liquid lubricants R_F–CO–Y–R.

No.	R_F	Y	R	mp/°C	Remark
1	C_7F_{15}	O	$C_{18}H_{37}$	29	
2	C_7F_{15}	O	$C_{18}H_{31}$	<20	
3	C_7F_{15}	NH	$C_{18}H_{37}$	91	
4	C_7F_{15}	NH	$C_{18}H_{31}$	48	
5	C_7F_{15}	$O^-H_3N^+$	$C_{18}H_{37}$	55	
6	C_7F_{15}	$O^-H_3N^+$	$C_{18}H_{31}$	<20	
7	C_9F_{19}	$O^-H_3N^+$	$C_{12}H_{25}$	61	
8	C_9F_{19}	$O^-H_3N^+$	$C_{14}H_{29}$	65	
9	C_9F_{19}	$O^-H_3N^+$	$C_{18}H_{37}$	71	
10	C_9F_{19}	$O^-H_3N^+$	$C_{24}H_{49}$	85	
11	C_9F_{19}	$O^-H_3N^+$	$C_{18}H_{35}$	26	
12	C_9F_{19}	$O^-H_3N^+$	$C_{18}H_{31}$	<20	
13	$-CF_2O-(CF_2O)_m-(CF_2CF_2O)_n-CF_2O-$	$O^-H_3N^+$	C_4H_9	<30	Z-lubricant
14	$-CF_2O-(CF_2O)_m-(CF_2CF_2O)_n-CF_2O-$	$O^-H_3N^+$	C_6H_{13}	<30	Z-lubricant
15	$-CF_2O-(CF_2O)_m-(CF_2CF_2O)_n-CF_2O-$	$O^-H_3N^+$	C_8H_{17}	<30	Z-lubricant
16	$-CF_2O-(CF_2O)_m-(CF_2CF_2O)_n-CF_2O-$	$O^-H_3N^+$	$C_{10}H_{21}$	<30	Z-lubricant
17	$-CF_2O-(CF_2O)_m-(CF_2CF_2O)_n-CF_2O-$	$O^-H_3N^+$	$C_{12}H_{25}$	<30	Z-lubricant
18	$-CF_2O-(CF_2O)_m-(CF_2CF_2O)_n-CF_2O-$	$O^-H_3N^+$	$C_{14}H_{29}$	<30	Z-lubricant
19	$-CF_2O-(CF_2O)_m-(CF_2CF_2O)_n-CF_2O-$	$O^-H_3N^+$	$C_{16}H_{33}$	< 30	Z-lubricant
20	$-CF_2O-(CF_2O)_m-(CF_2CF_2O)_n-CF_2O-$	$O^-H_3N^+$	$C_{18}H_{37}$	38–40	Z-lubricant
21	$-CF_2O-(CF_2O)_m-(CF_2CF_2O)_n-CF_2O-$	$O^-H_3N^+$	$C_{20}H_{41}$	58–61	Z-lubricant
22	$-CF_2O-(CF_2O)_m-(CF_2CF_2O)_n-CF_2O-$	$O^-H_3N^+$	$C_{18}H_{35}$	<30	Z-lubricant
23	$-CF_2O-(CF_2O)_m-(CF_2CF_2O)_n-CF_2O-$	$O^-H_3N^+$	$C_{18}H_{31}$	<30	Z-lubricant
24	$-CF_2O-(CF_2O)_m-(CF_2CF_2O)_n-CF_2O-$	$O^-H_2N^+(CH_3)$	$C_{18}H_{37}$	79–82	Z-lubricant
25	$-CF_2O-(CF_2O)_m-(CF_2CF_2O)_n-CF_2O-$	$O^-HN^+(CH_3)_2$	$C_{18}H_{37}$	55–57	Z-lubricant
26	$-CF_2O-(CF_2O)_m-(CF_2CF_2O)_n-CF_2O-$	$O^-H_2N^+(C_{18}H_{37})$	$C_{18}H_{37}$	51–55	Z-lubricant
27	$-CF_2O-(CF_2O)_m-(CF_2CF_2O)_n-CF_2O-$	$O^-H_2N^+(C_6H_5)$	$C_{18}H_{37}$	31–33	Z-lubricant
28	$F-(CF_2CF_2CF_2O)_n-CF_2CF_2-$	$O^-H_3N^+$	C_4H_9	<30	D-lubricant
29	$F-(CF_2CF_2CF_2O)_n-CF_2CF_2-$	$O^-H_3N^+$	C_6H_{13}	<30	D-lubricant
30	$F-(CF_2CF_2CF_2O)_n-CF_2CF_2-$	$O^-H_3N^+$	C_8H_{17}	<30	D-lubricant
31	$F-(CF_2CF_2CF_2O)_n-CF_2CF_2-$	$O^-H_3N^+$	$C_{10}H_{21}$	<30	D-lubricant
32	$F-(CF_2CF_2CF_2O)_n-CF_2CF_2-$	$O^-H_3N^+$	$C_{12}H_{25}$	<30	D-lubricant
33	$F-(CF_2CF_2CF_2O)_n-CF_2CF_2-$	$O^-H_3N^+$	$C_{14}H_{29}$	<30	D-lubricant
34	$F-(CF_2CF_2CF_2O)_n-CF_2CF_2-$	$O^-H_3N^+$	$C_{16}H_{33}$	<30	D-lubricant
35	$F-(CF_2CF_2CF_2O)_n-CF_2CF_2-$	$O^-H_3N^+$	$C_{18}H_{37}$	<30	D-lubricant
36	$F-(CF_2CF(CF_3)O)_n-CF_2-$	$O^-H_3N^+$	$C_{18}H_{37}$	<30	K-lubricant
37	$C_{17}H_{35}$	$O^-H_3N^+$	$C_{18}H_{37}$	92	
38	$8-C_{17}H_{33}$	$O^-H_3N^+$	$C_{18}H_{37}$	47	
39	$16-C_{17}H_{33}$	$O^-H_3N^+$	$C_{18}H_{37}$	84	
40	$8, 11, 14-C_{17}H_{29}$	$O^-H_3N^+$	$C_{18}H_{37}$	40	

TABLE 3: Structure of the lubricant.

	Molecular structure	mp/°C
Lubricant 7	$C_9F_{19} O^-H_3N^+ C_{12}H_{25}$	61
Lubricant 8	$C_9F_{19} O^-H_3N^+ C_{14}H_{29}$	65
Lubricant 9	$C_9F_{19} O^-H_3N^+ C_{18}H_{37}$	71
Lubricant 10	$C_9F_{19} O^-H_3N^+ C_{24}H_{49}$	85

lubricant (10) versus number of CSS operations is shown and the conventional functional PFPE (Z-DOL) shown for comparison. The relationship between the CSS durability and the molecular structure of the lubricant in terms of the polar group, chain length, and chain symmetry was investigated.

6.2.2. Magnetic Tapes. The frictional characteristic of the carboxylic acid ammonium salt coated on the magnetic tape by dip-coating is shown in Figure 6. The friction coefficient is shown as a function of the number of cycles of reciprocating motion over the cylinder. The frictional characteristic of a PFPE (Z-DOL) is shown for comparison. For the ammonium salt, the friction coefficient is low and remains at 0.18

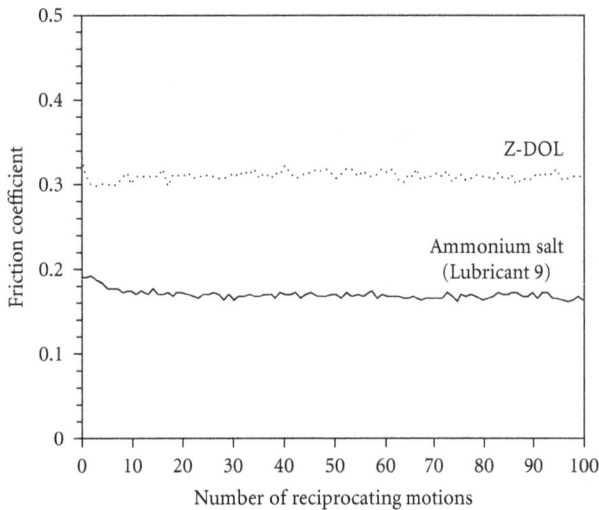

FIGURE 6: Friction coefficient of the carboxylic acid ammonium salt (Lubricant 9) and Z-DOL during friction test at 25°C and relative humidity of 60%.

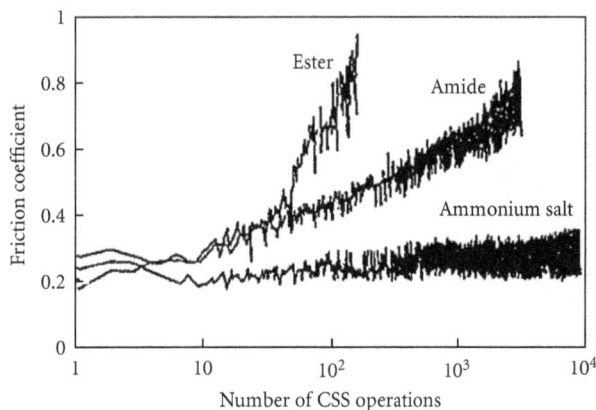

FIGURE 7: Friction coefficient of disk coated with ester lubricant (Lubricant 1), amide (Lubricant 3), and carboxylic acid ammonium salt (Lubricant 5) versus number of CSS operations.

even after 100 cycles of reciprocating motions, but it is more than 0.30 for the PFPE.

7. Effect of Molecular Structure on Friction

7.1. Hydrophilic Group

7.1.1. Effect of Hydrophilic Group on CSS Friction.
The CSS friction properties of three types of lubricants, that is, ester (Lubricant 1), amide (Lubricant 3), and carboxylic acid ammonium salt (Lubricant 5) are shown in Figure 7. These friction measurements of the synthesized lubricants revealed that the ester and the amide are far less durable than the comparable salt type. For the ester and amide lubricants, the friction coefficients (μ) are around 0.25 for the first ten CSS operations, but rise with the increasing number of CSS operations (n). Especially, for the ester lubricant, μ steeply increases after 20 operations, and the carbon protective layer

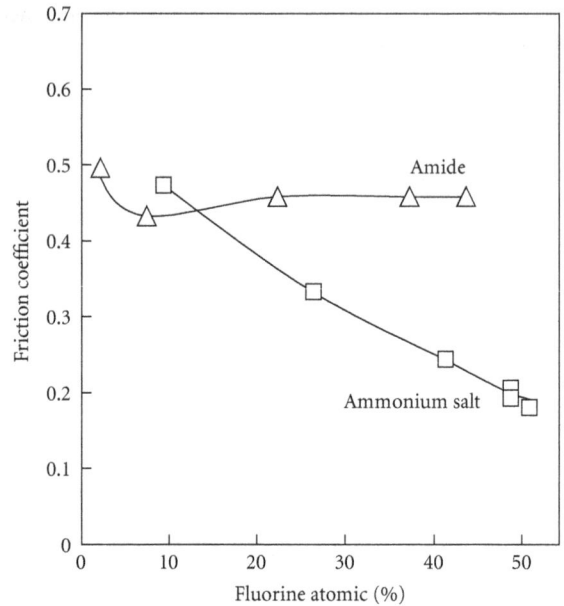

FIGURE 8: Relation between the relative intensity of fluorine atom on the tape surface and friction coefficient after 100 cycles of reciprocating motions.

gets damaged when the μ became over 0.90. The μ of the amide lubricant increased gradually and a wear scar occurred at 3279 operations. For the carboxylic acid ammonium salt lubricant, the μ value remained nearly constant at around 0.25 throughout the 10^4 CSS operations and the medium was scarcely damaged. The low initial value of μ, 0.2–0.3, indicates that there is sufficient lubricant film to protect the rubbing surface.

7.1.2. Frictional Tests for the Lubricant with a Different Polar Group for Magnetic Tapes.
The relation between the friction after 100 reciprocating cycles and the amount of the lubricant on the magnetic surface is shown in Figure 8. The amount of lubricant on the surface can be varied with the lubricant concentration of dip-coating solution. Clearly, the ammonium salt gives a better frictional characteristic than the corresponding amide. For the salt, the friction coefficient decreases with the increasing lubricant on the surface and reached 0.18. In contrast, the friction coefficient of the amide is almost independent of the amount of lubricant and is very high (approximately 0.45). These results revealed that the lubrication mechanism of the salt and the amide are different and depend on the polar group.

The frictional properties for the two lubricants depend on both the lubricant polar group and the surface concentration, but not the nonpolar hydrophobic groups, since the two had identical hydrophobic groups. Therefore, a comparison of the polar group effects on friction coefficient needs to be made with nearly the same amount of each lubricant on the surface. The fluorine content was measured by XPS and the film thickness was calculated [62–64]. The selected amount gave a relative intensity of the fluorine signal of about 40 atomic % in the XPS measurements,

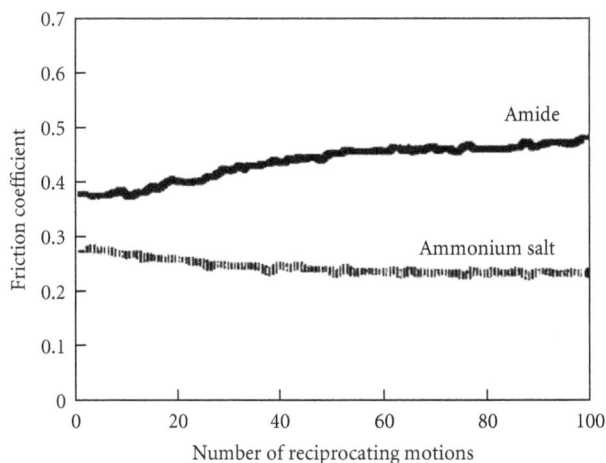

FIGURE 9: Comparison of frictional properties of ammonium carboxylate ionic liquid and the corresponding amide during a friction test with 40 atomic % fluorine.

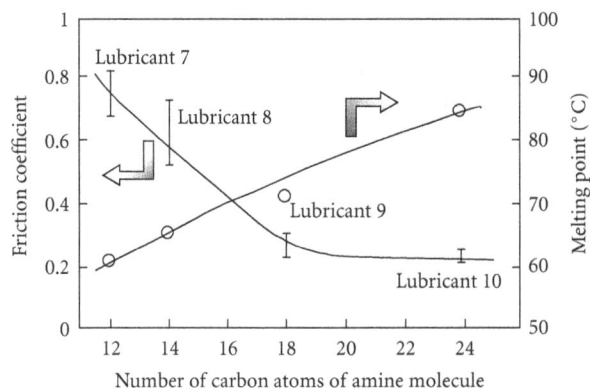

FIGURE 10: Friction coefficients of the salt type lubricants after 20,000 CSS operations.

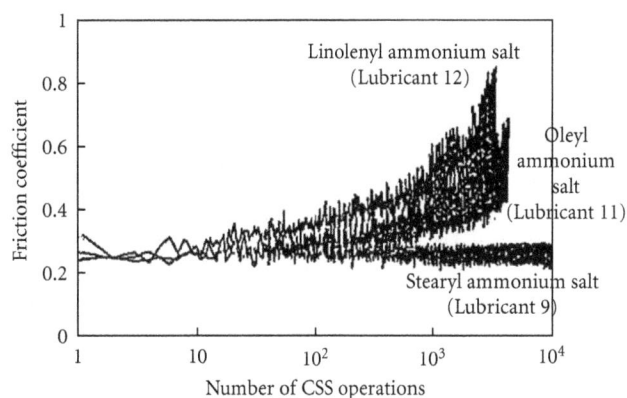

FIGURE 11: Friction properties of the salt type lubricant of long chain C-18 amine.

The shorter homologues showed an increase in μ as a result of the breakdown of their film [65].

For the given polar group, one of the key properties required for friction reduction is a high intermolecular cohesion energy (due to dispersive or van der Waal's interactions) between the hydrocarbon chains. Melting of the lubricant materials by heating involves disruption of the dispersive interactions between the hydrocarbon chains [66], therefore, the melting point of the lubricant should be related to the dispersive interactions of the hydrocarbon chain [64]. The melting points of the lubricants with different chain lengths listed in Table 2 are also plotted.

The melting point becomes higher with the increasing hydrocarbon chain length and μ decreases. Not only the polar group, but also the chain length due to dispersive interactions determined the durability.

7.2.2. Chain Symmetry (Double Bond Effect). The unsaturated oleyl (Lubricant 11) and linolenyl (Lubricant 12) ammonium salts used for a comparison with the saturated stearyl ammonium salt were synthesized, which have the same polar group and the same chain length. The oleyl amine has one double bond and the linolenyl amine has three. The CSS properties are shown in Figure 11.

Highly symmetrical (straight) molecules can be more readily arranged than the less symmetrically constituted (bent) molecules. Since high packaging in the lubricant layer is more favorable for highly symmetrical molecules, the symmetrical lubricants generally have higher cohesive interactions than their less symmetrical counterparts. As the salt type lubricants are strongly adsorbed on the carbon layer, we can show a model for the salt type lubricants in Figure 12. Saturated chains (e.g., stearyl ammonium salt) are linearly symmetrical and can efficiently pack. However, the unsaturated chains, particularly the cis-conformation chains (as in the oleyl and linolenyl ammonium salts), are bent, and less symmetrical, therefore, they do not pack well. Thus, the double bond ammonium salt has a low melting point and shows an increase in μ already mentioned.

since this is approximately the intensity of a monolayer for the salt prepared by the Langmuir-Blodget method [59]. This corresponds to the concentration of 0.26 mmol L^{-1} and 1.27 mmol L^{-1} of the chlorofluorocarbon solution for the salt and for the amide, respectively.

Figure 9 shows the friction coefficient variation versus the reciprocating cycles for the salt and for the amide at the above concentration. For the salt, the friction coefficient value remained low and steady at approximately 0.23 throughout the 100 reciprocating cycles. The friction coefficient with the amide increased with the number of cycles. The amide shows this increase in friction at all the concentration.

7.2. Hydrophobic Group

7.2.1. Effect of Alkyl Chain Length on CSS Friction. The hydrocarbon chain lengths of the salt type lubricant were changed and the CSS durability measured (Figure 10). The structures of the lubricant are shown in Table 3. As the number of carbon atoms in the amine molecule (n) increased, μ decreased until a nearly constant value of 0.24 was attained.

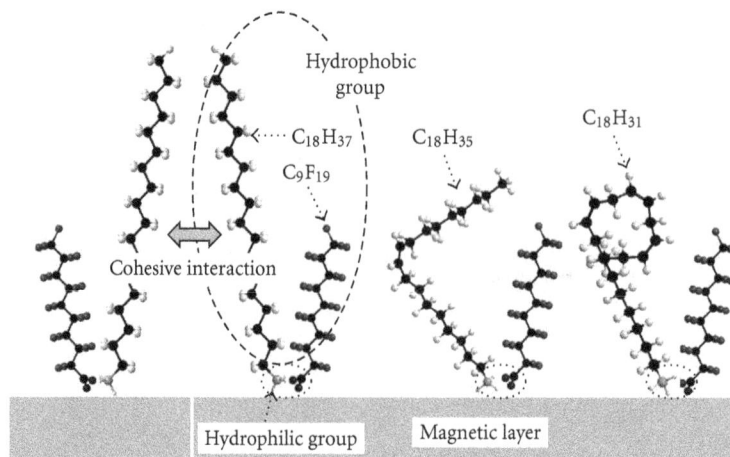

FIGURE 12: A model for the salt type lubricants adsorbed on the carbon layer.

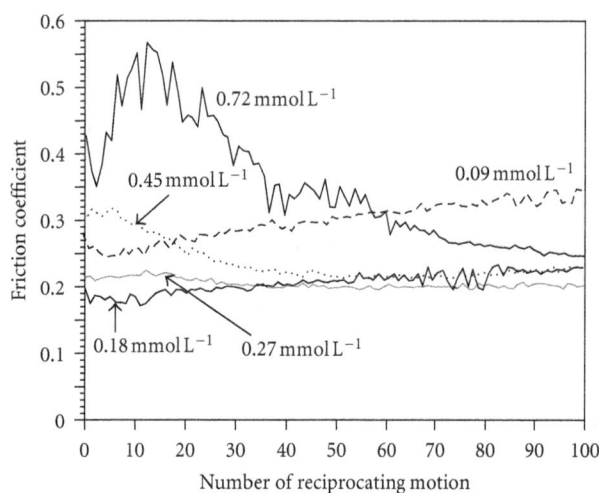

FIGURE 13: Concentration effect on friction of the lubricant solution.

7.3. Long Chain Hydrocarbon Carboxylic Acid Ammonium Salt

7.3.1. Effect of Concentration of Dip-Coating Solution on Friction. The frictional properties of the stearic acid stearyl ammonium salt (37) as a function of the dip-coating concentration are shown in Figure 13. The initial friction coefficient at the concentration of $0.09\,\text{mmol}\,\text{L}^{-1}$ is 0.25, and it increases with the number of reciprocating cycles. The lowest initial friction coefficient is observed at the concentration of $0.18\,\text{mmol}\,\text{L}^{-1}$ increased with the higher lubricant concentration, and it became very high at $0.72\,\text{mmol}\,\text{L}^{-1}$.

The concentration of $0.27\,\text{mmol}\,\text{L}^{-1}$ results in a thickness of approximately one monolayer for the lubricant which had the carboxylic acid ammonium salt as a polar group [46]. The increase in friction with the increasing reciprocating cycles seems to indicate that the surface of the magnetic layer is not sufficiently covered by the lubricant below that concentration.

On the contrary, since excess lubricant at the surface would result in a higher adhesional friction, the initial value became higher with the higher lubricant concentrations. According to the meniscus theory, friction should increase in the case of smooth surface with an increase in the lubricant thickness. The friction of lubricated media generally increases if the lubricant thickness is increased with respect to media roughness and creates menisci around individual asperity contacts [67]. A decrease in friction with the increasing reciprocating cycles reveals that an excess of lubricant transfers to the stainless steel cylinder counterface. For concentrations above $0.27\,\text{mmol}\,\text{L}^{-1}$ the friction coefficient approaches the value for the $0.27\,\text{mmol}\,\text{L}^{-1}$ concentration.

7.3.2. Number of Double Bond and Its Position Effect on Friction. In order to investigate the effect of the number of double bonds and their position in the molecular structure, the following three such lubricants were tested: 8-oleic acid (Lubricant 38) ammonium salt, 16-oleic acid (Lubricant 39) salt, and linolenic acid (Lubricant 40) salt which have three double bonds at positions 8, 11, and 14, respectively. Figure 14 reveals the friction coefficient variations versus reciprocating cycles for four lubricants at the concentration of $0.27\,\text{mmol}\,\text{L}^{-1}$. The introduction of a double bond causes an increase in friction with the number of cycles, and particularly for the linolenic acid ammonium salt, the friction coefficient steeply increases. For the oleic acid salt, the friction coefficient slightly increases with the number of double bonds in the hydrophobic group and the friction coefficient is higher for a double bond at the terminal position-16 compared to the center position-8.

The unsaturated chain, particularly the cis-conformation chains in the oleic acid and linolenic acid, are bent and, hence, they do not pack well. Therefore, with more double bonds, the melting point becomes lower, as demonstrated in Table 2, and also shows an increase in the friction coefficient. However, this does not explain two observations: (1) the friction coefficient increases with the number of reciprocating cycles, and (2) the lubricant with a double

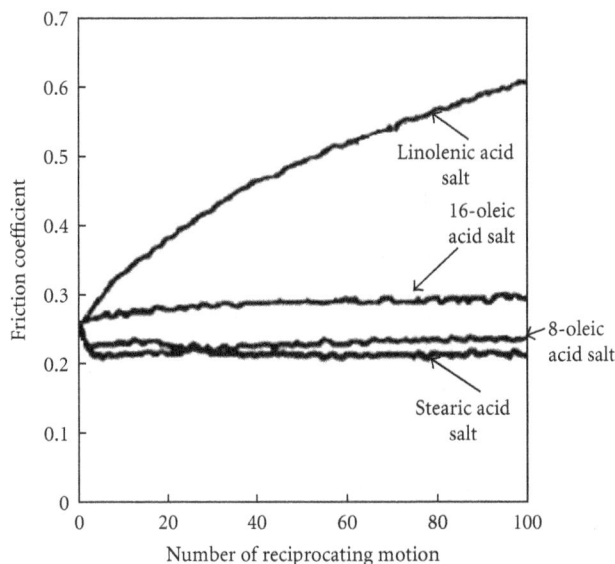

FIGURE 14: Effect of the number and position of the double bonds on the friction coefficient at a concentration of 0.27 mmol L^{-1}.

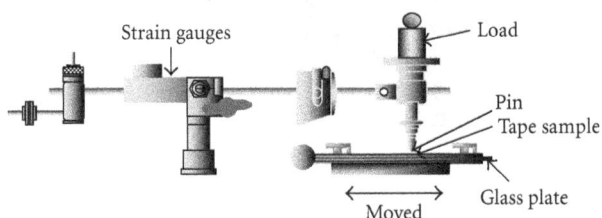

FIGURE 15: Schematic diagram of pin-on-flat friction tester.

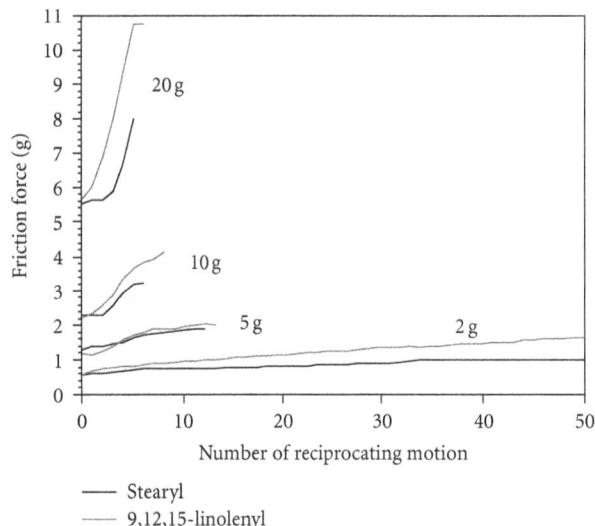

FIGURE 16: Change in friction force of stearic acid and linolenic acid salts in pin-on-flat tests.

bond at position 8 (the center) has a lower friction value than that at the terminal position 16.

In this case, the lubricant had a carboxylic acid ammonium salt as a polar group, which was strongly adsorbed on the tape surface compared to the double bond. This oleophobic group in the lubricant proved difficult to interact with the magnetic surface, but easily interacted with the stainless steel cylinder counterpart. This may help to explain why 16-oleic acid with a terminal double bond has a higher friction coefficient than the lubricant with a double bond at position 8.

7.3.3. Effect of Double Bond on Wear. The tapes treated with the linolenic and stearic acid ammonium salts were also tested by a pin-on-flat apparatus in Figure 15. The measured friction coefficient values versus the number of reciprocating cycles are shown in Figure 16. The test was terminated when the magnetic layer became damaged, except at the load of 2 g in which the tapes are scarcely damaged even after 50 reciprocating cycles. Again, all the lubricants with double bonds resulted in an increase in the friction, although the test geometry used here is much different from the tape tester used in the previously discussed results.

The lubricants having double bonds showed no obvious wear in the friction region where the tape coated with the nondouble bond lubricant was damaged. For example, the

nondouble bond lubricant required six reciprocating cycles at the load of 10 g to display damage and the frictional force of 3.1 g. However, the double bond lubricants had not been scarred until the frictional force reached 4.2 g at the eighth reciprocating cycle. This indicates that the double bond lubricant behaves like a wear protective film and has a higher load carrying capacity.

In order to characterize the surface of the magnetic layer from the tribological experiments, an FTIR-reflection-absorption spectroscopy (RAS) analysis was completed. The FTIR method allows chemical information to be obtained on the molecular level thickness, including the molecular structure.

The FTIR spectra in the 3000 cm^{-1} to 2800 cm^{-1} region before and after the friction tests are shown in Figure 17. The complete spectra of materials are not shown in this case. The peaks were shifted from 2916 cm^{-1} and 2850 cm^{-1} for the bulk (uncoated) lubricants to 2926 cm^{-1} and 2854 cm^{-1}, respectively, as shown in the spectra of Figures 17(a) and 17(c). For each lubricant film, these peaks are assigned to the CH$_2$ asymmetric and symmetric stretching vibrations. Nevertheless, the melting points of the bulk materials are higher than the ambient temperature, and this higher frequency shift reveals that the alkyl chain in the liquid phase or in a solution where the cohesive interaction between them is weak [68].

For the stearic acid salt film, the intensities of the spectra both before and after the friction test are similar, indicating that the rubbing motion did not cause a change in the microscopic structure of the film, such as the thickness and film formation. On the other hand, the intensities of the CH$_2$ stretching vibration became much weaker after the friction test for the linolenic acid film, which reveals that the thickness of the film was reduced due to rubbing. The decrease in film thickness resulted in a higher friction coefficient. For the pin-on-disk tests, the slight protective effect from double

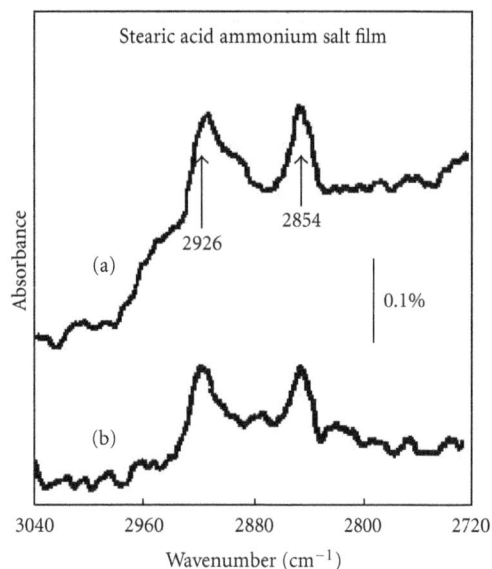

(a)

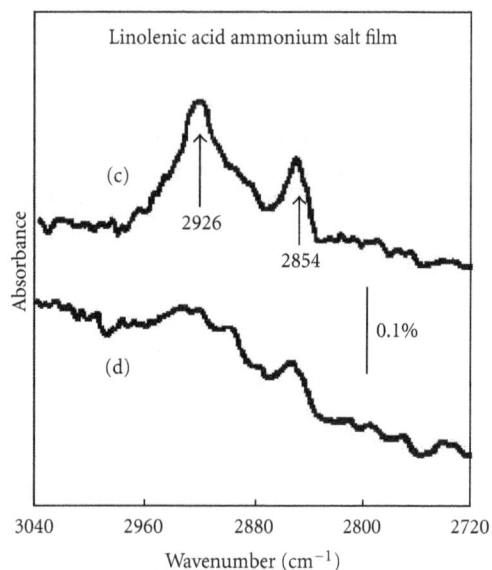

(b)

FIGURE 17: FTIR spectra of before and after the test. (a) RAS of stearic acid ammonium salt film, (b) RAS of film after the test. (c) RAS of linolenic acid ammonium salt film, (d) RAS of film after the test.

bond lubricants seems to reveal the existence of a polymer film, but this has not been proven.

7.4. Frictional Properties of Modified PFPEs on Magnetic Tapes

7.4.1. Performance of New Lubricant. The good viscosity characteristics, low melting point, low surface energy, low volatility, and good thermal stability of PFPE are among the important criteria for selecting a lubricant. By changing the perfluorocarboxylic acid with PFPE carboxylic acid, that is Z-lubricant in Scheme 3 and D- and K-lubricants in Scheme 2 of Figure 2, the modified PFPEs are expected to have lower

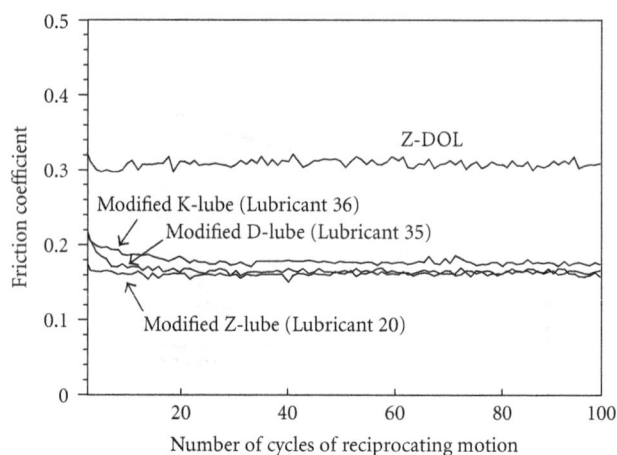

FIGURE 18: Frictional properties of the three types of the modified PFPEs using stearyl amine. The conventional Z-DOL is shown for comparison.

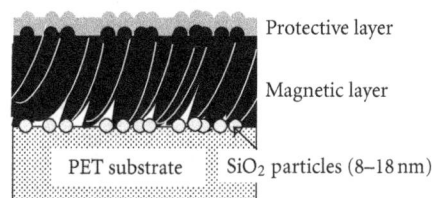

FIGURE 19: Schematic diagram of the cross section of a magnetic tape.

melting points and improved thermal properties compared to the corresponding perfluorocarboxylic acid homologue.

The frictional characteristics of the three types of modified PFPEs of the ammonium salt with carboxylate are shown in Figure 18. For the ammonium salt, it is low and approximately 0.17 even after 100 cycles of reciprocating motion, and is not dependent on the chain structure of the PFPE. However, it is over 0.30 for the conventional PFPE, and other types of end groups, such as the other hydroxyl and piperonyl, have a similar frictional coefficient of approximately equal to or greater than 0.30 (data not shown).

7.4.2. Friction on Smoother Surface. In order to design the surface morphology, the smaller spherical SiO_2 particles are coated onto the substrate with a polymer binder solution before the magnetic layer is deposited (Figure 19). The surface asperity can be controlled by changing the size of the SiO_2 particles to compromise the trade-off of electromagnetic characteristics and durability [69–71].

Most contacts in magnetic media are elastic; therefore, the kinetic frictions are expected to be higher for tapes with a lower surface roughness. The surface roughness (Ra) of the tapes with the surface SiO_2 particle diameters of 8, 12, and 18 nm were measured using an optical profiler, and are 3.3 nm, 1.7 nm, and 1.4 nm, respectively. The friction coefficients for the carbon coated tapes of different surface roughness are shown in Figures 20(a), 20(b), and 20(c).

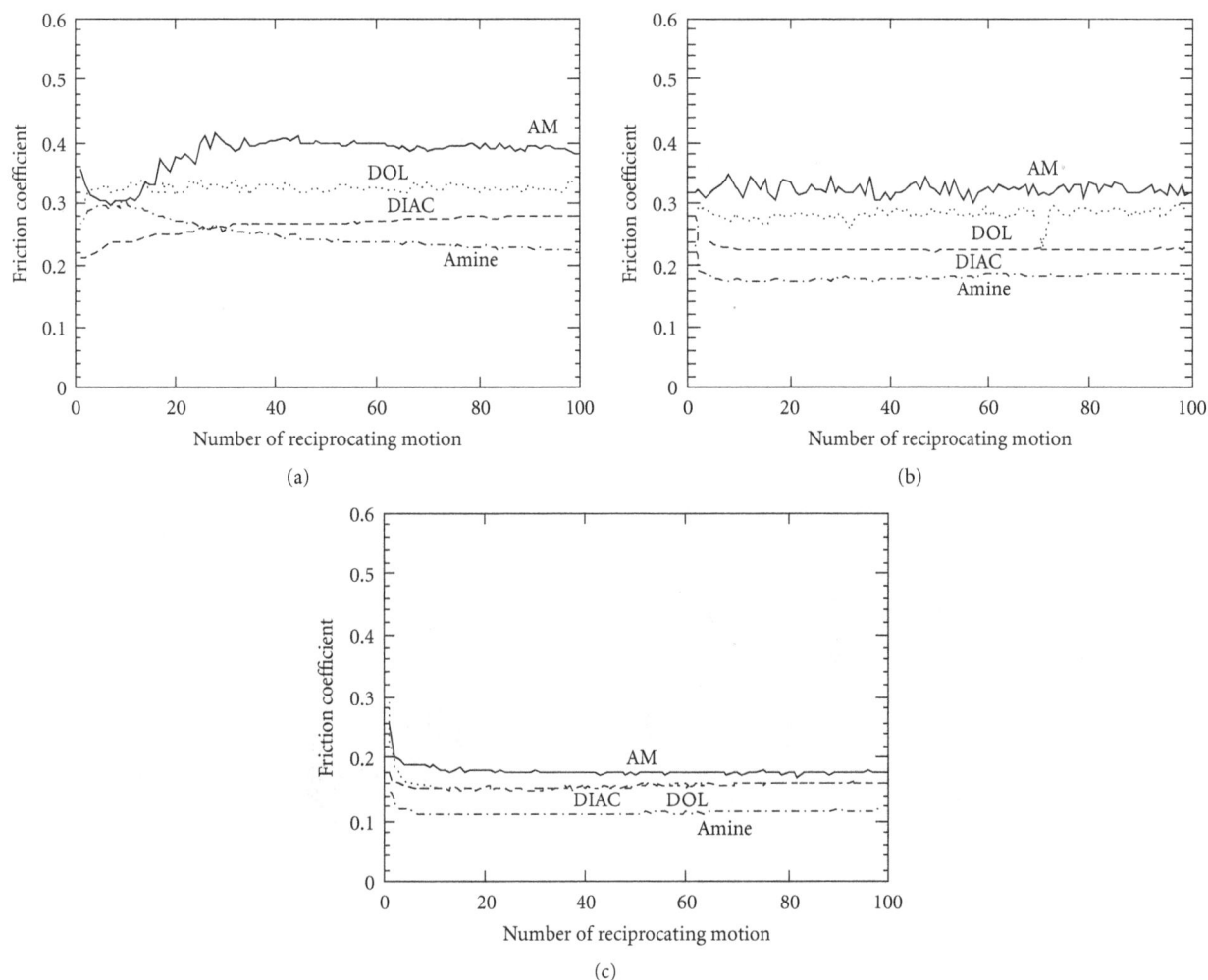

FIGURE 20: Frictional properties of each PFPE based on the different surface roughness using (a) 8 nm, (b) 12 nm, and (c) 18 nm particles.

The thickness of the lubricant was almost the same value of 1.2 nm for each PFPE tape. The friction coefficients were stable for the tapes with the 12 nm and 18 nm particles, but were unstable at the beginning (several ten reciprocating motions) for the smoother 8 nm tapes. The differences in friction for each PFPE tape were then compared: AM, Z-DOL, Z-DIAC, and ammonium salt (the friction coefficient decreases in that order). The difference is greater for the smoother surface.

Figure 21 shows the dynamic friction coefficient during the reciprocating operation for each PFPE in the case of the 12 nm tapes. The amplitude of the saw tooth pattern in the friction curve is significantly high for the AM and the Z-DOL tapes; these fluctuations in sliding resulted from the stick-slip process and are associated with squeal and chatter. However, the dynamic friction coefficient was relatively constant and the stick-slip phenomenon is only slightly observed for the Z-DIAC and the ammonium salt tapes.

7.4.3. Effect of Molecular Length of Amine. In order to examine the effect of the amine structure, the friction coefficients are also measured for the tapes versus the hydrocarbon chain

length of the amine. The relation between the number of carbon atoms and the frictional coefficients after 100 cycles of reciprocating motion for the modified Z- and D-lubricants is shown in Figure 22. As the number of carbon atoms in the amine molecule increases, the frictional coefficient decreases to a nearly constant value of 0.17 in both cases when the number of carbons exceeds 14.

7.4.4. Effect of Molecular Structure of Amine. Secondly, the molecular structure of the amine is changed using stearyl amine derivatives in order to fix the chain length of the longest substituent. The frictional results of the PFPE ammonium salts after 100 reciprocating motions and their melting points are summarized in Table 4. The introduction of a double bond into the hydrocarbon long chain, an oleyl with one double bond and a linolenyl with three, also makes the friction coefficient increase.

By replacing the hydrogen atoms of the amine group with alkyl and phenyl groups, a different series of salts would be obtained. This trend strongly suggests that replacing the hydrogen atom of the amine with an electron donating alkyl group increases the electrostatic interactions (including

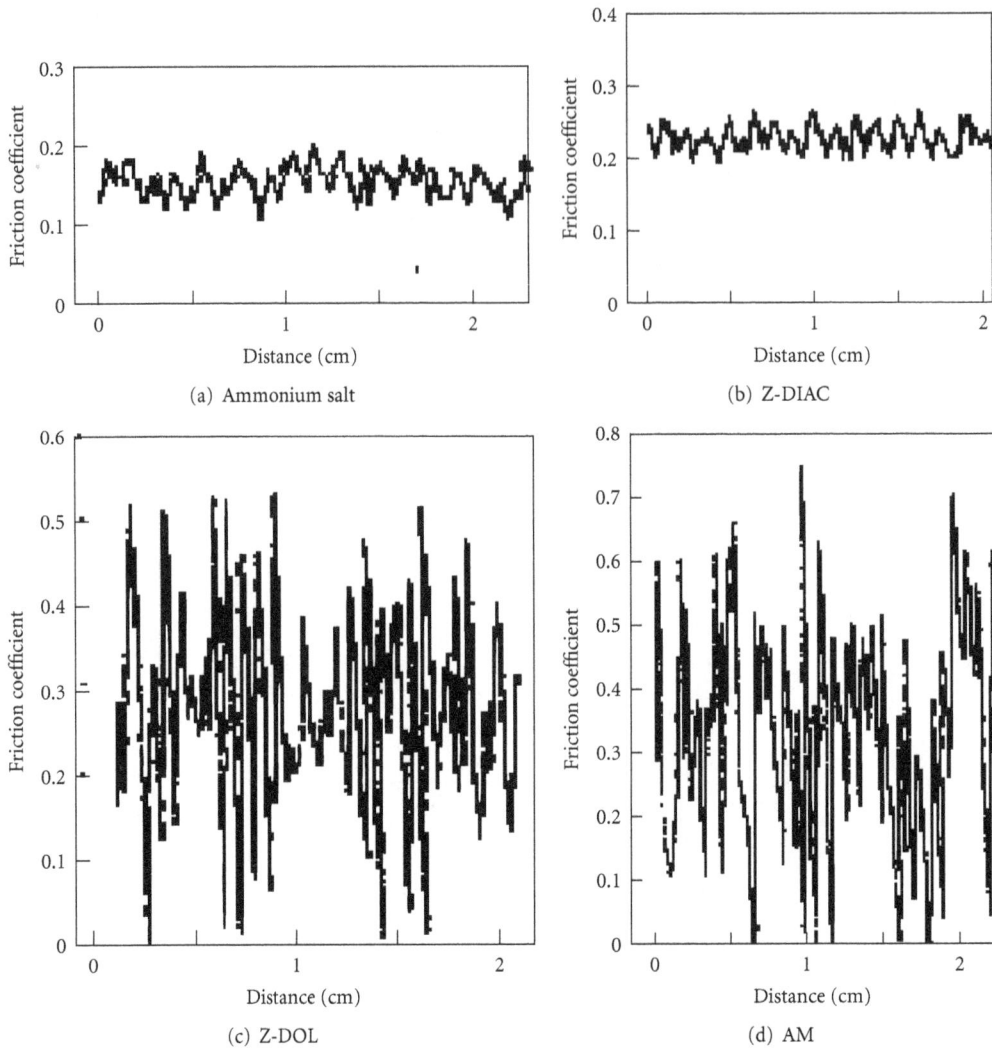

(a) Ammonium salt

(b) Z-DIAC

(c) Z-DOL

(d) AM

FIGURE 21: Change in dynamic friction coefficient during the reciprocating operation for the different PFPEs.

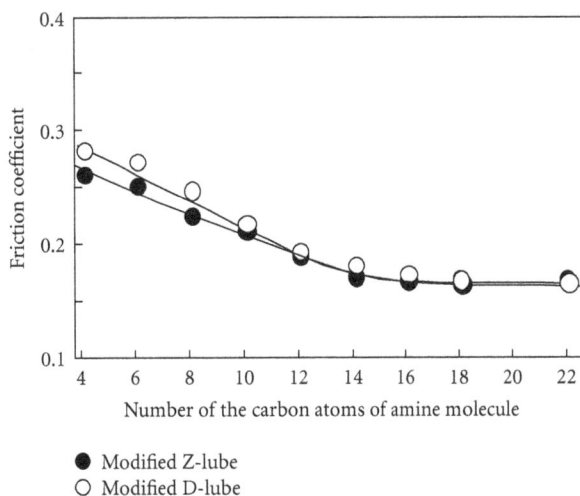

● Modified Z-lube
○ Modified D-lube

FIGURE 22: The relation between the number of carbon atoms and frictional coefficient after 100 cycles of reciprocating motion.

possible hydrogen bonding) between the cation and anion in the salts, which, in turn, raises the melting point. The alkyl substituted amine, for example, by methyl and stearyl groups, have higher melting points than the nonsubstituted stearyl amine, nevertheless, the friction coefficient was somewhat inferior to the stearyl amine. Exchanging the amine of the salts for the larger phenyl substituent generally produced a further decrease in the melting point. As a general trend, for each anion, the salts with cations of lower symmetry show a lower melting point than those with cations of higher symmetry. Also, adding the bulky phenyl group significantly increased the friction. These magnitudes of increase in the friction suggest a blocking effect by the large substituent attached to the amine nitrogen, which hinders adsorption of the polar group on the media surface. Steric hindrance by the polar group caused the high friction.

The cohesive energy density is normally lower for the fluorocarbon and ether group compared to the hydrocarbon. Cong et al. indicated that the film strength controlled by

TABLE 4: Fiction coefficients and melting points of the lubricants with the molecular structure of the amine having 18 carbon chains.

Lubricant number	Structure of lubricant	Friction coefficient	Melting point/°C
20 (stearyl)	$C_{18}H_{37}NH_2$	0.17	38–40
22 (oleyl)	$C_{18}H_{35}NH_2$	0.20	<30
23 (linolenyl)	$C_{18}H_{31}NH_2$	0.25	<30
24 (methyl stearyl)	$C_{18}H_{37}NHCH_3$	0.19	79–82
25 (dimethyl stearyl)	$C_{18}H_{37}N(CH_3)_2$	0.20	55–57
26 (distearyl)	$(C_{18}H_{37})_2NH$	0.21	51–55
27 (phenyl stearyl)	$C_{18}H_{37}NHC_6H_5$	0.30	31–33

intermolecular attractive forces is an important factor that affects the frictional properties of the monolayers, which is associated with a higher load-capacity [72]. The attractive forces between the fluorocarbon chains are lower than those between the hydrocarbon chains [73] and the presence of oxygen atoms in the PFPE backbone decreases the intermolecular attractive force [74–76]. For the ammonium salts of the PFPEs, the reason for the lower friction might be the incorporation of a hydrocarbon into the molecules. The longer the hydrocarbon chain length, the lower the friction coefficients become as shown in Figures 10 and 22. Therefore, it is expedient to increase the dispersive interaction by introduction of a hydrocarbon chain into a PFPE molecule without steric hindrance between the polar group of the lubricant and media surface. To be precise, a saturated straight long chain ammonium salt is the best selection.

The use of a conventional PFPE is limited by the solvent. However, since the modified PFPEs contain a hydrocarbon moiety and ammonium salt moiety, it is soluble in alcohols, and other conventional fluorinated solvents, which makes its practical use convenient. Figure 23 shows the consequences for friction of the stearyl ammonium salt of Z-lubricant by changing a thinner of dip-coating. In this case, ethanol, n-hexane-20% wt. ethanol, and a fluorinated solvent were used. It is evident that the frictional properties were independent of the dip-coating solvent.

8. Langmuir-Blodgett (LB) Films of the Salt Type Lubricant

In order to elucidate the molecular level structure of the spontaneously adsorbed layers, a comparative structural study with films prepared by the Langmuir-Blodgett (LB) method is useful [77]. Organized ultra thin films of controlled thickness are deposited on solid substrates by the LB technique. The effectiveness of LB films in protecting the magnetic thin film media was reported [78]. A stable and closed packed monolayer film can be obtained at the air-water interface using Joyce-Loebl trough. These films were transferred onto the magnetic media by vertical dipping method. Langmuir films (L films, i.e., monolayer on the water surface) and LB films of the ammonium salt lubricants were prepared and studied.

8.1. Basic Properties of the L Films and LB Films. The ammonium salt with perfluorocarboxylate (Lubricant 5), and the

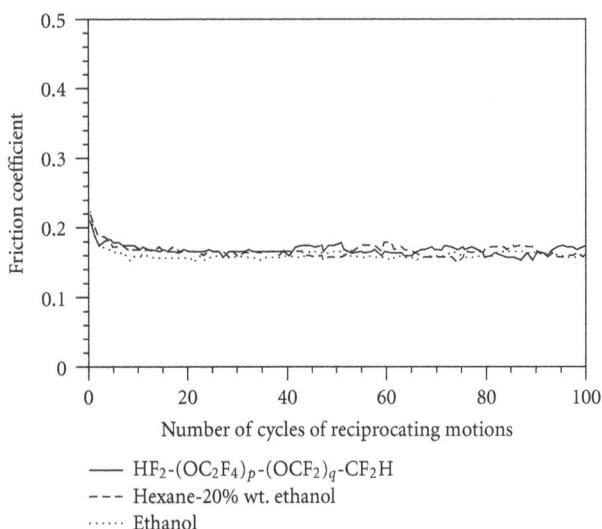

FIGURE 23: Influence of changing a thinner of lubricant on the friction coefficient.

— $HF_2-(OC_2F_4)_p-(OCF_2)_q-CF_2H$
--- Hexane-20% wt. ethanol
..... Ethanol

corresponding ester (Lubricant 1) and the amide (Lubricant 3) are compared. The salt forms a much more stable monolayer on the water surface than the other two ester and amide lubricants. Figure 24 shows the area decay curve of the L films of the salt and amide on the water surface when they are held at a surface pressure of 25 mNm^{-1}. The curve for the ester is unstable such that it collapsed as it was compressed. The area decay over 1 hour was less than 5% for the salt, whereas for the amide it was 45%, indicating a good stability for the salt monolayer. This result suggests a good balance between the polar and hydrophobic nature of the salt molecule, which is a necessary condition for producing a stable monolayer on the water surface.

Furthermore, the isotherm of the salt shown in Figure 25 suggests that the molecules of the salt were closely packed in the L film. From Figure 25, the area per molecule at 25 mNm^{-1} is about 0.6 nm^2. The areas occupied by an alkyl chain and a perfluoroalkyl chain are about 0.2 nm^2 and 0.4 nm^2, respectively, when their chain axes are perpendicular to the water surface. Therefore, it is suggested that the alkyl chains and perfluoroalkyl chains of the salt molecules are highly ordered and closely packed with their chain axes perpendicular to the water surface in the L film.

The closely packed monolayer is transferred onto the surface of the magnetic layer at 25 mNm^{-1} with a dipping

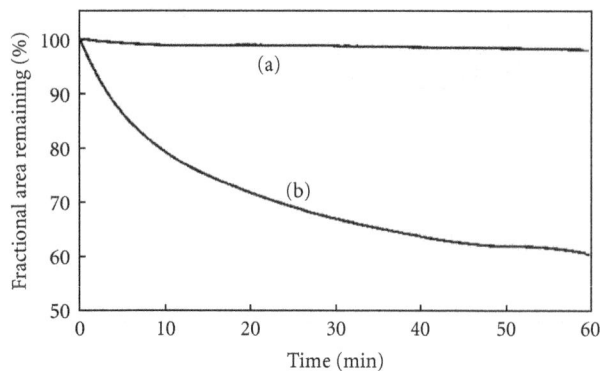

FIGURE 24: Area decay curve of the L films of the salt (a) and the amide (b) at a surface pressure of 25 mNm^{-1}.

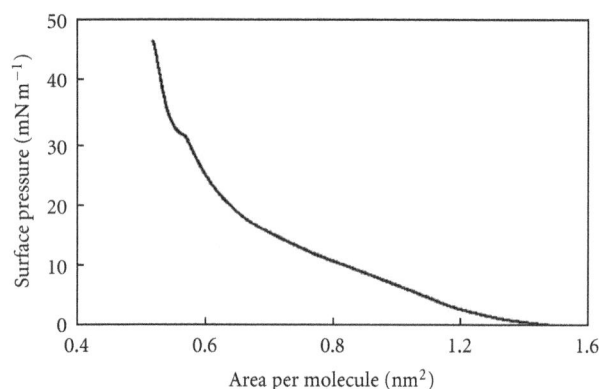

FIGURE 25: Isotherm of the salt.

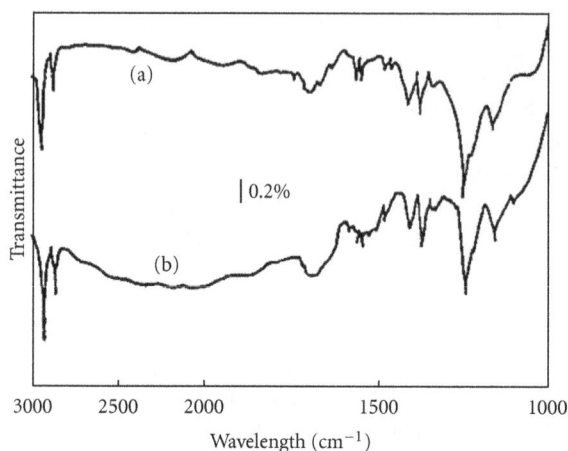

FIGURE 26: FTIR reflection adsorption spectra of the novel lubricant films prepared by (a) spontaneous adsorption method and (b) LB method on the surface of the magnetic layer.

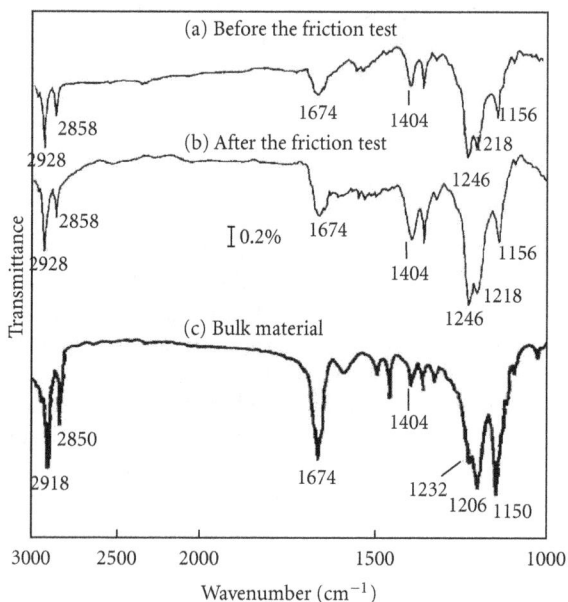

FIGURE 27: FTIR spectra of ammonium salt, C_7F_{15}-COO$^-$H$_3$N$^+$-$C_{18}F_{37}$ (Lubricant 5) (a) RAS spectra of adsorbed film (b) RAS spectra of adsorbed film after friction test (c) transmission spectrum of bulk material.

speed of 5 mm min^{-1} upon withdrawal from the water. The deposition trace suggested that there is an even deposition and the transfer ratio is about 0.75. Therefore, in the LB film monolayer the area occupied by one salt molecule was calculated to be about 0.8 nm^2. The difference in occupied area per molecule indicates that the chain leans slightly to the surface normal. The tilt of the chains in the LB monolayer is induced by the deposition process.

8.2. FTIR Study for LB Films. The friction properties may be enhanced by a microscopically smooth coverage of the lubricant film over the media surface. In the FTIR-RAS spectra, the components of the vibrational moments of the chemical bonds, which are parallel to the substrate normal, selectively appear [79]. Therefore, the FTIR-RAS spectra are useful for investigating the molecular orientation of a lubricant film on a substrate.

The salt-type lubricant films, which are prepared by the LB method and the spontaneous adsorption from the lubricant solution, are compared in Figure 26. The bands in the 3000–2800 cm^{-1} and 1370–1100 cm^{-1} regions are assigned to the CH and the CF stretching vibrations, respectively. The band at around 1674 cm^{-1} is assigned to the COO$^-$ antisymmetric vibration. The absolute intensities of these bands are very similar in both spectra, which suggest that the adsorbed layer of the salt is in fact a monolayer and that the degree

of orientation and packing of the alkyl and perfluoroalkyl chains closely resembles that in the LB monolayer.

9. Spectroscopy of Adsorbed Lubricant Film

9.1. FTIR Study for Lubricant Film of Different Hydrophilic Groups. Figures 27 and 28 compare the RAS of the lubricant film and transmission spectra of the bulk material for the salt and the amide, respectively. The mode assignment and peak positions for the RAS and for the transmission spectra are summarized in Table 5.

For the salt, the differences in the spectra between the RAS of the adsorbed lubricant film on the substrate

TABLE 5: Mode assignment and peak positions for the ammonium salt and amide in KBr and as adsorbed on a magnetic layer.

Peak position/cm^{-1}				
Amide		Ammonium salt		Vibration mode
Bulk	Film	Balk	Film	
2918	2928	2918	2918	CH$_2$ asymmetric stretching
2850	2856	2850	2858	CH$_2$ symmetric stretching
		1674	1674	COO$^-$ asymmetric stretching
		1404	1404	COO$^-$ symmetric stretching
1692				CO stretching
1232		1232	1246	CF$_2$ asymmetric stretching (E$_1$ symmetry)
1206		1206	1218	CF$_2$ asymmetric stretching (A$_2$ symmetry) + CF$_3$ stretching
1148		1150	1156	CF$_2$ symmetric stretching (E$_1$ symmetry)

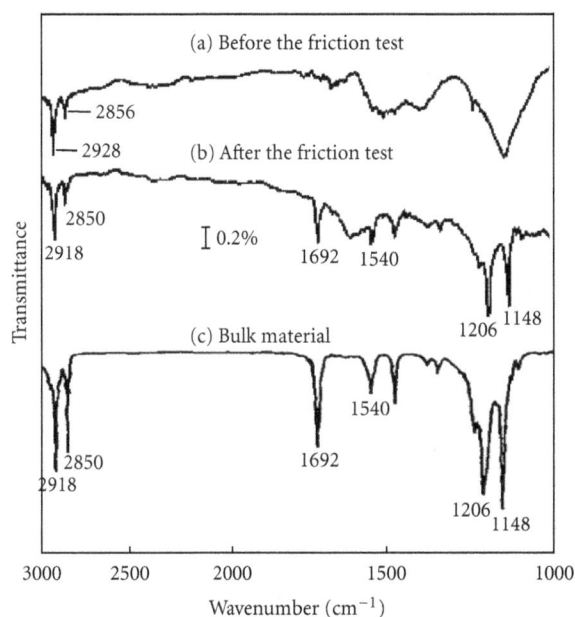

FIGURE 28: FTIR spectra of amide, C$_7$F$_{15}$-CONH-C$_{18}$F$_{37}$ (Lubricant 3) (a) RAS spectra of adsorbed film (b) RAS spectra of adsorbed film after friction test (c) transmission spectrum of bulk material.

(a) and the transmission spectra of the bulk materials (c) are described as follows.

(1) In the RAS spectrum of the adsorbed film, the bond assigned to the COO$^-$ asymmetric stretching vibration at 1674 cm^{-1} is much weaker than in the spectrum of the bulk material.

(2) The relative intensity in wavenumber of the CF$_2$ stretching vibrations in the 1250–1140 cm^{-1} region is changed and shifted to a higher frequency in the RAS spectrum.

(3) The band assigned to the CH$_2$ stretching vibration in the RAS spectrum in the vicinity of 2900 cm^{-1} is also shifted to a higher frequency than that in the spectrum of the bulk.

The weakness of the band at 1674 cm^{-1} suggests that asymmetric stretching of the COO$^-$ (hydrophilic) group is almost parallel to the substrate surface in the adsorbed film. The changes in the relative intensity and the higher frequency shift in the 1250 to 1140 cm^{-1} region are similar to those reported for a monolayer of perfluorocarboxylic acid [80]. These changes have been attributed to adsorption of the molecules with a preferential orientation and the perfluoroalkyl chains tilted in the adsorbed layer.

The higher frequency shift of the CH$_2$ stretching vibrations is often observed when the alkyl chains are in a liquid phase or in a solution, where the cohesive interaction between them is weak [66]. A probable cause of the higher frequency shift in the RAS spectrum is that the perfluoroalkyl chains hinder a cohesive interaction between the alkyl chains in the monolayer.

For the amide, the spectral pattern of the film in the region of the CF$_2$ and CF$_3$ stretching vibrations, 1250 to 1140 cm^{-1}, is different from that of the salt. However, the decrease in intensity of the stretching vibrations for C=O at 1692 cm^{-1} and the high frequency shift of the CH$_2$ stretching mode behaves similar to the carbonyl stretch and high frequency CH$_2$ shift for the salt.

The FTIR-RAS spectra of the lubricant film after the friction test are also shown in Figures 27(b) and 28(b). The absolute intensity of the CH$_2$ stretching region is similar in the spectra both before and after the friction test, indicating that sliding causes no substantial change in the surface concentration of the alkyl groups. For the salt, the spectra both before and after the test are similar, therefore, the rubbing motion caused no change in the microscopic structure of the lubricant film. On the other hand, there are several differences in the RAS spectrum of the adsorbed film of the amide before and after the test. These results are summarized as follows.

(1) The bands at approximately 1692 cm^{-1} and 1540 cm^{-1}, which are assigned to the C=O stretching vibration and the NH bending vibration, respectively, appear after the test.

(2) The pattern of the CF$_2$ and CF$_3$ stretching vibrations at 1250 cm^{-1} and 1140 cm^{-1} changes.

(3) The band of the CH$_2$ antisymmetrical and symmetrical stretching vibrations in the 3000 to 2800 cm^{-1} region shifts to a lower frequency.

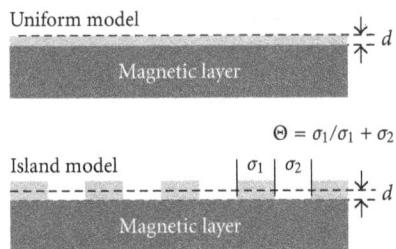

FIGURE 29: Lubricant coverage models. Θ is the lubricant coverage ratio in island model and d is the average thickness on magnetic layer.

Thus, the spectrum of the adsorbed amide film approaches the spectrum of the bulk material with sliding. This suggests rearrangement of the amide monolayer by the sliding process, perhaps into three-dimensional crystals or amorphous piles.

These results show that the layers of the amide were composed of orientated molecules, although they produced a higher friction coefficient than those of the salt. A possible cause for the higher friction coefficient is that a layer structure with a high degree of molecular orientation is less stable for the amide compared to the salt. This is demonstrated by the decay curve on the water surface in Figure 24.

The FTIR-RAS spectra showed that the polar group of both lubricants interacted with the magnetic surface before the friction test. For the ammonium salt, the friction coefficient is low and constant throughout the 100 reciprocating cycles (Figure 9). From a spectroscopic point of view, before and after the spectra of the friction test are similar, suggesting no change in the structure of the lubricant film. In contrast, for the amide, the friction increases with the number of cycles, thus the adsorbed lubricant was changed into bulk-like aggregates, which brings about the bare contact and leads to a higher friction.

9.2. Angle Resolved X-Ray Photoelectron Spectroscopy (ARXPS) Study for Microscopic Coverage.
The uniformity of the monolayer level lubricant on the magnetic thin film media has been investigated using ARXPS with the finding that the PFPE is a discontinuous film [50, 81]. Kimachi et al. proposed an island model to describe the coverage of conventional PFPEs on the magnetic recording media [82]. On the other hand, the modified PFPE has a good lubricity compared to the conventional ones, which implies a different film formation. In order to gain insight into this effect, microscopic coverage of the lubricant on the surface is investigated by ARXPS. The Z-lubricants with and without modification are deposited under the same conditions.

A lubricant layer completely covers the surface with a constant thickness of d in the uniform model. In the island model, the surface is discretely covered with lubricant islands having average the thickness of d (Figure 29). It is assumed in these models that the surface is flat and that elemental atoms are homogeneously distributed in the lubricant layer [83]. Based on these assumptions, the photo-electron intensity ratio, I_{lub}/I_{sub} is expressed as a function of the take-off angle,

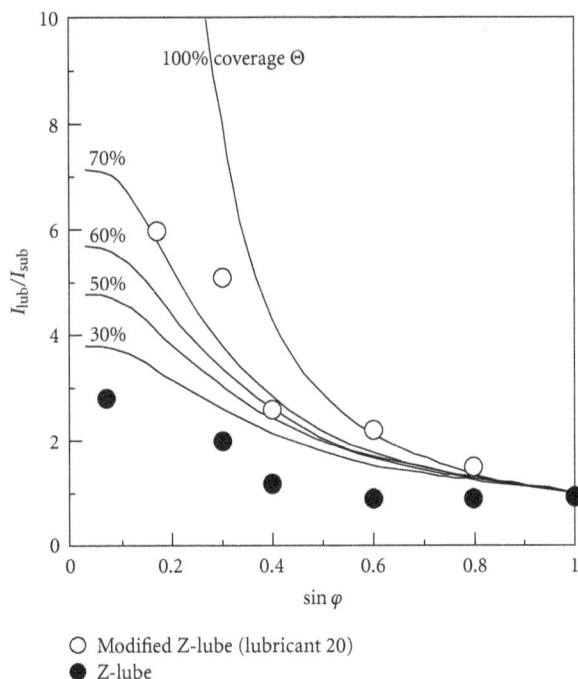

○ Modified Z-lube (lubricant 20)
● Z-lube

FIGURE 30: Coverage of the media surface with lubricant by ARXPS and the calculated curves (solid line) using island model shown for comparison.

φ, and the lubricant coverage ratio, Θ, on the surface of the magnetic layer. I_{lub} is the intensity of the photoelectron detected from the lubricant layer, and I_{sub} is the intensity from the under layer.

In Figure 30, I_{lub}/I_{sub} is plotted as a function of $\sin \varphi$, and the calculated model curves of Θ using the island model are also shown. Because the I_F of the modified PFPE is somewhat low, I_{lub}/I_{sub} is normalized in such a way that the I_{lub}/I_{sub} at 90 degrees is 1.

The coverage of the modified PFPE is greater than the conventional lubricant, despite the fact that the amount of lubricant on the surface is somewhat lower. The friction coefficient of the magnetic media coated with bonded lubricant film decreases linearly with increasing surface coverage [84]. The higher coverage reduces the dry contacts at the rubbing surface and minimizes friction.

It can be considered that two factors determine low friction, namely, better coverage and a strong interaction between the lubricant and the media surface. Better coverage is ascribed to the balance of the hydrophobic and hydrophilic properties of the lubricant. It is well balanced when the hydrocarbon chain is introduced. The polar group of the ammonium salt without steric hindrance has strong interactions at the surface.

10. Physicochemical Aspect

10.1. Surface Energy.
The surface energy is determined from the contact angle, θ, measurements using water and diiodomethane. The methodology for these measurements was

TABLE 6: Calculated dispersion and polar contribution to surface energy based upon contact angle measurements.

Lubricant	Contact angle/deg		γ^d/mJm^{-2}	γ^p/mJm^{-2}	γ^{total}/mJm^{-2}
	H$_2$O	CH$_2$I$_2$			
Stearyl (9)	83.8	64.7	21.4	6.7	28.1
Oleyl (11)	83.9	67.4	19.7	7.3	27.0
Linolenyl (12)	82.9	73.2	16.1	9.4	25.6

TABLE 7: Heat of adsorption and desorption of the lubricants on the surface.

Structure and lubricant number	Dielectric constant**	Heat of adsorption/erg cm^{-2}	Heat of desorption/erg cm^{-2}
$C_7F_{15}CO\ O\ C_{18}H_{31}$ (2)	2.01	2.1	1.0
$C_7F_{15}CO\ NH\ C_{18}H_{31}$* (4)	2.36	2.7	1.9
$C_7F_{15}\ O^-H_3N^+C_{18}H_{31}$ (6)	2.66	18.9	Not detected

*Stearyl amide is slightly soluble in the solvent, therefore, the double bonded lubricants are used. **Dielectric constant is measured at 3 MHz.

described by Kaeble [85]; therefore, an extensive reiteration is not necessary here.

The contact angle is related to the surface energy via Young's equation:

$$\gamma_S = \gamma_{SL} + \gamma_L \cos\theta, \qquad (1)$$

where γ_S is the surface energy of the solid, γ_L is the surface energy of the liquid, and γ_{SL} is the interfacial energy of the solid and liquid. When the reference liquid is capable of interacting with the surface through both dispersive and polar forces, the interfacial surface energy can be written as follows:

$$
\begin{aligned}
\gamma_{SL} &= \gamma_S + \gamma_L - 2\left(\gamma_S\gamma_L\right)^{1/2} \\
&= \gamma_S + \gamma_L - 2\left[\left(\gamma_S^d\gamma_L^d\right)^{1/2} + \left(\gamma_S^p\gamma_L^p\right)^{1/2}\right],
\end{aligned} \qquad (2)
$$

where γ_S^d and γ_S^p are the dispersive and polar components of the solid surface energy, respectively (Schrader [86]). Also, γ_L^d and γ_L^p are the liquid surface energies. The dispersive component is the London dispersion force contribution to the surface energy, and the polar component is a measure of the polar interactions such as hydrogen bonding or acid-base neutralization (Fowkes [87]). γ_S^d and γ_S^p are obtained by the substitution of (2) into (1).

The disorder of the hydrocarbon chains is also distinguishable by the wettability, which is determined by the nature of the outermost atomic group of the lubricant layer. Table 6 summarizes the surface properties calculated from the contact angle measurements.

The surface energies are evaluated by Kaelble's method [80]. As the adsorption model is shown in Figure 12, the unsaturated chains are bent and inefficiently packed, so that the perfluoroalkyl chains appear near the outside surface, and lower the surface energy value, which is mainly determined by the dispersive component.

10.2. Heat of the Preferential Adsorption. These three types of lubricants, the salt (Lubricant 5), the amide (Lubricant 3), and the ester (Lubricant 1), have the same hydrophobic groups, but have different friction coefficients, which is related to the polar group of the lubricant molecule. A plausible explanation is that the salt-type lubricant leads to greater adhesion than the ester and amide. As the surface of the sputtered carbon protective layer is classified as rather polar because it contains 5 and 7 atom % hydroxyl and carbonyl groups, respectively, a source of the attractive force at the surface may be the interactions between these polar groups [63]. In order to gain insight into this effect on the friction properties, the dielectric loss measurements are conducted.

The relative dielectric constant (ε) of the lubricants is measured at 3 MHz and summarized in Table 7. The relative dielectric constant, which is a parameter of the dipole moment, is 2.01 and 2.36 for the ester and the amide, respectively, but 2.66 for the salt-type. The higher the dielectric constant, the more strongly the lubricant seems to adsorb on the medium [88].

The formation of the lubricant film is a spontaneous process caused by a decrease in the free energy of the solid surfaces and lubricant molecule adsorption. The heat of adsorption of lubricants on a rubbing surface can be taken as a measure of the strength of attachment of the lubricant molecules to the surfaces and are also shown in Table 7.

For the ester and the amide-type, the heat of adsorption is small, and the heat of desorption is partially observed. However, for the salt-type, the heat of adsorption was very high compared to the ester and the amide, and the heat of desorption could not be detected. These results confirm that the ester and the amide were held by weak interactions on the carbon surface, whereas irreversible adsorption had taken place for the salt type. The high heat of adsorption of the salt type on the carbon surface results from these strong interactions and accounts for its low and steady friction. The low heat of adsorption for the ester and the amide, by contrast, apparently produces a film which is easily disrupted by sliding to give a rising friction coefficient with sliding.

11. Conclusions

The frictional properties of newly synthesized ionic liquid lubricants for magnetic media have been investigated.

A novel ionic liquid, which has an ammonium salt with a carboxylate as a hydrophilic group, has a lower frictional coefficient than the other conventional PFPE. It can be considered that two factors determined the low friction, namely, better coverage and strong interaction between the lubricant and the media surface.

These ionic lubricants invented around 1987 [26, 44, 89–91] have been used for magnetic tapes of 8 mm video, digital video cassette, the AIT system, and the broadcast application for about a quarter century because of their good lubricity and also from an environmental point of view.

The effects of the molecular structure of the modified PFPE on the frictional properties are summarized as follows.

(1) Stronger adsorption due to the adhesive interaction of the polar group, thus the novel carboxylic acid ammonium salt has a lower and more stable friction coefficient.

(2) Sufficient length and symmetry of the hydrocarbon chain cause extensive cohesive interactions, and these dispersive interactions compensate for the friction reduction.

(3) On the contrary, for the olefinic lubricant, the friction coefficient increased with the increasing number of reciprocating cycles because of the weaker cohesive interactions. The introduction of a double bond at the terminal position affected an increase in friction compared to the center position.

(4) The ammonium salt polar group can be introduced to the commercial PFPE, and the frictional coefficient is independent of the molecular structure of the PFPE backbone polymer and of the thinner dip-coating, which makes practical use convenient without any environmental problems.

From a microscopic point of view, the lubricant film coverage was also investigated as follows.

(1) A layer of the carboxylic acid ammonium salt lubricant film prepared by the spontaneous adsorption is highly ordered and closely packed as in the LB monolayer.

(2) The polar COO^- groups of the salt are adsorbed almost parallel to the surface, and the sliding scarcely changes the salt-type lubricant film, while the arrangement of the amide film into a bulk phase occurs due to the sliding process.

(3) The modified PFPE uniformly covers the magnetic surfaces; this is why it minimizes the friction.

References

[1] B. Bhushan, *Tribology and Mechanics of Magnetic Storage Devices*, Springer, NewYork, NY, USA, 1990.

[2] R. W. Wood, J. Miles, and T. Olson, "Recording technologies for terabit per square inch systems," *IEEE Transactions on Magnetics*, vol. 38, no. 4, pp. 1711–1718, 2002.

[3] M. S. Jhon and H. J. Cho, "Lubricants in future data storage technology," *Journal of Industrial and Engineering Chemistry*, vol. 7, no. 5, pp. 263–275, 2001.

[4] A. Maesaka and H. Ohmori, "Transmission electron microscopy analysis of lattice strain in epitaxial Co-Pd multilayers," *IEEE Transactions on Magnetics*, vol. 38, no. 5, pp. 2676–2678, 2002.

[5] S. Onodera, H. Kondo, and T. Kawana, "Materials for magnetic-tape media," *MRS Bulletin*, vol. 21, no. 9, pp. 35–41, 1996.

[6] Y. Shiraishi and A. Hirota, "Magnetic recording at video cassette recorder for home use," *IEEE Transactions on Magnetics*, vol. 14, no. 5, pp. 318–320, 1978.

[7] H. Naruse, K. Sato, H. Osaki, K. Chiba, T. Sasaki, and H. Yoshimura, "Advanced metal evaporated tape for consumer digital VCR'S (DV cassette)," *IEEE Transactions on Consumer Electronics*, vol. 42, no. 3, pp. 851–859, 1996.

[8] K. Kanota, H. Inoue, A. Uetake, M. Kawaguchi, K. Chiba, and Y. Kubota, "A high density recording technology for digital VCRs," *IEEE Transactions on Consumer Electronics*, vol. 36, no. 3, pp. 540–547, 1990.

[9] T. Ito, Y. Iwasaki, H. Tachikawa, Y. Murakami, and D. Shindo, "Microstructure of a Co-CoO obliquely evaporated magnetic tape," *Journal of Applied Physics*, vol. 91, no. 7, pp. 4468–4473, 2002.

[10] H. Tachikawa, Y. Murakami, T. Ito, Y. Iwasaki, and D. Shindo, "Microstructural analysis of obliquely evaporated Co-CoO tape using TEM and EELS," *Nippon Kinzoku Gakkaishi/Journal of the Japan Institute of Metals*, vol. 65, no. 5, pp. 349–355, 2001.

[11] Y. Kaneda, "Tribology of metal-evaporated tape for high-density magnetic recording," *IEEE Transactions on Magnetics*, vol. 33, no. 2, pp. 1058–1068, 1997.

[12] B. Xu, K. Motohashi, S. Onodera, and W. D. Doyle, "Magnetic characteristics and recording properties of thin Co-CoO metal evaporated tapes," *IEEE Transactions on Magnetics*, vol. 37, no. 4, pp. 1630–1633, 2001.

[13] S. Onodera, T. Takeda, and T. Kawana, "The archival stability of metal evaporated tape for consumer digital VCRs," *Journal of Applied Physics*, vol. 79, no. 8, pp. 4875–4877, 1996.

[14] S. Fukuda, T. Ozue, and S. Onodera, "Recording over 15 ktpi using multichannel heads in a tape system," *IEEE Transactions on Magnetics*, vol. 42, no. 2, pp. 182–187, 2006.

[15] N. Sekiguchi, K. Kawakami, T. Ozue, M. Yamaga, and S. Onodera, "Examination of newly developed metal particle media for >3 Gb/in^2 recording in GMR-based tape systems," *IEEE Transactions on Magnetics*, vol. 41, no. 10, pp. 3235–3237, 2005.

[16] K. Noma, M. Matsuoka, H. Kanal, Y. Uehara, K. Nomura, and N. Awaji, "Ultra-high magnetic moment films for write head," *IEEE Transactions on Magnetics*, vol. 42, no. 2, pp. 140–144, 2006.

[17] B. Bhushan, *Tribology and Mechanics of Magnetic Storage Devices*, Springer, New York, NY, USA, 2nd edition, 1996.

[18] S. Sato, Y. Arisaka, and S. Matsumura, "Surface design of aramid film for future me tapes," *IEEE Transactions on Magnetics*, vol. 35, no. 5, pp. 2760–2762, 1999.

[19] A. M. Homola, "Lubrication issues in magnetic disk storage devices," *IEEE Transactions on Magnetics*, vol. 32, no. 3, pp. 1812–1818, 1996.

[20] J. Lin and A. W. Wu, "Lubricants for magnetic rigid disks," in *Proceedings of International Tribology Conference*, pp. 599–604, Nagoya, Japan, October 1990.

[21] E. E. Klaus and and B. Bhushan, "Lubricants in magnetic media-a review," *ASLE*, vol. SP-19, pp. 7–15, 1985.

[22] B. Marchon, Q. Dai, F. Hendriks, and U. V. Nayak, "Disk drive having reduced variation of disk surface lubricant layer thickness," U.S. Patent 7002768, 2006.

[23] J. Liu, M. J. Stirniman, and J. Gui, "Lubricant for thin film storage media," U.S. Patent 6916531, 2005.

[24] M. Ishida, T. Nakakawaji, Y. Ito, H. Matsumoto, H. Tani, and H. Ishihara, "Lubricant, magnetic disk and magnetic disk apparatus," U.S. Patent 6869536, 2005.

[25] S. Gunsel, C. Venier, and I.-C. Chiu, "Lubricant for magnetic recording medium and use thereof," U.S. Patent 666728, 2003.

[26] J. Liu, M. J. Stirniman, and J. Gui, "Lubricant film containing additives for advanced tribological performance of magnetic storage medium," U.S. Patent 7060377, 2006.

[27] H. Kondo and T. Uchimi, "Novel Perfluoropolyether Derivatives Lubricants and Magnetic Recording Medium using the same," US Patent 5453539, 1992.

[28] B. Marchon, X.-C. Guo, T. Karis et al., "Fomblin multidentate lubricants for ultra-low magnetic spacing," *IEEE Transactions on Magnetics*, vol. 42, no. 10, pp. 2504–2506, 2006.

[29] H. Chiba, E. Yamasaka, T. Tokairin, Y. Oshikubo, and K. Watanabe, "Synthesis of multi-functional PFPE lubricant and its tribological characteristics," in *Proceedings of the 3rd World Tribology Congress*, Washington, DC, USA, September 2005, Paper WTC2005-63165.

[30] http://www.solvayplastics.com/sites/solvayplastics/EN/specialty_polymers/Fluorinated_Fluids/Pages/Fomblin_Functional_Fluids_PFPE.aspx.

[31] R. Sheldon, "Catalytic reactions in ionic liquids," *Chemical Communications*, vol. 23, pp. 2399–2403, 2001.

[32] H. Ohno, *Electrochemical Aspects of Ionic Liquids*, Wiley-Interscience, Hoboken, NJ, USA, 2005.

[33] A. E. Visser, R. P. Swatloski, W. M. Reichert et al., "Task-specific ionic liquids for the extraction of metal ions from aqueous solutions," *Chemical Communications*, no. 1, pp. 135–136, 2001.

[34] J. L. Anderson and D. W. Armstrong, "High-stability ionic liquids. A new class of stationary phases for gas chromatography," *Analytical Chemistry*, vol. 75, no. 18, pp. 4851–4858, 2003.

[35] C. M. Gordon, "New developments in catalysis using ionic liquids," *Applied Catalysis A*, vol. 222, no. 1-2, pp. 101–117, 2001.

[36] T. Welton, "Room-temperature ionic Liquids. Solvents for synthesis and catalysis," *Chemical Reviews*, vol. 99, no. 8, pp. 2071–2083, 1999.

[37] K. Binnemans, "Ionic liquid crystals," *Chemical Reviews*, vol. 105, no. 11, pp. 4148–4204, 2005.

[38] C. Ye, W. Liu, Y. Chen, and L. Yu, "Room temperature ionic liquids: a novel versatile lubricant," *Chemical Communications*, no. 21, pp. 2244–2246, 2001.

[39] W. Liu, C. Ye, Q. Gong, H. Wang, and P. Wang, "Tribological performance of room-temperature ionic liquids as lubricant," *Tribology Letters*, vol. 13, no. 2, pp. 81–85, 2002.

[40] J. Qu, J. J. Truhan, S. Dai, H. Luo, and P. J. Blau, "Ionic liquids with ammonium cations as lubricants or additives," *Tribology Letters*, vol. 22, no. 3, pp. 207–214, 2006.

[41] B. S. Phillips and J. S. Zabinski, "Ionic liquid lubrication effects on ceramics in a water environment," *Tribology Letters*, vol. 17, no. 3, pp. 533–541, 2004.

[42] R. A. Reich, P. A. Stewart, J. Bohaychick, and J. A. Urbanski, "Base oil properties of ionic liquids," *Lubrication Engineering*, vol. 59, no. 7, pp. 16–21, 2003.

[43] P. Iglesias, M. D. Bermúdez, F. J. Carrión, and G. Martínez-Nicolás, "Friction and wear of aluminium-steel contacts lubricated with ordered fluids-neutral and ionic liquid crystals as oil additives," *Wear*, vol. 256, no. 3-4, pp. 386–392, 2004.

[44] B. A. Omotowa, B. S. Phillips, J. S. Zabinski, and J. M. Shreeve, "Phosphazene-based ionic liquids: synthesis, temperature-dependent viscosity, and effect as additives in water lubrication of silicon nitride ceramics," *Inorganic Chemistry*, vol. 43, no. 17, pp. 5466–5471, 2004.

[45] H. Kondo, "Magnetic Thin Film Media," Japanese Patent 2581090, 1987.

[46] H. Kondo, J. Seto, and K. Haga S. Ozawa, "Novel lubricant for magnetic thin film media," *Journal of the Magnetics Society of Japan*, vol. 13, supplement 1, pp. 213–218, 1989.

[47] H. Kondo, A. Seki, H. Watanabe, and J. Seto, "Frictional properties of novel lubricants for magnetic thin film media," *IEEE Transactions on Magnetics*, vol. 26, no. 5, pp. 2691–2693, 1990.

[48] H. Kondo, A. Seki, and A. Kita, "Comparison of an amide and amine salt as friction modifiers for a magnetic thin films medium," *Tribology Transactions*, vol. 37, no. 1, pp. 99–104, 1994.

[49] H. Kondo, "Effect of double bonds on friction in the boundary lubrication of magnetic thin film media," *Wear*, vol. 202, no. 2, pp. 149–153, 1997.

[50] H. Kondo, A. Seki, and A. Kita, "The effect of molecular structure and microscopic coverage of lubricants on the frictional properties," *Journal of the Surface Science Society of Japan*, vol. 14, no. 6, pp. 331–335, 1993.

[51] H. Kondo, Y. Hisamichi, and T. Kamei, "Lubrication of modified perfluoropolyether on magnetic media," *Journal of Magnetism and Magnetic Materials*, vol. 155, no. 1–3, pp. 332–334, 1996.

[52] H. Kondo and Y. Kaneda, "Development of modified perfluoropolyether tape," in *Proceedings of the International Tribology Conference (AUSTRIB '94)*, pp. 415–420, Perth, Australia, 1994.

[53] H. Kondo, "Protic ionic liquids with ammonium salts as lubricants for magnetic thin film media," *Tribology Letters*, vol. 31, no. 3, pp. 211–218, 2008.

[54] T. L. Greaves and C. J. Drummond, "Protic ionic liquids: properties and applications," *Chemical Reviews*, vol. 108, no. 1, pp. 206–237, 2008.

[55] W. Xu and C. A. Angell, "Solvent-free electrolytes with aqueous solution-like conductivities," *Science*, vol. 302, no. 5644, pp. 422–425, 2003.

[56] M. Yoshizawa, W. Xu, and C. A. Angell, "Ionic liquids by proton transfer: vapor pressure, conductivity, and the relevance of ΔpK_a from aqueous solutions," *Journal of the American Chemical Society*, vol. 125, no. 50, pp. 15411–15419, 2003.

[57] T. L. Greaves, A. Weerawardena, C. Fong, I. Krodkiewska, and C. J. Drummond, "Protic ionic liquids: solvents with tunable phase behavior and physicochemical properties," *Journal of Physical Chemistry B*, vol. 110, no. 45, pp. 22479–22487, 2006.

[58] F. Kohler, H. Atrops, H. Kalali et al., "Molecular interactions in mixtures of carboxylic acids with amines. 1. Melting curves and viscosities," *Journal of Physical Chemistry*, vol. 85, no. 17, pp. 2520–2524, 1981.

[59] F. Kohler, R. Gopal, G. Götze et al., "Molecular interactions in mixtures of carboxylic acids with amines. 2. Volumetric, conductimetric, and NMR properties," *Journal of Physical Chemistry*, vol. 85, no. 17, pp. 2524–2529, 1981.

[60] A. Seki and H. Kondo, "FTIR reflection absorption spectra of novel lubricant layers on magnetic thin film media," *Journal of*

the Magnetics Society of Japan, vol. 15, supplement 2, pp. 745–749, 1991.

[61] H. Kondo and A. Seki, "Relation between the microscopic structure of the lubricant film and the frictional properties," *Journal of Japanese Society of Tribologists*, vol. 38, no. 1, pp. 40–45, 1993.

[62] R. E. Linder and P. B. Mee, "ESCA determination of fluorocarbon lubricant film thickness on magnetic disk media," *IEEE Transactions on Magnetics*, vol. 18, no. 6, pp. 1073–1076, 1982.

[63] M. F. Toney, C. Mathew Mate, and D. Pocker, "Calibrating ESCA and ellipsometry measurements of perfluoropolyether lubricant thickness," *IEEE Transactions on Magnetics*, vol. 34, no. 4, pp. 1774–1776, 1998.

[64] H. Kondo and Y. Nishida, "Quantitative analysis of surface functional groups on the amorphous carbon in magnetic media with XPS preceded by chemical derivatization," *Bulletin of the Chemical Society of Japan*, vol. 80, no. 7, pp. 1405–1412, 2007.

[65] M. Beltzer and S. Jahanmir, "Role of dispersion interactions between hydrocarbon chains in boundary lubrication," *ASLE Transactions*, vol. 30, no. 1, pp. 47–54, 1987.

[66] K. S. Markley, *Fatty Acids and Their Chemistry, Properties, Production, and Uses Part 1*, John Wiley & Sons, New York, NY, USA, 2nd edition, 1968.

[67] B. Bhushan and M. T. Dugger, "Liquid-mediated adhesion at the thin film magnetic disk/slider interface," *Journal of Tribology*, vol. 112, no. 2, pp. 217–223, 1990.

[68] I. M. Asher and I. W. Levin, "Effects of temperature and molecular interactions on the vibrational infrared spectra of phospholipid vesicles," *Biochimica et Biophysica Acta*, vol. 468, no. 1, pp. 63–72, 1977.

[69] K. Chiba, K. Sato, Y. Ebine, and T. Sasaki, "Metal evaporated tape for high band 8mm video system," *IEEE Transactions on Consumer Electronics*, vol. 35, no. 3, pp. 421–428, 1989.

[70] H. Osaki, K. Fukushi, and K. Ozawa, "Wear mechanisms of metal-evaporated magnetic tapes in helical scan videotape recorders," *IEEE Transactions on Magnetics*, vol. 26, no. 6, pp. 3180–3185, 1990.

[71] H. Osaki, "Role of surface asperities on durability of metal-evaporated magnetic tapes," *IEEE Transactions on Magnetics*, vol. 29, no. 1, pp. 11–20, 1993.

[72] P. Cong, T. Igari, and S. Mori, "Effects of film characteristics on frictional properties of carboxylic acid monolayers," *Tribology Letters*, vol. 9, no. 3-4, pp. 175–179, 2001.

[73] R. M. Overney, E. Meyer, J. Frommer et al., "Force microscopy study of friction and elastic compliance of phase-separated organic thin films," *Langmuir*, vol. 10, no. 4, pp. 1281–1286, 1994.

[74] G. Marchionni, G. Ajroldi, M. C. Righetti, and G. Pezzin, "Molecular interactions in perfluorinated and hydrogenated compounds: linear paraffins and ethers," *Macromolecules*, vol. 26, no. 7, pp. 1751–1757, 1993.

[75] K. Paserba, N. Shukla, A. J. Gellman, J. Gui, and B. Marchen, "Bonding of ethers and alcohols to a-CN$_x$ films," *Langmuir*, vol. 15, no. 5, pp. 1709–1715, 1999.

[76] N. Shukla, A. J. Gellman, and J. Gui, "Interaction of CF$_3$CH$_2$OH and (CF$_3$CF$_2$)$_2$O with amorphous carbon films," *Langmuir*, vol. 16, no. 16, pp. 6562–6568, 2000.

[77] G. Robert, *Langmuir-Blodgett Films*, Plenum Publishing Corporation, New York, NY, USA, 1990.

[78] J. Seto, T. Nagai, C. Ishimoto, and H. Watanabe, "Frictional properties of magnetic media coated with Langmuir-Blodgett films," *Thin Solid Films*, vol. 134, no. 1–3, pp. 101–108, 1985.

[79] R. J. H. Clark, *Spectroscopy of Surface*, John Wiley & Sons, New York, NY, USA, 1988.

[80] L. K. Chau and M. D. Porter, "Composition and structure of spontaneously adsorbed monolayers of *n*-perfluorocarboxylic acids on silver," *Chemical Physics Letters*, vol. 167, no. 3, pp. 198–204, 1990.

[81] J. F. Moulder, J. S. Hammond, and K. L. Smith, "Using angle resolved ESCA to characterize Winchester disks," *Applied Surface Science*, vol. 25, no. 4, pp. 446–454, 1986.

[82] Y. Kimachi, F. Yoshimura, M. Hoshino, and A. Terada, "Uniformity quantification of lubricant layer on magnetic recording media," *IEEE Transactions on Magnetics*, vol. 23, no. 5, pp. 2392–2394, 1987.

[83] C. S. Fadley, "Instrumentation for surface studies: XPS angular distributions," *Journal of Electron Spectroscopy and Related Phenomena*, vol. 5, no. 1, pp. 725–754, 1974.

[84] J. Choi, M. Kawaguchi, and T. Kato, "The surface coverage effect on the frictional properties of patterned PFPE nanolubricant films in HDI," *IEEE Transactions on Magnetics*, vol. 39, no. 5, pp. 2492–2494, 2003.

[85] D. H. Kaelble, "Dispersion-polar surface tension properties of organic solids," *Journal of Adhesion*, vol. 2, pp. 66–81, 1970.

[86] M. E. Schrader, "Contact angle and vapor adsorption," *Langmuir*, vol. 12, no. 15, pp. 3728–3732, 1996.

[87] F. M. Fowkes, "Attractive forces at interfaces," *Industrial Engineering Chemistry*, vol. 56, no. 12, pp. 40–52, 1964.

[88] Y. B. Yang, T. Bang, A. Mochizuki, and S. Kobayashi, "The surface anchoring dependence of the dielectric constant of an FLC material showing electroclinic effect," *Ferroelectrics*, vol. 121, no. 1, pp. 113–125, 1991.

[89] S. Haga and H. Kondo, "Magnetic recording medium," Japanese patent 2590482, 1987.

[90] H. Kondo and S. Haga, "Magnetic recording medium," Japanese patent 2629725, 1987.

[91] H. Kondo, "Magnetic recording medium," US Patent 5741593, 1992.

16

Tribological Studies on AISI 1040 with Raw and Modified Versions of Pongam and Jatropha Vegetable Oils as Lubricants

Y. M. Shashidhara and S. R. Jayaram

Department of Mechanical Engineering, Malnad College of Engineering, Karnataka state, Hassan 573 201, India

Correspondence should be addressed to Y. M. Shashidhara, shashi.yms@gmail.com

Academic Editor: Arvind Agarwal

The friction and wear tests on AISI 1040 are carried out under raw, modified versions of two nonedible vegetable oils Pongam (*Pongamia pinnata*) and Jatropha (*Jatropha curcas*) and also commercially available mineral oil using a pin-on-disc tribometer for various sliding distances and loads. A significant drop in friction and wear for AISI 1040 is observed under Pongam and Jatropha raw oil compared to mineral oil, for the complete tested sliding distance and load, increasing the potential of vegetable oil for tribological applications. Stribeck curves are also drawn to understand the regimes of lubrication. Both the vegetable oils showed a clear reduction in the boundary lubrication regimes, leading to an early start of full film lubrication.

1. Introduction

Escalating prices, rapid depletion of fossil reserves as well as the stringent legislations enforced by the International authorities have enlarged the scope around the globe to explore alternative ecofriendly lubricants [1, 2]. Oils of plant origin are promising, renewable, environmental the friendly, and nontoxic fluids, pose no work place health hazards and are readily biodegradable. Vegetable oils exhibit, number of performance blessings such as natural high viscosity (30 to 80% higher than mineral oil) and excellent lubricity due to their ester functionality. These are over 95% bio degradable and thus degrade much faster than mineral oils (20 to 30%), thereby offering greater potential when it comes to reducing the cost of disposal [3]. Another important property of vegetable oils is their highflash points; for instance the flash point of Soybean oil is 326°C while mineral oils have approximately 200°C, thus reducing emissions and work place pollution. Further, biobased lubricants help to provide energy security for countries who use them and create local and regional economic development opportunities.

Vegetable-based oils are finding their applications in various sectors of industry. For instance, the automotive lubricants derived from rapeseed, soy, and sunflower oils are finding good markets in European countries [4]. Also,

some of the vegetable oilbased lubricants and greases are finding their opportunities in specific applications like metal forming and working, food processing industry, machine elements, marine, locomotives, and so forth [5, 6]. Even though, the vegetablebased oils are finding their scope in various types of applications, it is meaningful to select oil for a particular application only after ascertaining the properties and behaviour. Otherwise, oil can also be selected based on the task the oil is to perform. In this context, the selected lube is expected to satisfy the basic requirements under two broad areas, namely, tribological behaviour and stability. The triglyceride structure of a vegetable oil provides these desirable qualities due to their long and polar fatty acid chains [7, 8]. It also generates high-strength lubricant films that interact strongly with metallic surfaces, reducing both friction and wear [9].

Many contributions comparing the vegetable oil performance with mineral oil performance are reported. Canola oil with boric acid showed about 30% lower friction coefficients compared to mineral oil, tested under pin-on-disc tribometer [10]. Lower magnitudes (about 35%) of wear are reported under electronized vegetable oil, tested with steel [11].

Soybean and sunflower oils showed about 25% drop in friction and wear rate when compared to mineral oil

TABLE 1: Fatty acid composition of Pongam and Jatropha oil.

Compound name	Pongama raw oil (%)	Jotrapha raw oil (%)
Lauric acid (C20:0)	00.30	00.40
Myristic acid (C14:0)	00.00	00.00
Palmitic acid (C16:0)	12.00	14.60
Stearic acid (C18:0)	06.70	06.60
Arachidic acid (C20:4)	01.60	00.22
Palmitoleic acid (C16:1)	00.00	01.20
Oleic acid (C18:1)	53.20	40.60
Linoleic acid (C18:2)	20.80	36.20
Linolenic acid (C18:3)	04.00	00.30

[9]. 30 percent reduction in coefficient of friction [12] is reported under sunflower oil as lubricant with coated tools compared to mineral oil. Phosphate esters with rapeseed oil exhibit about 40% reduction in friction and wear values [13, 14]. About 80% reductions in wear rate is reported for soybean and with amine phosphate as lubricant tested under tribology test rigs [15]. Soybean, canola, and sunflower oils with various oleic acid compositions, showed lower friction rates [16]. 37% drop in wear scar diameter size and about 15% decline in friction are reported when coconut oil blended with AW/EP additive [17].

Pongam and Jatropha are nonedible and plentifully available vegetable oils in Indian scenario. Currently, the oils are being used for the production of biodiesel. Due to their higher monosaturates (oleic acid) in their fatty acid composition (Table 1) and higher thermal and oxidative stability, they are projected as potential lubricants for wide range of operations. Oxidative stability is an important property to be considered for the consistent formation of oil layer at operating temperature for the whole life of the oil. These facts highlight the potential of vegetable oils as good lubricants.

The steel, AISI 1040, is a proven material for the manufacture of different components of machine and plentifully used in many other manufacturing industries also.

The motive of the present work is to bring out the enormous potential of vegetable oils to be used in manufacturing sector as straight cutting oils or lubricants. This has gained more importance in the light of the recent restrictions made by world leaders like OSHA, HOSH, EPA, and so forth, where in, they have suggested to come out with replacements for mineral oils, which are most environmentally friendly and also are depleting. Also, there is a large consumption of cutting oils/lubricants for the manufacturing sector.

In the present study, the two raw oils and their chemically modified versions are tested for physicochemical properties. These raw, modified versions of the vegetable oils and two commercially available mineral oils are used as lubricants to conduct friction and wear tests on AISI 1040 using pin-on-disc tribometer. Experiments are conducted for different loads and sliding distances. The tribological performance of the two materials with vegetable oils and with mineral oils is

compared. Further, the Stribeck curves are also drawn with all these oils to analyse the onset of full fluid film.

2. Methodology

2.1. Oil Modification. The raw vegetable oils have certain limitations like low thermo-oxidative stability [18]. This problem is addressed by various methods, namely, reformulation of additives, chemical modification, and genetic modification of the oil seed [2]. In the present work, chemical modification methods such as epoxidation [19] and transesterification [20] are used to modify the structure of two raw oils. After the modifications, their polyunsaturated C=C bonds are eliminated in the oil structure and the thermo-oxidative stabilities are enhanced.

Pongam raw oil (PRO) is modified into Pongam methyl ester (PME) and epoxidized Pongam raw oil (EPRO). Similarly, Jatropha raw oil (JRO) is altered to Jatropha methyl ester (JME) and epoxidized Jatropha raw oil (EJRO). Further, Pongam methyl ester (PME) and Jatropha methyl ester (JME) are modified into epoxidized Pongam methyl ester (EPME) and epoxidized Jatropha methyl ester (EJME), respectively.

2.2. Physicochemical Properties. The physicochemical properties of mineral, raw, and modified vegetable oils such as viscosity (ASTM D 445), viscosity index, (ASTM D 2270), flash point (ASTM D 92), pour point (ASTM D97), iodine values, and others are tested as per the standards (Table 2).

The result show that, the viscosity of epoxidized Pongam and Jatropha oils is increased by about 20% compared to their raw versions. This is attributed to high molecular weight and more polar structure in the epoxidized oils than their raw versions [19]. A marginal increase in viscosity index is seen under EPRO and EJRO. However, about 30% and 37% increase in viscosity index under EPME and EJME, respectively, are observed compared to their raw versions. Further, about 30% drop in flash point under both EPME and EJME are seen. The modification slightly improved the low temperature property (pour point) of the oils.

The epoxidation of the two oils is also confirmed by the reduction in iodine values, an indicator of unsaturation. EPRO, EPME, EJRO, and EJME show about 75%–80% drop in iodine values compared to their raw versions. The material composition of AISI 1040 is also ascertained (Table 3).

2.3. Friction and Wear Tests. Experiments are conducted on AISI 1040 steel pins of 8 mm diameter under the mineral, raw, and modified versions of the two vegetable oils using a pin-on-disc tribometer. Friction and wear tests are conducted with the wear track diameter of 100 mm and sliding speed of 4.2 m/s and for the loads 70 N, 100 N, 150 N, and 200 N. The volumetric wear and coefficient of friction for various distances of 1.25 km, 2.50 km, 3.75 km, 5.00 km, 6.25 km, and 7.5 km and loads are determined. The variation of coefficient of friction as well as wear as a function of sliding distance for different loads and oils are drawn.

TABLE 2: Physicochemical properties of raw and modified oils.

Properties	Pongam raw oil	Epoxidized Pongam raw oil	Epoxidized Pongam methyl ester	Jatropha raw oil	Epoxidized Jatropha raw oil	Epoxidized Jatropha methyl ester	Mineral oil (MRO)
Kinematic viscosity at 40°C (cSt)	53.65	65.41	20.15	35.36	63.65	11.22	33.00
Kinematic viscosity at 100°C (cSt)	13.90	20.17	11.50	11.70	19.97	07.07	12.00
Viscosity index	163.58	171.08	219.00	177.23	178.10	225.36	185.21
Total acid value (mg KOH g^{-1})	00.73	00.18	00.13	05.91	01.28	00.05	1.75
Flash point (0°C)	268.00	270.00	190.00	284.00	310.00	180.00	160
Pour point (0°C)	02.00	02.00	−02.00	−05.00	−06.00	−03.00	00.00
Iodine value (mg l g^{-1})	80.44	20.21	21.41	101.44	20.00	22.00	06.50

TABLE 3: Compositions of AISI 1040.

Contents	C	Si	Mn	P	S
Percentage	0.444	0.130	0.702	0.026	0.017

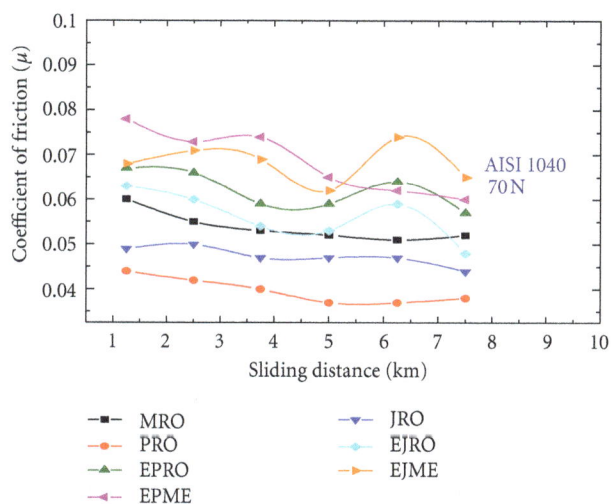

FIGURE 1: Variation of coefficient of friction with sliding distance for 70 N.

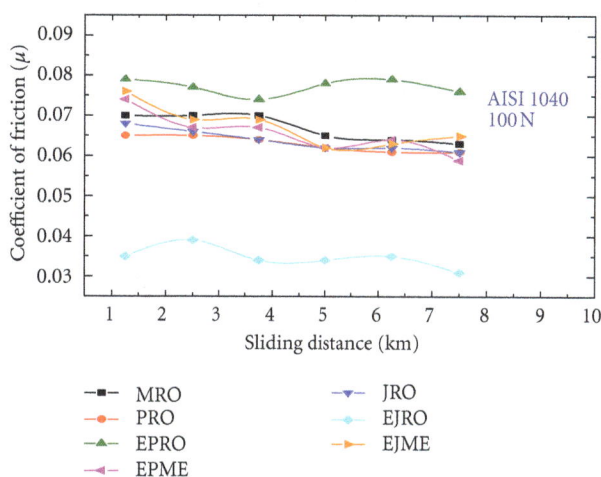

FIGURE 2: Variation of coefficient of friction with sliding distance for 100 N.

3. Analysis of Results

3.1. Friction Behaviour of AISI 1040. Figure 1 represents the variation of coefficient of friction with sliding distance for 70 N load. It is seen that about 50% drop in co-efficient of friction values under Pongam raw oil for the complete range of sliding distance tested when compared to mineral oil. However, under its epoxidized versions, EPRO and EPME, about 10% and 15% increase in friction respectively is observed compared to mineral oil. Further, under Jatropha raw, about 15% drop in the friction coefficient is seen compared to petroleum oil. On the other hand, marginal increase in friction is observed under EJRO and about 10% increase is noticed under EJME compared to mineral oil.

Low friction coefficients are seen under Pongam, Jatropha raw, and their versions of oils except EPRO for 100 N

load (Figure 2). About 10% drop is seen under all versions except EJRO and EPRO. EJRO show about 40% lower friction compared to mineral oil. However, EPRO show marginally higher values of friction.

PRO shows about 60% drop in friction for the load 150 N (Figure 3) compared to mineral oil. Also, EPRO and EPME exhibit about 35% and 40% lower friction, respectively, compared to under mineral oil. Further, JRO, EJRO, and EJME follow similar trend with 35% drop in friction compared to petroleum oil.

Figure 4 shows the variation of coefficient of friction with sliding distance for 200 N. About 20% and 13% drop in friction is seen under PRO and EPRO, respectively, compared to mineral oil. On the other hand, EPME exhibits about 30% lower friction. Further, JRO and EJRO show the trend similar to PRO and EPRO. However, under EJME, marginally lower frictional values are seen compared to mineral oil.

The lower friction values under the two vegetable raw oils are attributed to their polar nature and viscosity properties. Significant reductions under vegetable oils can be due to the fact that, the thin surface film that develops in boundary lubrication is formed by the adsorption of polar compounds

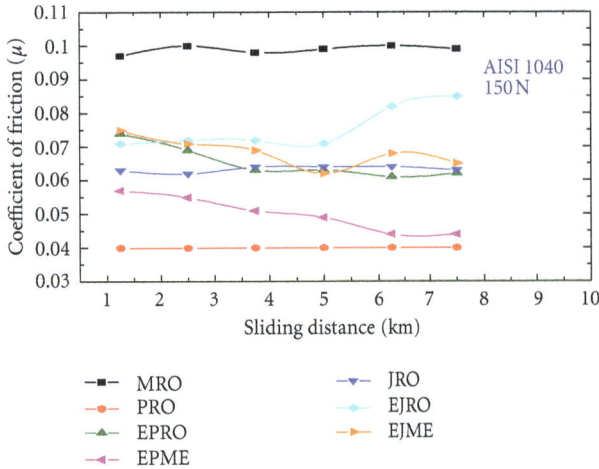

FIGURE 3: Variation of coefficient of friction with sliding distance for 150 N.

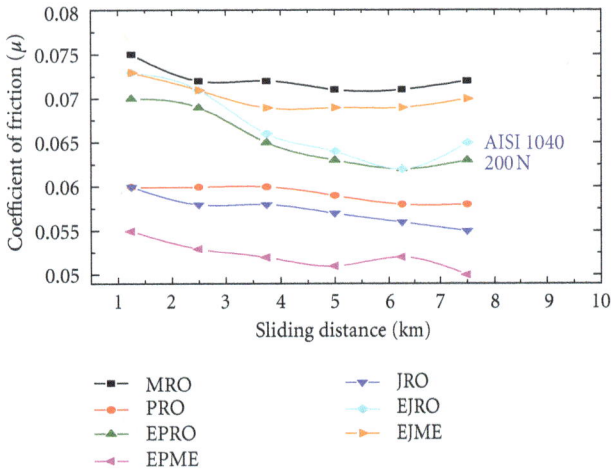

FIGURE 5: Variation of wear with sliding distance for 70 N.

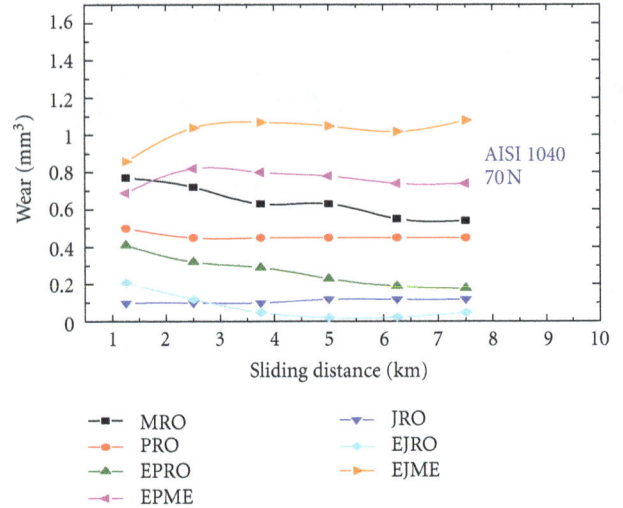

FIGURE 4: Variation of coefficient of friction with sliding distance for 200 N.

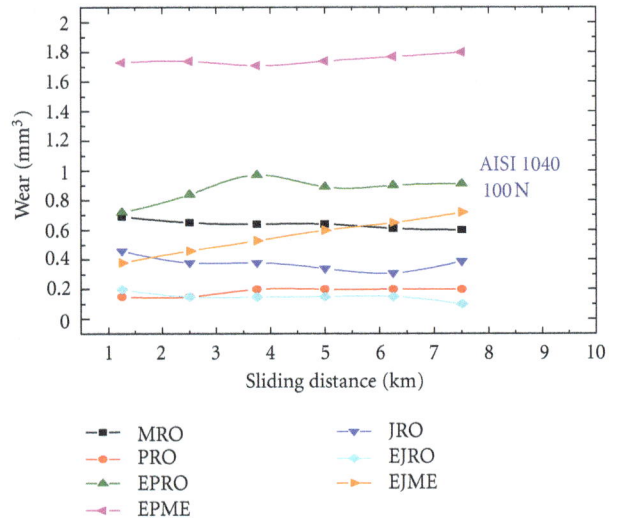

FIGURE 6: Variation of wear with sliding distance for 100 N.

at the metal surface of the mating pair or by chemical reaction of the lubricant at the surface. Since boundary lubrication by fatty acids is associated with the adsorption of the acid by dipolar attraction at the surface, they are capable of reducing the friction between the surfaces. Further, the oleic acid composition play, a vital role in reducing the friction coefficient largely when compared to linoleic and linolenic acids. Higher oleic acid composition is desirable for lower friction [21].

PRO show lower frictional values due to its polar in nature and presence of little higher oleic acid composition in the structure compared to JRO and its versions. However, epoxidized versions of Pongam oil have higher viscosity, leading to higher friction at low loads. On the other hand, as the load and temperature increases, a tribochemical reaction film could be formed on the frictional surface. The three member oxirane ring could form polyester or polyether material due to tribo-polymerization, which is tribologically effective to reduce friction [22].

3.2. Wear Behaviour of AISI 1040. Figures 5, 6, 7, and 8 show the variation of wear with sliding distance. The two raw oils show lower wear values compared to mineral oil for the complete tested sliding distance irrespective of load.

On the contrary, their epoxidized ester versions exhibit higher wear values, particularly at lower loads. The wear patterns observed here follow the same trend under friction measurement.

PRO shows 65%–98% drop in wear for various loads. EPRO exhibits 65%–70% lower wear at lower loads and 35%–95% lower values of wear at higher loads. EPME show 20%–90% higher wear under lower loads and contrast results are seen under higher loads.

JRO shows 50%–98% reduction in wear for the tested loads. EJRO exhibits maximum reduction in wear in the range 75%–98% for the tested loads. Further, under EJME, the material wears more at lower loads but, reduction in wear is seen at higher loads.

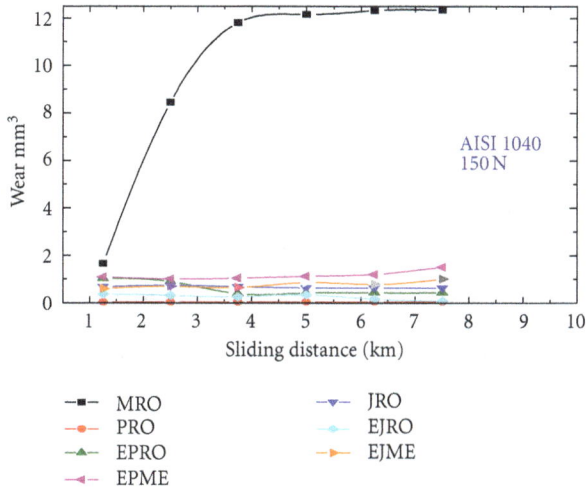

FIGURE 7: Variation of wear with sliding distance for 150 N.

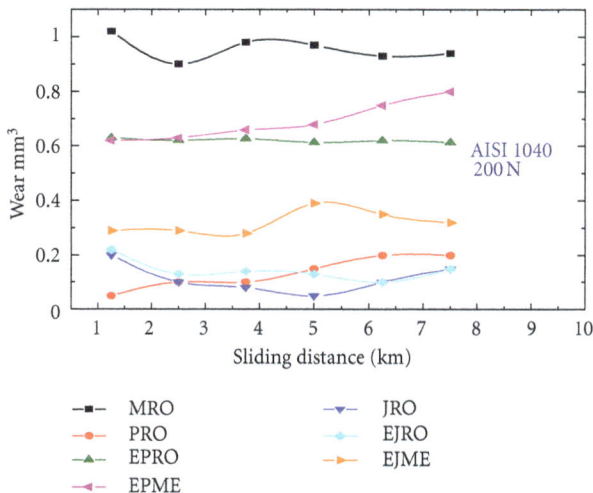

FIGURE 8: Variation of wear with sliding distance for 200 N.

More wear resistance under Pongam and Jatropha oils is observed. It is attributed to high oleic acid composition in the oils, which generates stronger adsorption on metal surfaces and produces greater lateral interaction between the ester chains [23]. Stronger adsorption capability is reported with high oleic acid content in the fatty acid composition of a vegetable oil [24]. Further, epoxidized Pongam and Jatropha forms produce polyester material due to tribo-polymerization, which reduces wear significantly, specifically at higher loads.

3.3. Stribeck Curve. Stribeck curves represent the regimes of lubrication and would throw some light on the possibility of expanding the borders of the film lubrication ranges (when moving from the right to the left over the Stribeck curve), the film lubrication will stay longer and support higher loads and in unloading cycle (moving from left to the right), the change from boundary lubrication to film lubrication will occur earlier [11]. Stribeck curves are drawn for various coefficient

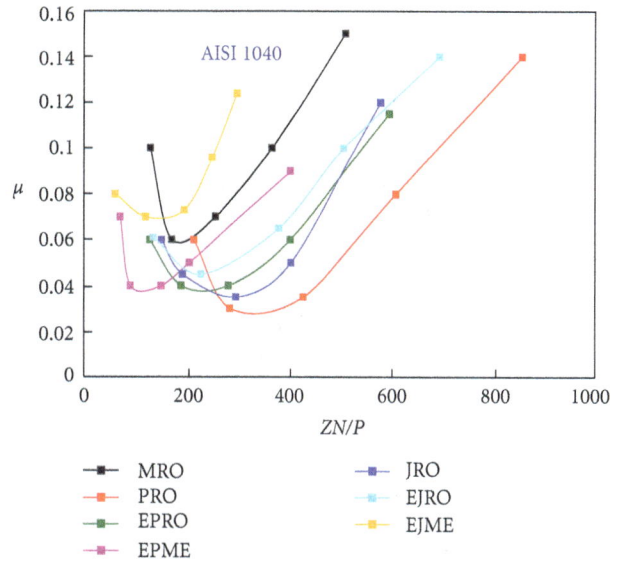

FIGURE 9: Stribeck curves for AISI 1040.

of friction (μ) and loads (P), keeping speed (N) and viscosity (Z) constant; Figure 9 represents the Stribeck curves for AISI 1040 under various lubricant modes.

Early full film formation is seen under both the vegetable oils and their versions compared to petroleum oil. This is clearly observed as the Stribeck curves for the vegetable oils are shifted towards the left. The Stribeck curve under EPRO and EPME show about 30% shift towards left and similar trends are seen under Jatropha and its versions.

Pongam and its epoxidized versions of oils are found to be better for AISI 1040 among the two vegetable oils as they generate early full film and produce low friction.

4. Conclusions

A comparative study reveals a significant reduction in friction and wear for AISI 1040 under Pongam and Jatropha raw oils compared to mineral oil. However, their modified versions of both oils exhibit slightly higher friction at low-load operations.

Pongam, Jatropha and epoxidized Jatropha raw oils offer more wear resistance for the tested sliding distances and loads compared to petroleum oil.

Both Pongam and Jatropha oils show a promising trend of extension of film lubrication with reduction in boundary lubrication regimes compared to mineral oil.

References

[1] S. Z. Erhan, B. K. Sharma, and J. M. Perez, "Oxidation and low temperature stability of vegetable oil-based lubricants," *Industrial Crops and Products*, vol. 24, no. 3, pp. 292–299, 2006.

[2] N. J. Fox and G. W. Stachowiak, "Vegetable oil-based lubricants—a review of oxidation," *Tribology International*, vol. 40, no. 7, pp. 1035–1046, 2007.

[3] L. Pop, C. Puşcaş, G. Bandur, G. Vlase, and R. Nuţiu, "Basestock oils for lubricants from mixtures of corn oil and synthetic

diesters," *Journal of the American Oil Chemists' Society*, vol. 85, no. 1, pp. 71–76, 2008.

[4] Y. M. Shashidhara and S. R. Jayaram, "Vegetable oils as a potential cutting fluid—an evolution," *Tribology International*, vol. 43, no. 5-6, pp. 1073–1081, 2010.

[5] S. Z. Erhan and S. Asadauskas, "Lubricant basestocks from vegetable oils," *Industrial Crops and Products*, vol. 11, no. 2-3, pp. 277–282, 2000.

[6] D. Hörner, "Recent trends in environmentally frinedly lubricants," *Journal of Synthetic Lubrication*, vol. 18, no. 4, pp. 327–347, 2002.

[7] H. Wagner, R. Luther, and T. Mang, "Lubricant base fluids based on renewable raw materials: their catalytic manufacture and modification," *Applied Catalysis A*, vol. 221, no. 1-2, pp. 429–442, 2001.

[8] M. A. Maleque, H. H. Masjuki, and S. M. Sapuan, "Vegetable-based biodegradable lubricating oil additives," *Industrial Lubrication and Tribology*, vol. 55, no. 2-3, pp. 137–143, 2003.

[9] M. T. Siniawski, N. Saniei, B. Adhikari, and L. A. Doezema, "Influence of fatty acid composition on the tribological performance of two vegetable-based lubricants," *Journal of Synthetic Lubrication*, vol. 24, no. 2, pp. 101–110, 2007.

[10] M. A. Kabir, "A pin-on-disc experimetal study on a green particulare-fluid lubricant," *J. Tribology*, vol. 130, pp. 01–06, 2008.

[11] R. Michel and T. M. Elektrionized, "Vegetable oils as lubricity componets in metal working lubricants," *FME Transactions*, vol. 36, pp. 133–138, 2008.

[12] M. Kalin and J. Vižintin, "A comparison of the tribological behaviour of steel/steel, steel/DLC and DLC/DLC contacts when lubricated with mineral and biodegradable oils," *Wear*, vol. 261, no. 1, pp. 22–31, 2006.

[13] L. Jiusheng, R. Wenqi, R. Tianhui, F. Xingguo, and L. Weimin, "Tribological properties of phosphate esters as additives in rape seed oil," *Journal of Synthetic Lubrication*, vol. 20, no. 2, pp. 151–158, 2003.

[14] W. Huang, B. Hou, P. Zhang, and J. Dong, "Tribological performance and action mechanism of S-[2-(acetamido)thiazol-1-yl] dialkyl dithiocarbamate as additive in rapeseed oil," *Wear*, vol. 256, no. 11-12, pp. 1106–1113, 2004.

[15] W. Castro, D. E. Weller, K. Cheenkachorn, and J. M. Perez, "The effect of chemical structure of basefluids on antiwear effectiveness of additives," *Tribology International*, vol. 38, no. 3, pp. 321–326, 2005.

[16] A. M. Petlyuk and R. J. Adams, "Oxidation stability and tribological behavior of vegetable oil hydraulic fluids," *Tribology Transactions*, vol. 47, no. 2, pp. 182–187, 2004.

[17] N. H. Jayadas, K. Prabhakaran Nair, and A. G, "Tribological evaluation of coconut oil as an environment-friendly lubricant," *Tribology International*, vol. 40, no. 2, pp. 350–354, 2007.

[18] B. K. Sharma, A. Adhvaryu, and S. Z. Erhan, "Friction and wear behavior of thioether hydroxy vegetable oil," *Tribology International*, vol. 42, no. 2, pp. 353–358, 2009.

[19] X. Wu, X. Zhang, S. Yang, H. Chen, and D. Wang, "Study of epoxidized rapeseed oil used as a potential biodegradable lubricant," *Journal of the American Oil Chemists' Society*, vol. 77, no. 5, pp. 561–563, 2000.

[20] R. A. Holser, "Transesterification of epoxidized soybean oil to prepare epoxy methyl esters," *Industrial Crops and Products*, vol. 27, no. 1, pp. 130–132, 2008.

[21] S. Bhuyan, S. Sundararajan, L. Yao, E. G. Hammond, and T. Wang, "Boundary lubrication properties of lipid-based compounds evaluated using microtribological methods," *Tribology Letters*, vol. 22, no. 2, pp. 167–172, 2006.

[22] B. K. Sharma, K. M. Doll, and S. Z. Erhan, "Oxidation, friction reducing, and low temperature properties of epoxy fatty acid methyl esters," *Green Chemistry*, vol. 9, no. 5, pp. 469–474, 2007.

[23] A. Adhvaryu, S. Z. Erhan, and J. M. Perez, "Tribological studies of thermally and chemically modified vegetable oils for use as environmentally friendly lubricants," *Wear*, vol. 257, no. 3-4, pp. 359–367, 2004.

[24] A. Adhvaryu, S. Z. Erhan, Z. S. Liu, and J. M. Perez, "Oxidation kinetic studies of oils derived from unmodified and genetically modified vegetables using pressurized differential scanning calorimetry and nuclear magnetic resonance spectroscopy," *Thermochimica Acta*, vol. 364, no. 1-2, pp. 87–97, 2000.

17

Non-Newtonian Effects on the Squeeze Film Characteristics between a Sphere and a Flat Plate: Rabinowitsch Model

Udaya P. Singh and Ram S. Gupta

Department of Applied Science and Humanities, Kamla Nehru Institute of Technology, Sultanpur 228118, India

Correspondence should be addressed to Udaya P. Singh, journals4phd@gmail.com

Academic Editor: J. Paulo Davim

The use of additives (polyisobutylene, ethylene-propylene, lithium hydroxy stearate, hydrophobic silica, etc.) changes lubricants' rheology due to which they show pseudoplastic and dilatant nature, which can be modelled as cubic stress fluid model (Rabinowitsch fluid model). The present theoretical analysis investigates the effects of non-Newtonian pseudoplastic and dilatant lubricants on the squeezing characteristics of a sphere and a flat plate. The modified Reynolds equation has been derived and an asymptotic solution for film pressure is obtained. The results for the film pressure distribution, load carrying capacity, and squeezing time characteristics have been calculated for various values of pseudoplastic parameter and compared with the Newtonian results. These characteristics show a significant variation with the non-Newtonian pseudoplastic and dilatant behavior of the fluids.

1. Introduction

Squeeze film between a sphere and plate is observed in various machine elements such as ball bearings, cam and followers, and gears. The mechanical action (squeezing, shearing, etc.) leading to the generation of high pressure at the contacts [1–4] changes the rheology of the lubricants such as viscosity and density which account for the performance characteristics of machine elements. Dowson [5], Wada and Hayashi [6], and Yadav and Kapur [7] emphasized the variation of viscosity and density with temperature and pressure and reported significant changes in bearing characteristics. Denn [8], Rajagopal [9], and Renardy [10] indicated that in high pressure lubrication applications, the variation of viscosity becomes more important than the density. Variation of the viscosity also causes the instability of the lubricants' nature by changing its shearing stress-strain rate relation due to which the estimated characteristics of lubricated contacts such as sphere-plate contacts (point contacts) may deviate from the desired value. This situation is avoided by enhancing the efficiency of stabilizing properties of lubricants by the addition of additives (polyisobutylene, ethylene propylene, etc.). The use of additives minimizes the

sensitivity of the lubricant to the change in the shearing strain rate and the lubricants behave like non-Newtonian pseudoplastic, dilatant, and viscoplastic fluids depending on the nature and quantity of the additives. To account for the effects of lubricant additives on the performance characteristics of lubricated point contacts, various non-Newtonian fluid models like power law, micropolar, and couple stress fluid models have been studied by researchers from time to time [11–14]. Among these fluid models, Rabinowitsch fluid model [6] is an established model to predict the effects of additives on the performance characteristics of the lubricated bearings. The shearing stress-strain relation in this model for one-dimensional fluid flow is given by

$$\bar{\tau}_{rz} + \bar{\kappa}\bar{\tau}_{rz}^3 = \bar{\mu}\frac{\partial \bar{u}}{\partial \bar{z}}, \tag{1}$$

where $\bar{\mu}$ is the initial viscosity of lubricant and $\bar{\kappa}$ is the nonlinear factor responsible for the non-Newtonian effects of the fluid which will be referred to as the coefficient of pseudoplasticity. This model can be applied to Newtonian, dilatant, and pseudoplastic lubricants for $\bar{\kappa} = 0$, $\bar{\kappa} < 0$, and $\bar{\kappa} > 0$, respectively. The advantage of this model lies in the fact that the theoretical analysis for the present model

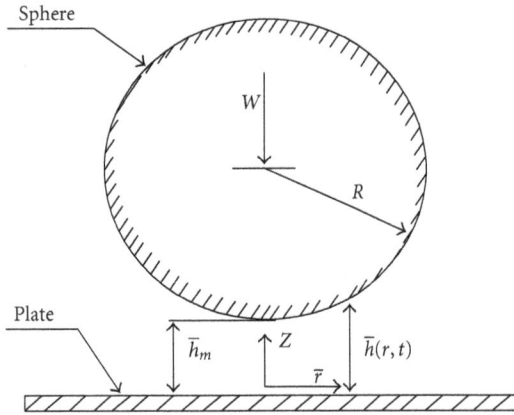

FIGURE 1: Schematic diagram of squeeze film between a sphere and a plate.

was verified with the experimental justification by Wada and Hayashi [6]. They used spindle oil as the base lubricant and concluded a decrease of dimensionless film pressure with the increase of additive (polyisobutylene). Afterwards, the theoretical study of bearing performance with non-Newtonian lubricants using this models was done by Bourgin and Gay [15] on journal bearing, Hashimoto and Wada [16] on circular plates bearing, and Lin [17] on parallel circular plates. Recently Singh et al. [18–21] used this model to study the performance characteristics of hydrostatic thrust bearings and slider bearings.

The objective of this paper is to extend the results [18–21] to squeeze film characteristics between a sphere and a plate by introducing a quantitative analysis using Rabinowitsch fluid model, which accounting for the effect of additives in the lubricant. The modified Reynolds equation governing the squeeze film pressure is derived. Squeeze film characteristics such as film pressure, load-carrying capacity, and squeezing time are presented. The importance of the present analysis lies in that the earlier theoretical investigations on sphere-plate squeezing [11–14] are based on couple stress or micropolar fluid models, which suffer the scarcity of experimental verification.

2. Constitutive Equations and Boundary Conditions

The physical configuration of a sphere-plate system is shown in Figure 1. The sphere is approaching towards the plate with a normal velocity $(-d\bar{h}/d\bar{t})$, separated by a lubricant thin film. The lubricant in the system is taken as a non-Newtonian Rabinowitsch fluid. The body forces and body couples are assumed to be absent.

Under the assumptions of hydrodynamic lubrication applicable to thin film as considered by Dowson [5], the field equations governing the one-dimensional motion of an incompressible non-Newtonian Rabinowitsch fluid model in polar coordinates (r, θ, z) system [18] are

$$\frac{1}{\bar{r}}\frac{\partial}{\partial \bar{r}}(\bar{r}\bar{u}) + \frac{\partial \bar{w}}{\partial \bar{z}} = 0, \quad (2)$$

$$\frac{\partial \bar{p}}{\partial \bar{r}} = \frac{\partial \bar{\tau}_{rz}}{\partial \bar{z}}, \quad (3)$$

$$\frac{\partial \bar{p}}{\partial \bar{z}} = 0 \quad (4)$$

which are solved under the no-slip boundary conditions:

$$\bar{u} = 0, \quad \bar{w} = 0, \quad \text{at } \bar{z} = 0,$$
$$\bar{u} = 0, \quad \bar{w} = -\frac{\partial \bar{h}}{\partial \bar{t}}, \quad \text{at } \bar{z} = \bar{h}, \quad (5)$$

where $\bar{u}$ and $\bar{w}$ are the velocity components in $\bar{r}$ and $\bar{z}$ directions, respectively, and $\bar{h}$ is the film thickness between the sphere and plate.

3. Analysis

Integrating (3) with respect to $\bar{z}$ under boundary conditions (5) and using (1), the expression for velocity $\bar{u}$ is obtained as

$$\bar{u} = \frac{1}{2\bar{\mu}}\left[\frac{\partial \bar{p}}{\partial \bar{r}}\bar{z}(\bar{z}-\bar{h}) + \bar{\kappa}\left(\frac{\partial \bar{p}}{\partial \bar{r}}\right)^3 \right.$$
$$\left. \times \left(\frac{1}{2}\bar{z}^4 - \bar{z}^3\bar{h} + \frac{3}{4}\bar{z}^2\bar{h}^2 - \frac{1}{4}\bar{z}\bar{h}^3\right)\right]. \quad (6)$$

Integrating (2) with respect to $\bar{z}$ under the relevant boundary conditions (5) for $\bar{w}$ and using (6), the modified Reynolds equation is obtained as

$$\frac{1}{\bar{r}}\frac{\partial}{\partial \bar{r}}\left[\bar{r}\left\{\bar{h}^3\frac{\partial \bar{p}}{\partial \bar{r}} + \frac{3\bar{\kappa}\bar{h}^5}{20}\left(\frac{\partial \bar{p}}{\partial \bar{r}}\right)^3\right\}\right] = -12\bar{\mu}\frac{\partial \bar{h}}{\partial \bar{t}}. \quad (7)$$

In the limiting case of $\kappa \to 0$, (7) reduces to the Newtonian form of Reynolds equation obtained by Conway and Lee [22]:

$$\frac{1}{\bar{r}}\frac{\partial}{\partial \bar{r}}\left[\bar{r}\bar{h}^3\frac{\partial \bar{p}}{\partial \bar{r}}\right] = -12\bar{\mu}\frac{\partial \bar{h}}{\partial \bar{t}}. \quad (8)$$

The expression for film thickness between the sphere and plate at a time $\bar{t}$ is taken of the form [1]

$$\bar{h} = \bar{h}_m + \frac{\bar{r}^2}{2R}, \quad (9)$$

where R denote the radius of the sphere. In case of squeezing between two spheres, the value of the radius R can be taken as (i) $R^{-1} = R_1^{-1} + R_2^{-1}$ for external contact and (ii) $R^{-1} = R_1^{-1} - R_2^{-1}$ for internal contact, where R_1 and R_2 are the radii of the spheres.

The modified Reynolds equation (7) takes the dimensionless form:

$$\frac{1}{r}\frac{d}{dr}\left[r\left\{h^3\frac{dp}{dr} + \frac{3\alpha h^5}{20}\left(\frac{dp}{dr}\right)^3\right\}\right] = -\frac{12}{\beta}, \quad (10)$$

where $p = \bar{h}_{mo}^2\bar{p}/\bar{\mu}R(d\bar{h}_m/d\bar{t})$ is the dimensionless pressure, $\alpha = \bar{\kappa}(\bar{\mu}R(d\bar{h}_m/d\bar{t})/\bar{h}_{mo}^2)^2$ is the parameter of pseudoplasticity, $\beta = \bar{h}_{mo}/R$ is the sphere parameter, $h_m = \bar{h}_m/\bar{h}_{mo}$ is the minimum film thickness at a time t, and $\overline{h}_{mo}$ is the minimum film thickness at $\overline{t} = 0$. The value of the pseudoplastic coefficient $\overline{\kappa}$ depends on the type of and the quantity of additives which can be determined experimentally [6]. Thus, the values of R, $\overline{h}_{mo}$, and $\overline{\mu}$ being known for a particular bearing and lubricant and the vales of α can be calculated with the appropriate value of $\overline{\kappa}$. However, for the validity of the present analysis, the value of α is restricted to $|\alpha| < 0.01$.

As (10) is a nonlinear equation in p, it is not easy to solve it using analytical methods. Therefore, the classical perturbation method is used to solve it. The perturbation series for p can be expressed in the form:

$$p = p_o + \alpha p_1 + \alpha^2 p_2 + \cdots. \qquad (11)$$

For $\alpha \ll 1$, it is sufficient, for analysis, to consider the first order term in α as follows:

$$p = p_o + \alpha p_1. \qquad (12)$$

For the higher values of α, second and higher order terms can be considered to increase the accuracy of the results. However, for the higher values of α, it will be more appropriate to adopt a numerical solution procedure such as the finite element method to solve the Reynolds equation.

Substituting (12) in (10), the perturbation equations are obtained as

$$\frac{1}{r}\frac{d}{dr}\left[rh^3\frac{dp_o}{dr}\right] = -\frac{12}{\beta},$$

$$\frac{1}{r}\frac{d}{dr}\left[r\left\{h^3\frac{dp_1}{dr} + \frac{3h^5}{20}\left(\frac{dp_o}{dr}\right)^3\right\}\right] = 0. \qquad (13)$$

Solving (13) under the boundary conditions

$$\frac{dp}{dr} = 0 \quad \text{at } r = 0, \qquad p = 0 \quad \text{at } r = 1, \qquad (14)$$

the dimensionless pressure developed in the film region is:

$$p = 12\beta^2\left[\frac{1}{(1+2h_m\beta)^2} - \frac{1}{(r^2+2h_m\beta)^2}\right]$$
$$- \frac{3456}{25}\alpha\beta^4\left[\frac{3+h_m\beta}{(1+2h_m\beta)^6} - \frac{3r^2+h_m\beta}{(r^2+2h_m\beta)^6}\right]. \qquad (15)$$

3.1. Load Carrying Capacity. The load carrying capacity can be obtained by integrating the film pressure over the squeezing film area as follows:

$$\overline{w} = 2\pi\int_0^R \overline{p}\overline{r}d\overline{r} \qquad (16)$$

which takes the dimensionless form:

$$w = \int_0^1 prdr, \qquad (17)$$

where

$$w = \overline{w}\left[\frac{\overline{h}_{mo}^2}{\left(2\pi\overline{\mu}R^3\left(d\overline{h}_m/d\overline{t}\right)\right)}\right]. \qquad (18)$$

3.2. Squeezing Time. The squeezing time can be calculated by integrating (18) with respect to t under the condition that $h_m = 1$ at $t = 0$ as follows:

$$t = \int_{h_f}^1 wdh = \int_{h_f}^1\int_0^1 rpdrdh_m, \qquad (19)$$

where

$$t = \frac{\overline{w}\overline{h}_{mo}}{2\pi\overline{\mu}R^3}\overline{t}. \qquad (20)$$

4. Results and Discussions

Based on the Rabinowitsch fluids model, the effects of non-Newtonian rheology on the squeeze-film characteristics between a sphere and a plate are investigated using a dimensionless parameter α which accounts for the non-Newtonian nature of the lubricant, that is, for the induced nature due to the use of additives. The parameters $\alpha = 0$, $\alpha < 0$, and $\alpha > 0$ describe the Newtonian, dilatant, and pseudoplastic lubricants, respectively. For the validity of the analysis, the numerical results for non-Newtonian lubricants are compared with the Newtonian results [12].

In order to analyze the non-Newtonian effects of fluids on the squeeze-film performance of sphere-plate system, various squeeze-film characteristics are presented with the following values:

(i) pseudoplastic parameter $\alpha = -0.01$ to 0.01, [2, 18];

(ii) sphere parameter $\beta = 0.03, 0.05$, [12, 14].

Figure 2 shows the variation of dimensionless film pressure (p) with respect to the dimensionless coordinate (r). It is clear from the figure that the pressure is the maximum at $r = 0$, that is, at the minimum film thickness and decreases towards the outer of the sphere and hence, the analysis obeys the basic theory of film pressure in sphere-plate system. Again, the pressure for dilatant lubricants ($\alpha < 0$) is higher than the pressure for Newtonian lubricants whereas the pressure for pseudoplastic lubricants ($\alpha > 0$) is lower than the pressure for Newtonian lubricants. Further, the pressure increases as α decrease from 0.01 to -0.01. It is observed that the effect of non-Newtonian pseudoplastic and dilatant lubricants produces a remarkable change in the film pressure near $r = 0$, and it decreases towards periphery $r = 1$. For design parameters $\beta = 0.05$ and $h_m = 0.7$, a small value of pseudoplastic parameter $\alpha = -0.01$ (dilatant fluids) increases the film pressure by nearly 25% at $r = 0.1$ and 3% at $r = 0.5$. For the same design parameters, the value of pseudoplastic parameter $\alpha = 0.01$ (pseudoplastic fluids) decreases the film pressure by nearly 30% at $r = 0.1$ and 2% at $r = 0.5$. It shows that the pseudoplastic and dilatant lubricants produce larger effects with the higher film pressure.

Figure 3 shows the variation of dimensionless maximum film pressure ($p_{\max}$) with respect to the dimensionless minimum film thickness (h_m). The effect of dilatant lubricants is observed to increase the maximum film pressure from

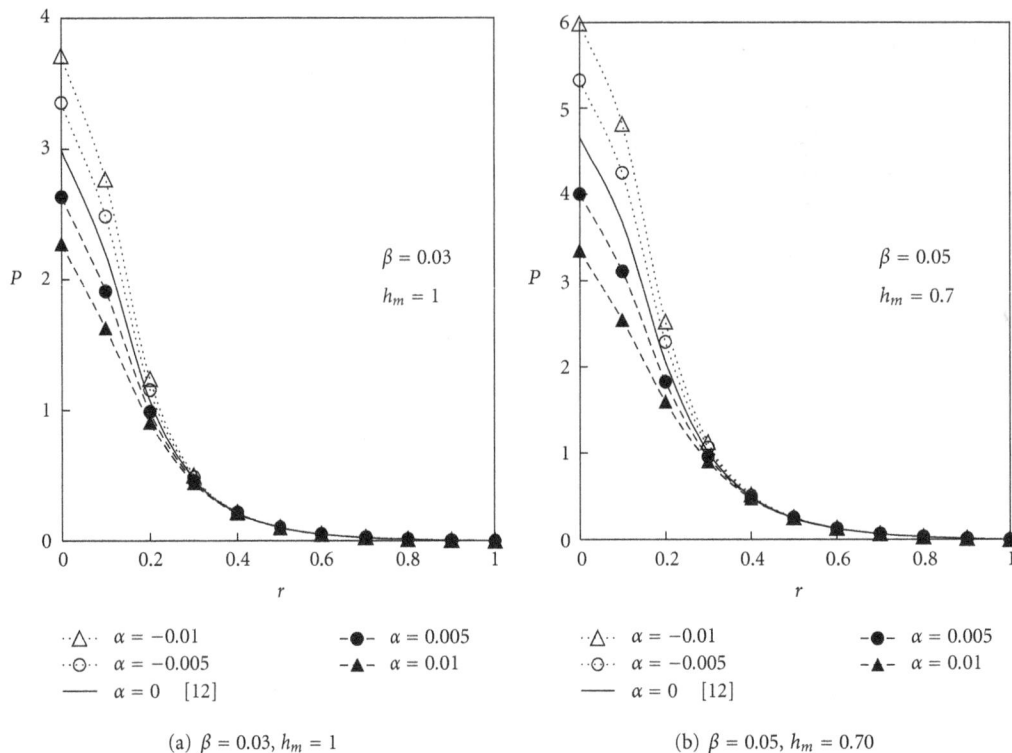

(a) $\beta = 0.03$, $h_m = 1$

(b) $\beta = 0.05$, $h_m = 0.70$

FIGURE 2: Variation of dimensionless pressure with respect to the dimensionless radius r for different values of pseudoplastic parameter α.

its value in Newtonian case whereas the effect of pseudoplastic lubricant decreases the maximum film pressure from its value in the Newtonian case. Furthermore, the maximum pressure increases with the decrease of pseudoplastic parameter α from 0.01 to −0.01. There is also a relative change in the maximum pressure for pseudoplastic and dilatant lubricants, which increases with the decrease of film thickness. For the sphere parameter $\beta = 0.05$, a small value of pseudoplastic parameter $\alpha = -0.01$ increases the maximum film pressure by nearly 12% to 25% as the minimum film thickness h_m decreases from 1 to 0.7. For the same value of β and h_m, the value of pseudoplastic parameter $\alpha = 0.01$ (pseudoplastic fluids) decreases the film pressure nearly 15% to 30%. Therefore, it can be safely said that the higher the film pressure, the greater the change produced by pseudoplastic and dilatant lubricants is.

Figure 4 shows the variation of dimensionless load carrying capacity (W) of the system with respect to the dimensionless minimum film thickness (h_m). It is observed that the load capacity obtained with dilatant lubricants is higher than that with Newtonian lubricants, and the load capacity obtained with pseudoplastic lubricants is lower than its value obtained with Newtonian lubricants. Furthermore, the load capacity increases with the decrease of pseudoplastic parameter α from 0.01 to −0.01. It is also observed that there is a relative change in load capacity obtained with the different values of pseudoplastic parameter, which increases with the decrease of film thickness. For the sphere parameter $\beta = 0.05$, the effect of dilatant lubricant $\alpha = -0.01$ increases the load capacity by 10% to 15% as the minimum film

thickness h_m decreases from 1 to 0.7 and for the same value of β and h_m, the effect of pseudoplastic lubricants $\alpha = 0.01$ decreases the film pressure from 15% to 20%.

Figure 5 shows the time (t) elapsed in squeezing the film from its initial thickness $h_m = 1$ to a final thickness $h_m = h_f$. It is observed that for each value of h_f and β, the squeeze time for dilatant lubricants is longer than that with the Newtonian lubricants, whereas, the squeeze time for pseudoplastic lubricants is less than its value in the Newtonian case. Further, the squeeze time increases with the decrease of pseudoplastic parameter from 0.01 to −0.01. This phenomenon can be interpreted as a result of the increase in the film pressure from pseudoplastic to dilatant lubricants. For the sphere parameter $\beta = 0.03$, the time to squeeze the film to $h_f = 0.7$ is increased by nearly 23% with dilatant lubricant $\alpha = -0.01$ in comparison with the Newtonian case, whereas the same is decreased by nearly 18% for pseudoplastic lubricants. For the sphere parameter $\beta = 0.05$, the time to squeeze the film from $h = 1$ to $h = 0.7$ is increased by nearly 14% with dilatant lubricant $\alpha = -0.01$ in comparison with the Newtonian case. For the same value of β and h_f, the squeeze time is reduced by nearly 16% with pseudoplastic lubricants $\alpha = 0.01$. Thus, dilatant lubricants increase and pseudoplastic lubricants reduce the squeeze time of the bearing.

5. Conclusions

Based on the Rabinowitsch fluid model (cubic stress model) for non-Newtonian pseudoplastic and dilatant fluids, the

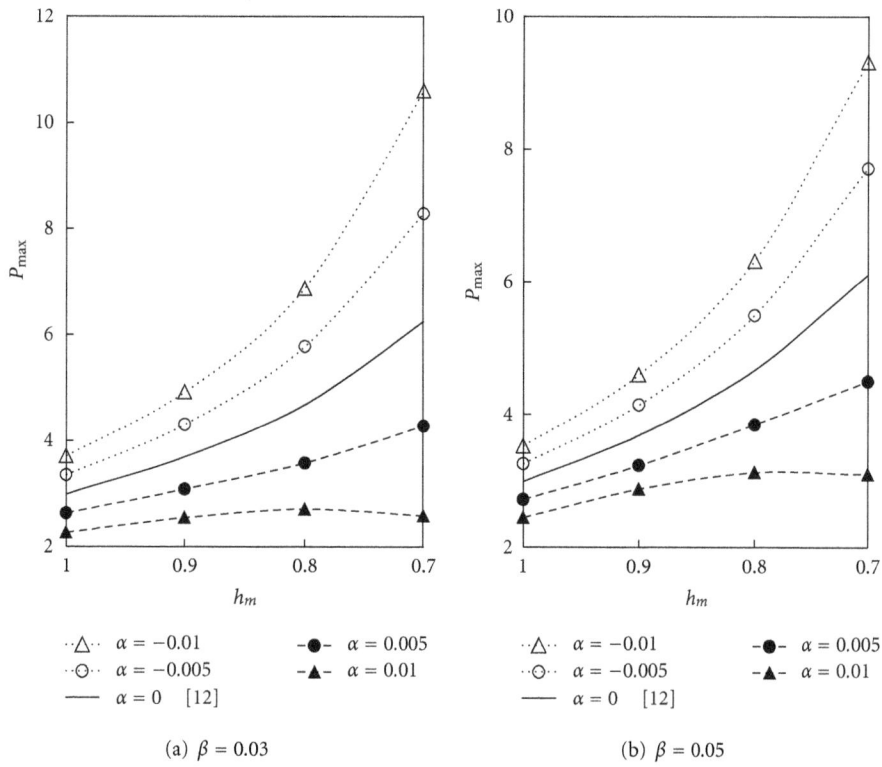

FIGURE 3: Variation of the dimensionless maximum film pressure with respect to the minimum film thickness h_m for different values of parameter α.

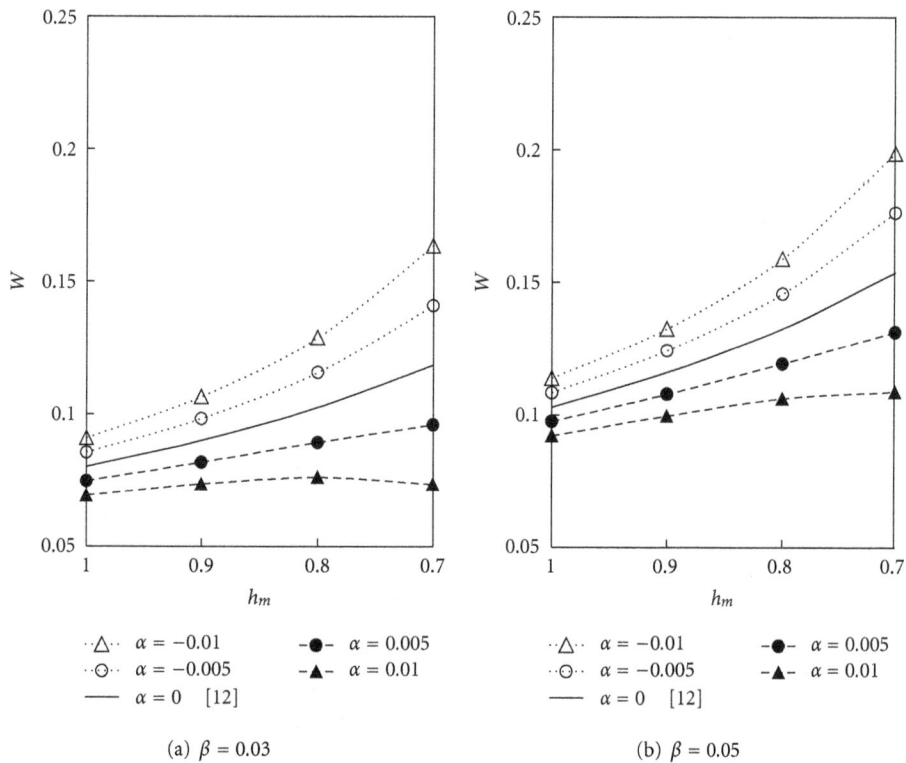

FIGURE 4: Variation of the dimensionless load capacity with respect to the dimensionless minimum film thickness h_m for values of parameter α.

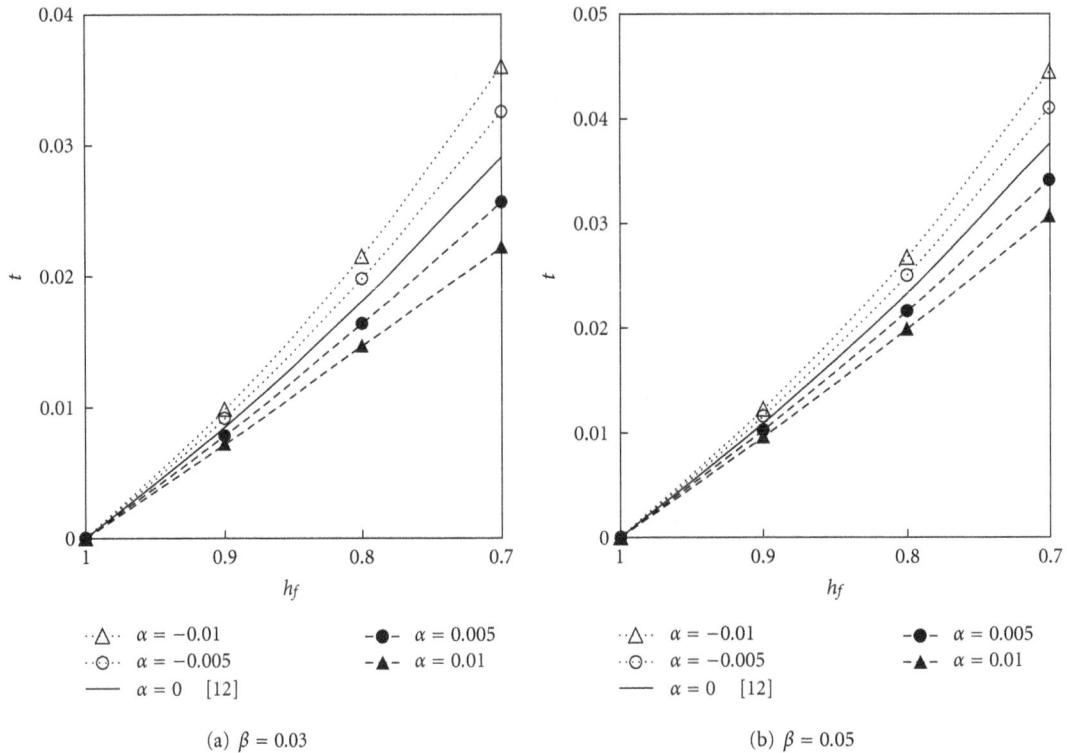

(a) $\beta = 0.03$

(b) $\beta = 0.05$

Figure 5: Variation of dimensionless squeeze time with respect to the squeezed film thickness h_f for different values of parameter α.

effects of lubricant additives on the performance characteristics of squeezing film between a sphere and a plate are presented avoiding the inertia and cavitation effects. The analytical solution for pressure distribution is obtained using a classical perturbation technique. Based on the present theoretical analysis, the following results have been drawn.

(1) Dilatant lubricants increase the pressure and load carrying capacity significantly, whereas the case is reversed with the pseudoplastic lubricants.

(2) On comparing with the Newtonian case, dilatant lubricants increase the squeeze time, whereas the pseudoplastic lubricants decrease it.

(3) As the squeezing time of the sphere-plate system is significantly increased with the dilatant lubricants, it is expected that the use of additives can reduce the vibration in the sphere-plate systems.

Thus, the present analysis can also provide a guideline to control the vibration in the system. Hence, the results are expected to be more helpful for better bearing performance and stability. However, an experimental validation of these results is required at laboratory level.

Nomenclature

$-$:	Bar denotes the dimensional quantities
$\bar{h}, h$:	Film thickness defined in (9), $h = \bar{h}/h_{mo}$
$\bar{h}_m, h_m$:	Minimum film thickness, $h_m = \bar{h}_m/\bar{h}_{mo}$
$\bar{h}_{mo}$:	Initial minimum film thickness
$\bar{p}, p$:	Film pressure, $p = \bar{h}_{mo}^2 \bar{p}/\bar{\mu}R(d\bar{h}_m/d\bar{t})$
p_o, p_1:	Dimensionless perturbed film pressures
$\bar{r}, r$:	Radial coordinate, $r = \bar{r}/R$
R:	Radius of sphere
$\bar{t}, t$:	Time, $t = (\overline{W}\bar{h}_{mo}/2\bar{\mu}BR^2)\bar{t}$
$\bar{u}, \bar{w}$:	Components of velocity
$\overline{W}, W$:	Load capacity, $W = (h_{mo}^2/(2\bar{\mu}BR^2(d\bar{h}_m/d\bar{t})))\overline{W}$
α:	$\bar{\kappa}[\bar{\mu}R(d\bar{h}_m/d\bar{t})/\bar{h}_{mo}^2]^2$
β:	Design parameter $(\bar{h}_{mo}/R)$
$\bar{\kappa}$:	Coefficient pseudoplasticity
$\bar{\mu}$:	Viscosity of lubricant
$\bar{\tau}_{rz}$:	Stress component.

Acknowledgments

The authors, hereby, thank Dr. M. Fillon (Director of Research, Centre National de la Recherche Scientifique, University of Poitiers) and Dr. V. K. Kapur (Former Professor and Chairman, KNIT, Sultanpur, India) for providing useful materials and guidelines to enhance the content of the paper.

References

[1] M. D. Pascovici, C. S. Popescu, and V. G. Marian, "Impact of a rigid sphere on a highly compressible porous layer imbibed with a Newtonian liquid," *Proceedings of the Institution of Mechanical Engineers J*, vol. 224, no. 8, pp. 789–795, 2010.

[2] L. E. Goodman and G. E. Bowie, "Experiments on damping at contacts of a sphere with flat plates—Test results for the constant normal force/varying tangential force problem are compared with theoretical predictions," *Experimental Mechanics*, vol. 1, no. 2, pp. 48–54, 1961.

[3] H. A. Kalameh, A. Karamali, C. Anitescu, and T. Rabczuk, "High velocity impact of metal sphere on thin metallic plate using smooth particle hydrodynamics (SPH) method," *Frontiers of Structural and Civil Engineering*, vol. 6, no. 2, pp. 101–110, 2012.

[4] G. H. Meeten, "Squeeze flow between plane and spherical surfaces," *Rheologica Acta*, vol. 40, no. 3, pp. 279–288, 2001.

[5] D. Dowson, "Inertia effects in hydrostatic thrust bearings," *Journal of Basic Engineering*, vol. 83, no. 2, pp. 227–234, 1961.

[6] S. Wada and H. Hayashi, "Hydrodynamic lubrication of journal bearings by pseudo- plastic lubricants," *Bulletin of JSME*, vol. 14, no. 69, pp. 279–286, 1971.

[7] J. S. Yadav and V. K. Kapur, "On the viscosity variation with temperature and pressure in thrust bearing," *International Journal of Engineering Science*, vol. 19, no. 2, pp. 269–277, 1981.

[8] M. M. Denn, *Polymer Melt Processing*, Cambridge University Press, Cambridge, UK, 2008.

[9] K. R. Rajagopal, "On implicit constitutive theories for fluids," *Journal of Fluid Mechanics*, vol. 550, pp. 243–249, 2006.

[10] M. Renardy, "Parallel shear flows of fluids with a pressure-dependent viscosity," *Journal of Non-Newtonian Fluid Mechanics*, vol. 114, no. 2-3, pp. 229–236, 2003.

[11] J. R. Gregory, "Squeeze film between two spheres in a power-law fluid," *Journal of Non-Newtonian Fluid Mechanics*, vol. 63, no. 2-3, pp. 141–152, 1996.

[12] J. R. Lin, L. J. Liang, and L. M. Chu, "Effects of non-Newtonian micropolar fluids on the squeeze-film characteristics between a sphere and a plate surface," *Proceedings of the Institution of Mechanical Engineers J*, vol. 224, no. 8, pp. 825–832, 2010.

[13] N. B. Naduvinamani, P. S. Hiremath, and G. Gurubasavaraj, "Effect of surface roughness on the couple-stress squeeze film between a sphere and a flat plate," *Tribology International*, vol. 38, no. 5, pp. 451–458, 2005.

[14] A. A. Elsharkawy and K. J. Al Fadhalah, "Squeeze film characteristics between a sphere and a rough porous flat plate with micropolar fluids," *Lubrication Science*, vol. 23, no. 1, pp. 1–18, 2011.

[15] P. Bourgin and B. Gay, "Determination of the load capacity of a finite width journal bearing by a finite-element method in the case of a non-newtonian lubricant," *Journal of Tribology*, vol. 106, no. 2, pp. 285–290, 1984.

[16] H. Hashimoto and S. Wada, "Effects of fluid inertia forces in parallel circular squeeze film bearings lubricated with pseudo-plastic fluids," *Journal of Tribology*, vol. 108, no. 2, pp. 282–287, 1986.

[17] J. R. Lin, "Non-newtonian effects on the dynamic characteristics of one-dimensional slider bearings: rabinowitsch fluid model," *Tribology Letters*, vol. 10, no. 4, pp. 237–243, 2001.

[18] U. P. Singh, R. S. Gupta, and V. K. Kapur, "On the steady performance of hydrostatic thrust bearing: rabinowitsch fluid model," *Tribology Transaction*, vol. 54, no. 5, pp. 723–729, 2011.

[19] U. P. Singh, R. S. Gupta, and V. K. Kapur, "Effects of inertia in the steady state pressurised flow of a non-newtonian fluid between two curvilinear surfaces of revolution: rabinowitsch fluid model," *Chemical and Process Engineering*, vol. 32, no. 4, pp. 333–349, 2011.

[20] U. P. Singh, R. S. Gupta, and V. K. Kapur, "On the steady performance of annular hydrostatic thrust bearing: rabinowitsch fluid model," *Journal of Tribology*, vol. 134, no. 4, Article ID 044502, 5 pages, 2012.

[21] U. P. Singh, R. S. Gupta, and V. K. Kapur, "On the performance of pivoted curved slider bearings: rabinowitsch fluid model," *Tribology in Industry*, vol. 34, no. 3, pp. 127–136, 2012.

[22] H. D. Conway and H. C. Lee, "Impact of a lubricated surface by a sphere," *Journal of Tribology*, vol. 97, no. 4, pp. 613–615, 1975.

18

Optimum Groove Location of Hydrodynamic Journal Bearing Using Genetic Algorithm

Lintu Roy and S. K. Kakoty

Indian Institute of Technology Guwahati, Guwahati 781309, India

Correspondence should be addressed to S. K. Kakoty; sashin@iitg.ernet.in

Academic Editor: Michel Fillon

This paper presents the various arrangements of grooving location of two-groove oil journal bearing for optimum performance. An attempt has been made to find out the effect of different configurations of two groove oil journal bearing by changing groove locations. Various groove angles that have been considered are 10°, 20,° and 30°. The Reynolds equation is solved numerically in a finite difference grid satisfying the appropriate boundary conditions. Determination of optimum performance is based on maximization of nondimensional load, flow coefficient, and mass parameter and minimization of friction variable using genetic algorithm. The results using genetic algorithm are compared with sequential quadratic programming (SQP). The two grooved bearings in general have grooves placed at diametrically opposite directions. However, the optimum groove locations, arrived at in the present work, are not diametrically opposite.

1. Introduction

Journal bearings are used extensively in rotating machines because of their low wear and good damping characteristics. Fluid-film journal bearings are available to support a rotating shaft in a turbo machinery system. A full circular journal bearing has a much simple configuration but exhibits instability at higher rotational speeds. It is relatively less expensive compared to the multilobe bearings. It is well known that whirl instability occurs at high speed in oil journal bearing. Present day bearings, at over increasing speeds and loads, confront the engineer with many new problems. Excessive power losses reduce the efficiency of the engine, and high bearing temperature poses a danger to material of the bearing as well as the lubricant. Instability arising mainly in the form of oil whip may ruin not only the bearing but the machine itself. New bearing designs are sought to meet the new requirements. A journal bearing fed by two axial grooves has a wide practical application due to its good load carrying capacity and ability to operate when reversal of shaft rotation occurs [1]. These bearing usually have the grooves positioned orthogonal to the predominant load direction. Among the previous works on two axial

groove oil journal bearings; Klit and Lund [2] used finite element method to find dynamic coefficients of plain circular bearing with two 20° axial grooves. Gethin and Deihi [3] studied the effect of loading direction on the performance of a twin-axial groove cylindrical bore bearing. It has been anticipated that, if the bearing is loaded into the groove, its load carrying ability will be diminished, but the effect on hydrodynamic lubricant flow and power loss is not so obvious. If the positions of the grooves are arranged for carrying a relatively higher load, then the likelihood of bearing instability reduces, since the journal will run more eccentrically. Again hydrodynamic leakage and friction are affected by the direction of loading. So a question arises: where the position of the groove should lie so as to give the optimum load capacity, flow, friction, and critical speed. A new technique for optimizing hybrid journal bearings was presented by Rowe and Koshal [4]. The method involved the comparison of the bearings to be optimized with a reference bearing on the basis of load/total power, load/pumping power, and load/flow. Lin and Noah [5] used genetic algorithm to optimize the performance of a hydrodynamic journal bearing. Hashimoto and Matsumoto [6] described the optimum design methodology for improving operating characteristics of hydrodynamic journal bearings. The hybrid optimization technique combining the direct search method and the successive quadratic programming has been applied to find the optimum design of elliptical journal bearings. Boedo and Eshkabilov [7] described the implementation of a genetic algorithm suitable for the optimal shape design of finite-width, isoviscous, fluid film journal bearings under steady load and steady journal rotation. Hirani [8] formulated a problem to minimize temperature rise, power loss, and oil flow. An evolution-based optimization methodology for cylindrical journal bearings had been applied for journal bearings.

David et al. [9] in their paper presented the basic concepts of traditional genetic algorithm, its advantages with variety of applications. The paper also pointed out advanced features and future directions. McCall [10] presented genetic algorithms (GAs), a heuristic search and optimization technique inspired by natural evolution. GAs have been successfully applied to a wide range of real-world problems of significant complexity. When there are hundreds of publications on application of GAs, only couple of representative publications are cited here.

It has been observed that GAs have been successfully applied for optimizing bearing performance. However, the performance of two-groove journal bearing has not been optimized pertaining to location of groove positions with multiple objectives. In view of this, an attempt has been made in this paper to obtain an optimum configuration of the two grooves positions around the circumference of the hydrodynamic journal bearing for maximum oil flow, minimum friction loss, maximum load bearing capacity, and maximum critical speed vis-à-vis mass parameter, a function of speed.

1.1. Oil Flow. The oil flow rate depends on several factors, such as the viscosity of the lubricant, the geometry (length, diameter, and radial clearance) of the bearing, operating eccentricity, the inlet oil pressure, the arrangement of feeding sources, and groove location of the bearing. The pressure developed in the film due to journal motion also contributes to the flow. An adequate oil flow takes away frictional heat and does not allow rapid rise in temperature.

1.2. Friction Loss. The calculation of friction loss within a bearing oil film is an integral part of the design of the bearing. The friction loss appears as heat, raises the temperature of the lubricant and lowers its viscosity, which is a key parameter of the bearing analysis. Therefore, the accurate prediction of friction loss is desired. The friction force is calculated by integrating shear stress over the journal surface. It is desired to keep the friction loss at minimum.

1.3. Load Carrying Capacity. The load carrying capacity of the bearing within a bearing is developed due to pressure developed in the film. For a more accurate analysis, careful consideration of film extent needs to be included. This is expected to influence hydrodynamic leakage significantly and load carrying ability under some circumstances. If the feeding groove (in which pressure is zero) falls in the load carrying film, this part of the bearing makes no contribution to the load-carrying ability. Thus the location of the groove plays a role in determining the load carrying ability of the bearing.

1.4. Critical Speed of Instability. Plain circular bearing is mostly replaced by some other bearings, as plain bearing does not suit the stability requirements of high-speed machines and precision machine tools. Grooved circular bearings and multilobe bearings with two lobes, three lobes, and four lobes are commonly used. The critical mass parameter (a measure of stability) is a function of speed. The higher the critical speed is, the higher the stability limit is. The larger the eccentricity ratio is, the more stable the shaft is. If the eccentricity ratio is larger than 0.8, in particular, the shaft is always stable. In engineering analysis it is essential to know the critical speed at which oil whirl occurs and avoid it during operation. It has been found that severe whirl occurs when the shaft speed is approximately twice the bearing critical frequency.

1.5. Selection Procedure. To facilitate the optimum bearing design in the present paper, the nondimensional values of flow coefficient, load, and mass parameters along with friction variables for different configurations in groups are estimated. The optimum performance is determined on the basis of maximization of flow, load, mass parameter, and minimization of friction variable.

2. Theory

The Reynolds equation in two dimensions for an incompressible fluid is the governing equation. It can be written in a dimensionless form as

$$\frac{\partial}{\partial \theta}\left(\bar{h}^3 \frac{\partial \bar{p}}{\partial \theta}\right) + \left(\frac{D}{L}\right)^2 \left(\bar{h}^3 \frac{\partial^2 \bar{p}}{\partial \bar{z}^2}\right) = \frac{\partial \bar{h}}{\partial \theta} + 2\lambda \frac{\partial \bar{h}}{\partial \tau}, \quad (1)$$

where,

$$\theta = \frac{x}{R}, \qquad \bar{z} = \frac{z}{L/2}, \qquad \bar{h} = \frac{h}{C}, \qquad \bar{p} = \frac{pC^2}{6\eta UR},$$

$$\tau = \omega_p t, \qquad \lambda = \frac{\omega_p}{\omega}. \quad (2)$$

The pressure and film thickness can be expressed for small amplitude of vibration as

$$\bar{p} = \bar{p}_0 + \varepsilon_1 e^{i\tau} \bar{p}_1 + \varepsilon_0 \phi_1 e^{i\tau} \bar{p}_2,$$

$$\bar{h} = \bar{h}_0 + \varepsilon_1 e^{i\tau} \cos \theta + \varepsilon_0 \phi_1 e^{i\tau} \sin \theta. \quad (3)$$

Substitution of (3) into (1) and retaining the first order terms and by equating the coefficients of ε_0, $\varepsilon_1 e^{i\tau}$, and $\varepsilon_0 \phi_1 e^{i\tau}$, three differential equations in $\overline{p}_0$, $\overline{p}_1$, and $\overline{p}_2$ are obtained as shown in the following:

$$\frac{\partial}{\partial \theta}\left(\overline{h}_0^3 \frac{\partial \overline{p}_0}{\partial \theta}\right) + \left(\frac{D}{L}\right)^2 \frac{\partial}{\partial \overline{z}}\left(\overline{h}_0^3 \frac{\partial \overline{p}_0}{\partial \overline{z}}\right) = \frac{\partial \overline{h}_0}{\partial \theta}, \quad (4)$$

$$\frac{\partial}{\partial \theta}\left(\overline{h}_0^3 \frac{\partial \overline{p}_1}{\partial \theta}\right) + \left(\frac{D}{L}\right)^2 \frac{\partial}{\partial \overline{z}}\left(\overline{h}_0^3 \frac{\partial \overline{p}_1}{\partial \overline{z}}\right)$$
$$+ 3\frac{\partial}{\partial \theta}\left(\overline{h}_0^2 \frac{\partial \overline{p}_0}{\partial \theta}\cos\theta\right) + \left(\frac{D}{L}\right)^2 \frac{\partial}{\partial \overline{z}}\left(\overline{h}_0^2 \frac{\partial \overline{p}_0}{\partial \overline{z}}\cos\theta\right)$$
$$= -\sin\theta + i2\lambda\cos\theta, \quad (5)$$

$$\frac{\partial}{\partial \theta}\left(\overline{h}_0^3 \frac{\partial \overline{p}_2}{\partial \theta}\right) + \left(\frac{D}{L}\right)^2 \frac{\partial}{\partial \overline{z}}\left(\overline{h}_0^3 \frac{\partial \overline{p}_2}{\partial \overline{z}}\right)$$
$$+ 3\frac{\partial}{\partial \theta}\left(\overline{h}_0^2 \frac{\partial \overline{p}_0}{\partial \theta}\sin\theta\right) + \left(\frac{D}{L}\right)^2 \frac{\partial}{\partial \overline{z}}\left(\overline{h}_0^2 \frac{\partial \overline{p}_0}{\partial \overline{z}}\sin\theta\right)$$
$$= -\cos\theta + i2\lambda\sin\theta. \quad (6)$$

Boundary conditions used for the steady state pressure and dynamic pressure distribution are as follows:

$$\frac{\partial \overline{p}_i}{\partial \theta} = 0, \quad \overline{p}_i = 0 \text{ at } \theta = \theta_r,$$
$$\overline{p}_i(\theta, \overline{z}) = 0, \quad \text{when } \theta_s \le \theta \le \theta_e, \quad (7)$$

where, $\overline{p}_i = \overline{p}_0$, $\overline{p}_1$, $\overline{p}_2$ and θ_s: starting angle of the groove with respect to the vertical axis, θ_e: angle at which the groove ends with respect to the vertical axis, and θ_r: angle at which the film cavitates with respect to the vertical axis.

The nondimensional steady state load components as well as the nondimensional steady state load are given by

$$\overline{W}_{X_0} = \int_{\theta_1}^{\theta_2}\int_0^1 \overline{p}_0 \cos\theta \, d\theta \, d\overline{z}, \quad (8)$$

$$\overline{W}_{Z_0} = \int_{\theta_1}^{\theta_2}\int_0^1 \overline{p}_0 \sin\theta \, d\theta \, d\overline{z}, \quad (9)$$

$$\overline{W}_0 = \sqrt{\overline{W}_{X_0}^2 + \overline{W}_{Z_0}^2}. \quad (10)$$

Equation (4) is solved for the steady state pressure distribution ($\overline{p}_0$), discretizing in a finite difference grid of size 88×14 and using Gauss-Seidel method with successive overrelaxation (SOR) technique satisfying the boundary conditions. The convergence criterion adopted for pressure calculation is $|1 - \sum \overline{P}_{\text{old}}/\sum \overline{P}_{\text{new}}| \le 10^{-5}$. Chosen bearing eccentricity and arbitrary attitude angle picked at random result in magnitude of forces generated due to pressure wedge in the bearing. The attitude angle is changed till the horizontal force component ($\overline{W}_{Z_0}$) in the pressure wedge becomes zero. This eventually locates the attitude angle. For this equilibrium position the vertical force ($\overline{W}_{X_0}$) gives the load capacity, $\overline{W}_0$. The Sommerfeld number is given by $S = 1/\pi\overline{W}_0$.

The flow coefficient in the dimensionless form can be written as

$$\overline{q}_Z = \frac{1}{2}\left(\frac{D}{L}\right)^2 \int_0^{2\pi} \overline{h}_0^3 \frac{\partial \overline{p}_0}{\partial \overline{z}} d\theta. \quad (11)$$

The friction variable is given by

$$\overline{\mu} = \mu\left(\frac{R}{C}\right) = \frac{\int_0^{2\pi}\left(3\overline{h}(\partial \overline{p}_0/\partial \theta) + 1/\overline{h}\right)d\theta}{6\overline{W}}. \quad (12)$$

Equations (5) and (6) for $\overline{p}_1$ and $\overline{p}_2$ are solved satisfying the boundary conditions and known values of $\overline{p}_0$ using the same procedure used for calculating steady state pressure. Dynamic loads due to $\overline{p}_1$ and $\overline{p}_2$ are given by

$$\overline{W}_{X1} = \int_{\theta_1}^{\theta_2}\int_0^1 \overline{p}_1 \cos\theta \, d\theta \, d\overline{z},$$

$$\overline{W}_{Z1} = \int_{\theta_s}^{\theta_e}\int_0^1 \overline{p}_1 \sin\theta \, d\theta \, d\overline{z},$$

$$\overline{W}_{X2} = \int_{\theta_1}^{\theta_2}\int_0^1 \overline{p}_2 \cos\theta \, d\theta \, d\overline{z}, \quad (13)$$

$$\overline{W}_{Z2} = \int_{\theta_s}^{\theta_e}\int_0^1 \overline{p}_2 \sin\theta \, d\theta \, d\overline{z}.$$

It is found that the fluid film, which supports the bearing, is equivalent to a spring mass damping system. Since the journal executes small harmonic oscillations about its steady state position, the dynamic load carrying capacity can be expressed as a spring and a viscous damping force. The stiffness and damping coefficients are given by

$$\overline{K}_{XX} = -\text{Re}(\overline{W}_{X1}); \quad \overline{K}_{ZX} = -\text{Re}(W_{Zt1});$$
$$\overline{K}_{XZ} = -\text{Re}(\overline{W}_{Xt2}); \quad \overline{K}_{ZZ} = -\text{Re}(\overline{W}_{Zt2}),$$
$$\overline{C}_{XX} = -\text{Im}(\overline{W}_{Xt1}); \quad \overline{C}_{ZX} = -\text{Im}(\overline{W}_{Zt1}); \quad (14)$$
$$\overline{C}_{XZ} = -\text{Im}(\overline{W}_{Xt2}); \quad \overline{C}_{ZZ} = -\text{Im}(\overline{W}_{Zt2}).$$

2.1. Mass Parameter and Whirl Ratio. The nondimensional linearised equations of journal motion can be written as [11]

$$\overline{M}\Delta\ddot{\overline{X}}_0 + \overline{K}_{XX}\Delta\overline{X} + \overline{K}_{XZ}\Delta\overline{Z} + \overline{C}_{XX}\Delta\dot{\overline{X}} + \overline{C}_{XZ}\Delta\dot{\overline{Z}} = 0,$$
$$\overline{M}\Delta\ddot{\overline{Z}}_0 + \overline{K}_{ZX}\Delta\overline{X} + \overline{K}_{ZZ}\Delta\overline{Z} + \overline{C}_{ZX}\Delta\dot{\overline{X}} + \overline{C}_{ZZ}\Delta\dot{\overline{Z}} = 0, \quad (15)$$

where $\overline{M} = mC\omega^2/W$, a nondimensional mass parameter.

Now, $(\overline{x}_0, \overline{z}_0)$ is the steady state equilibrium position (nondimensional) of the journal. $(\Delta\overline{X}, \Delta\overline{Z})$ is the perturbed amount from this position at a nondimensional time "τ". The instantaneous positions are given by

$$\overline{x} = \overline{x}_0 e^{i\lambda\tau}, \quad \overline{z} = \overline{z}_0 e^{i\lambda\tau},$$
$$\dot{\overline{x}} = i\lambda\overline{x}_0 e^{i\lambda\tau}, \quad \dot{\overline{z}} = i\lambda\overline{z}_0 e^{i\lambda\tau}, \quad (16)$$
$$\ddot{\overline{x}} = -\lambda^2\overline{x}_0 e^{i\lambda\tau}, \quad \ddot{\overline{z}} = -\lambda^2\overline{z}_0 e^{i\lambda\tau}.$$

Substituting the above nondimensional terms in the equations of motion (see (15)), a characteristic equation is formed to find a non-trivial solution. Solving the characteristic equation, the following expressions for the mass parameter, $\overline{M}$, and the whirl ratio, λ, are arrived at

$$\lambda^2 \overline{M} = \frac{\overline{K}_{XX}\overline{C}_{ZZ} + \overline{K}_{ZZ}\overline{C}_{XX} - \left(\overline{K}_{XZ}\overline{C}_{ZX} + \overline{K}_{ZX}\overline{C}_{XZ}\right)}{\overline{C}_{XX} + \overline{C}_{ZZ}} = k_0.$$
(17)

So,

$$\lambda^2 = \frac{\left(\overline{K}_{XX} - k_0\right)\left(\overline{K}_{ZZ} - k_0\right) - \overline{K}_{XZ}\overline{K}_{ZX}}{\overline{C}_{XX}\overline{C}_{ZZ} - \overline{C}_{XZ}\overline{C}_{ZX}}, \qquad \overline{M} = \frac{k_0}{\lambda^2}.$$
(18)

3. Optimization Techniques

It has been found that the location of the groove has an influence on flow ($\overline{q}_Z$), frictional variable ($\overline{\mu}$), load carrying capacity ($\overline{W}$), and mass parameter ($\overline{M}$). Genetic algorithm (GA) is the most popular stochastic method used to find the optimum solution for all kinds of problems. The most striking difference between GAs and many traditional optimization methods is that GAs work with a population of points instead of a single point. On the other hand, since GAs require only function values at various discrete points, a discrete or discontinuous function can be handled with no extra burden. This allows GAs to be applied to a wide variety of problems. Another advantage with a population-based search algorithm is that multiple optimal solutions can be captured in the population easily, thereby reducing the effort to use the same algorithm many times. Genetic algorithms perform a multiple directional search by maintaining a population of potential solutions. The population-to-population approach attempts to make the search escape from local optima [6]. GAs are very helpful when the developer does not have precise domain expertise because GAs possess the ability to explore and learn from their domain.

3.1. Multiobjective Problem Formulation.
The problem is framed with four objectives. The variables used in the problem are in case-I starting angle of first groove (θ_1), starting angle of second groove (θ_2). The optimum configurations obtained for an eccentricity ratio range from 0.1 to 0.9 in this case. In case-II, the eccentricity ratio (ε), starting angle of first groove (θ_1), and starting angle of second groove (θ_2) are variables and act as chromosome, the groove angles being $10°$ in both cases. It has been found that for $10°$ groove angle the pressure development as well as load carrying capacity is higher in comparison with $20°$ and $30°$. The objectives are minimization of friction variable ($\overline{\mu}$), Equation (12), maximization of load capacity ($\overline{W}$), Equation (10), flow coefficient ($\overline{q}_Z$), Equation (11), maximization of mass parameter ($\overline{M}$), and Equation (17); objective function framing is same for both cases, and variable bounds are shown in Table 1.

TABLE 1: Variable bounds for the bearing problem.

Case	Variable	Lower bound	Upper bound
I	Starting angle of first groove	$0°$	$180°$
	Starting angle of second groove	$170°$	$350°$
II	ε	0.1	0.9
	Starting angle of first groove	$0°$	$180°$
	Starting angle of second groove	$170°$	$350°$

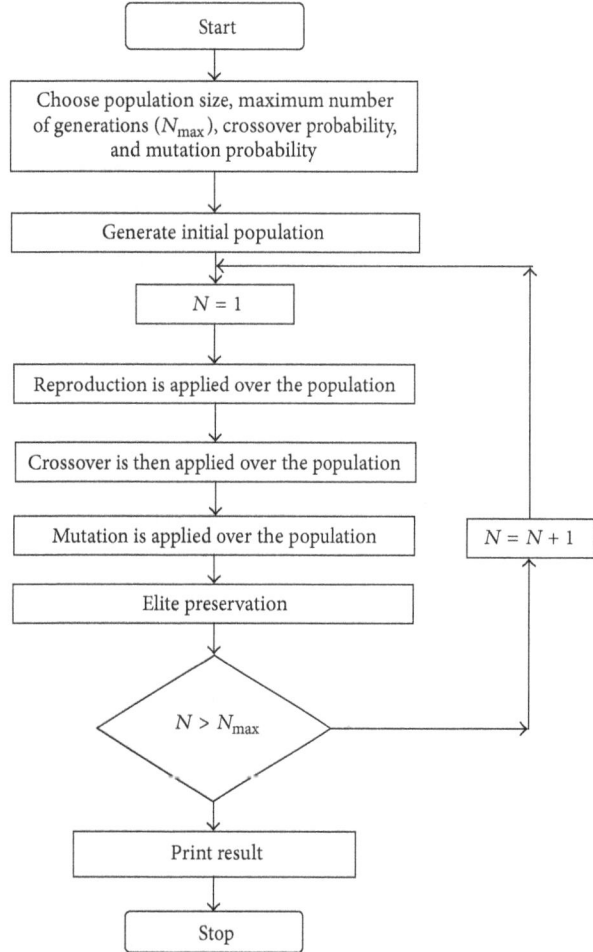

FIGURE 1: Flow chart for real-coded genetic Algorithm.

3.2. Real-Coded Genetic Algorithm Computational Procedure.
In this problem three variables called genes will form a chromosome. A set of chromosome is called population. With uniform probability distribution all chromosomes in the population are initialized. The population of each generation will have feasible design variables (chromosome) in terms of their allowable ranges but may be infeasible otherwise. The main steps involved in the genetic algorithm are discussed below and shown in flow chart (Figure 1).

Real-oded GA comprises of mainly six steps as follows.

Step 1. There are mainly four user-defined parameters in the program, population size, maximum number of generation, cross over probability, and mutation probability. The best value of population size is 50. It is found that the program is converging very fast with these values. Cross over probability and mutation probability are more sensitive parameters for this program.

Step 2. Second stage of program is to initialize the population size. So, 50 chromosomes are initialized using random probability for each variable span.

Step 3. The selection operator involves randomly choosing members of the population to enter a mating pool. The operator is carefully formulated to ensure that better members of the population (with higher fitness) have a greater probability of being selected for mating, but that worse members of the population still have a small probability of being selected. Having some probability of choosing worse members is important to ensure that the search process is global and does not simply converge to the nearest local optimum. Selection is one of the important aspects of the GA process, and there are several ways for the selection.

Step 4. Recombination is carried out through crossover and mutation operation in GA. The crossover operator is a method for sharing information between chromosomes. It ensures that the probability of reaching any point in the search space is never zero. The crossover operator is the main search operator in the GA. The search power of a crossover operator is defined as a measure of how flexible the operator is to create an arbitrary point in the search space. Crossover is useful in problems where building block exchange is necessary. It has been found that GAs may work well with large crossover probability and with a small mutation probability. A single point crossover preserves the structure of the parent string to the maximum. From a set of crossover operator, linear, blended crossover, and simulated binary crossover operators, it is found that, from trial run, the simulated binary crossover gives better convergence in limited time.

Step 5. From biological view, mutation is any change of DNA material that can be reproduced. From computer science view, mutation is a genetic operator that follows crossover operator. It usually acts on only one individual chosen based on a probability or fitness function. One or more genetic components of the individual are scanned. And this component is modified based on some user-definable probability or condition. Without mutation, offspring chromosomes would be limited to only the genes available within the initial population. Mutation should be able to introduce new genetic material as well as modify the existing one. With these new gene values, the genetic algorithm may be able to arrive at a better solution than was previously possible. Mutation operator prevents premature convergence to local optima by randomly sampling new points in the search space. There are many types of mutation, and these types depend on the representation itself. Random mutation finds a better suitability with the existing problem.

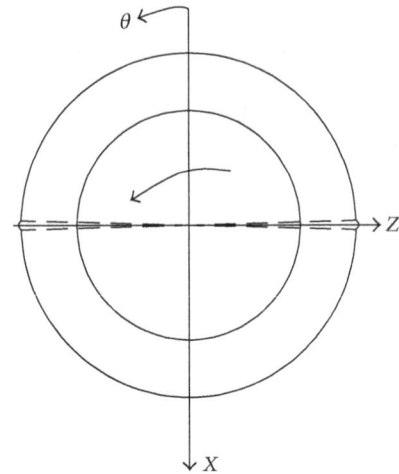

FIGURE 2: Hz-Hz configuration of two-groove oil journal bearing.

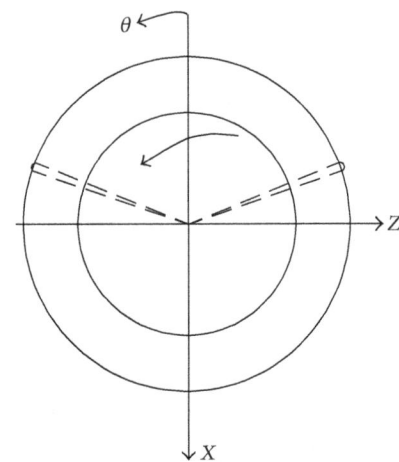

FIGURE 3: Up-Up configuration of two-groove oil journal bearing.

Step 6. Elite preservation forms a new population from the initial population and mutated one. This operator is responsible for convergence of the fitness by allowing better value to pass to the next generation.

4. Results and Discussion

The groove position located around the circumference is grouped as follows.

Group-I: Hz-Hz configuration: grooves are placed in a horizontal position 180° apart, that is, diametrically opposite to each other (Figure 2).

Group-II: Up-Up configuration: both grooves are placed (5° to 80°) above the horizontal position (Figure 3), and groove position is varied at 5° interval. Up-Up-10 configuration means that both grooves are 10° above the horizontal as shown in Figure 3.

Group-III: Up-Hz configuration: the left groove is (5° to 80°) above the horizontal position, and the other groove is in horizontal position (Figure 3). Groove

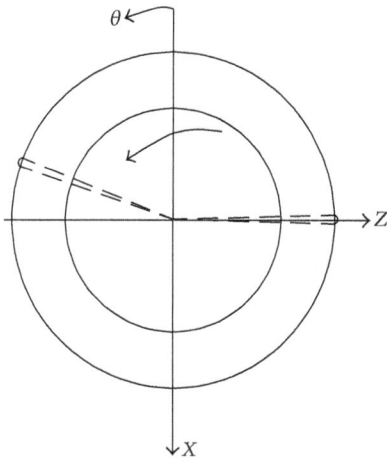

FIGURE 4: Up-Hz configuration of two-groove oil journal bearing.

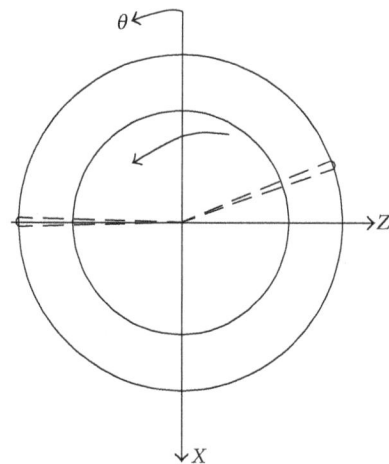

FIGURE 6: Hz-Up configuration of two-groove oil journal bearing.

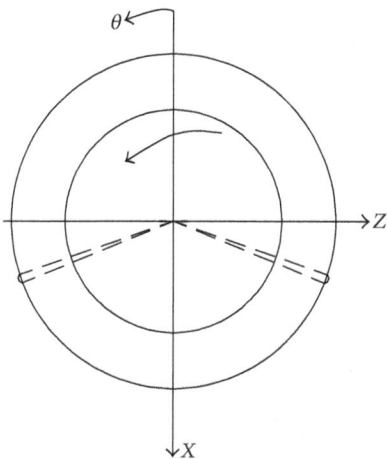

FIGURE 5: Dn-Dn configuration of two-groove oil journal bearing.

position is varied at 5° interval (Figure 4). Up-10 Hz configuration means that the left groove is 10° above the horizontal and the right groove is horizontal as shown in Figure 4.

Group-IV: Dn-Dn configuration: both grooves are placed (5° to 80°) below the horizontal position (Figure 5), and groove position is varied at 5° interval. Dn-Dn-10 configuration means that both grooves are 10° below the horizontal as shown in Figure 5.

Group-V: Hz-Up configuration: the left groove is in horizontal position, and the other groove is (5° to 80°) above the horizontal position (Figure 6) groove position is varied at 5° interval (Group-V). Hz-Up-10 configuration means that the left groove is horizontal and the right one is 10° above the horizontal as shown in Figure 6.

Group-VI: Dn-Up configuration: one of the grooves is (5° to 80°) below the horizontal position (the left one), and the other groove (the right one) is (5° to 80°) above the horizontal position; groove position is

varied at 5° interval (Figure 7). Dn-10-Up-10 configuration means that the left one is 10° down and the right groove is 10° above the horizontal as shown in Figure 7.

Group-VII: Dn-Hz configuration: the left groove is (5° to 80°) below the horizontal position, and the other groove is in horizontal position (Figure 8); groove position is varied at 5° interval (Figure 8). Dn-10 Hz configuration means that the left groove is 10° below the horizontal and the right one is horizontal as shown in Figure 8.

Group-VIII: Hz-Dn configuration: the left groove is in horizontal position, and the other groove is (5° to 80°) below the horizontal position; groove position is varied at 5° interval (Figure 9). Hz-Dn-10 configuration means that the left groove is horizontally placed while the right one is 10° below the horizontal as shown in Figure 9.

Group-IX: Up-Dn configuration: the left groove is (5° to 80°) above horizontal position, and the other groove is (5° to 80°) below the horizontal position; groove position is varied at 5° interval (Figure 10). Up-10-Dn-10 configuration means that the left groove is 10° above the horizontal and the right one is 10° below the horizontal as shown in Figure 10.

To ascertain the size of the groove for better performance, a comparison of nondimensional load is made for different groove angles as shown in Table 2. It has been observed that the load carrying capacity is slightly higher with 10° groove angles in comparison with 20° and 30° groove angles (Table 2) in case of two axial groove bearings. Therefore, 10° groove angles are considered throughout the analysis.

A code has been developed to calculate the steady state and dynamic characteristics for given values of L/D ratios and groove locations (group-I to group-IX), which is subsequently used for obtaining optimum groove locations for different objective functions. An optimum groove location has been obtained depending on maximization of load, flow

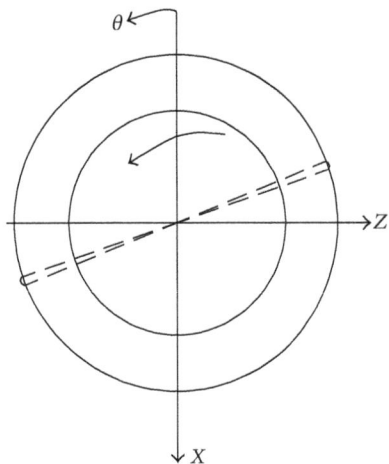

FIGURE 7: Dn-Up configuration of two-groove oil journal bearing.

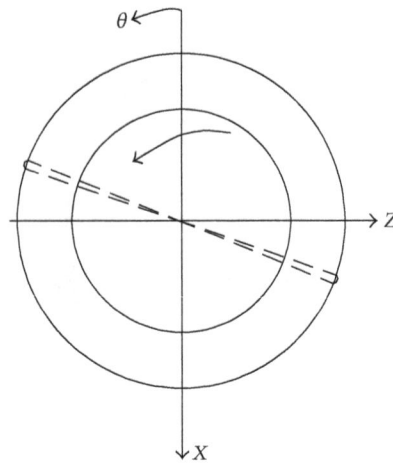

FIGURE 8: Dn-Hz configuration of two-groove oil journal bearing.

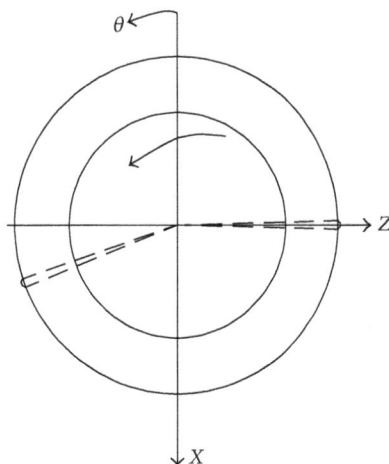

FIGURE 9: Hz-Dn configuration of two-groove oil journal bearing.

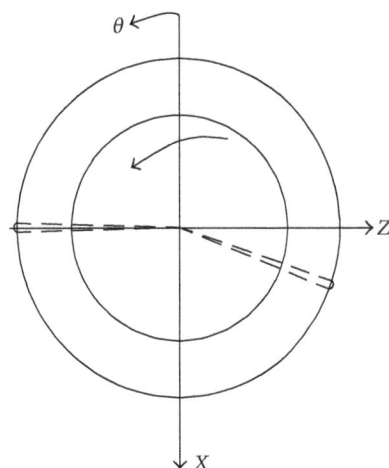

FIGURE 10: Up-Dn configuration of two-groove oil journal bearing.

TABLE 2: Comparison of nondimensional load values using 10°, 20°, and 30° groove angles.

ε	W		
	10° groove	20° groove	30° groove
0.200	0.077	0.074	0.0715
0.400	0.1865	0.181	0.175
0.600	0.406	0.399	0.389
0.800	1.135	1.123	1.107

TABLE 3: Comparison of GA and SQP results.

ε	Objective function value (minimum friction variable)	
	GA results	SQP results
0.100	25.841	25.841
0.200	12.575	12.575
0.300	7.991	7.991
0.400	5.603	5.603
0.500	4.050	4.050
0.600	3.023	3.023
0.700	2.146	2.146
0.800	1.501	1.501
0.900	0.358	0.358

and mass parameter, and minimization of friction with the help of Genetic Algorithm (GA) toolbox of MatLab. The obtained results from (GA) have been compared with the results obtained using sequential quadratic programming (SQP).

The optimum value of fitness function obtained corresponding to minimization of friction variable has been tabulated for both GA and SQP in Table 3.

Similarly maximum load, maximum flow, and maximum mass parameter values are also found to match both methods. It has been observed as stated above that the results using both methods are found to be the same. However, GA has been used in this work as GA, being a heuristic search and optimization technique inspired by natural evolution, has

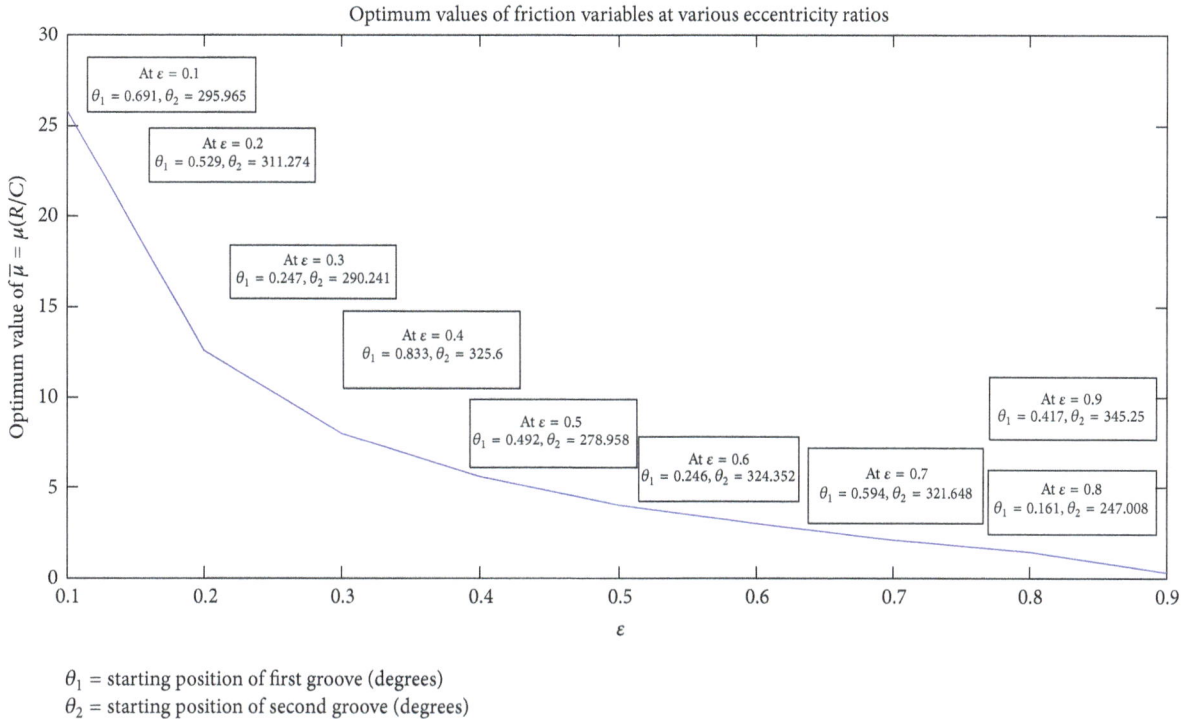

Optimum values of friction variables at various eccentricity ratios

At $\varepsilon = 0.1$
$\theta_1 = 0.691, \theta_2 = 295.965$

At $\varepsilon = 0.2$
$\theta_1 = 0.529, \theta_2 = 311.274$

At $\varepsilon = 0.3$
$\theta_1 = 0.247, \theta_2 = 290.241$

At $\varepsilon = 0.4$
$\theta_1 = 0.833, \theta_2 = 325.6$

At $\varepsilon = 0.5$
$\theta_1 = 0.492, \theta_2 = 278.958$

At $\varepsilon = 0.6$
$\theta_1 = 0.246, \theta_2 = 324.352$

At $\varepsilon = 0.7$
$\theta_1 = 0.594, \theta_2 = 321.648$

At $\varepsilon = 0.8$
$\theta_1 = 0.161, \theta_2 = 247.008$

At $\varepsilon = 0.9$
$\theta_1 = 0.417, \theta_2 = 345.25$

θ_1 = starting position of first groove (degrees)
θ_2 = starting position of second groove (degrees)

FIGURE 11: Variation of friction variable at optimum grooving location for different ε.

been successfully applied to a wide range of real-world problems of significant complexity [3, 9]. It has been suggested that heuristic optimization provides a robust and efficient approach for solving complex real-world problems [5].

Initially a single objective function has been taken up. The generic algorithm convergence rate to true optima depends on the probability of crossover and mutation, on one hand, and the maximum generation, on the other hand. In order to preserve a few very good strings and reject low-fitness strings, a high crossover probability is preferred. The mutation operator helps to retain the diversity in the population but disrupts the progress towards a converged population and interferes with beneficial action of the selection and crossover. Therefore, a low probability, 0.001–0.1, is preferred. The genetic algorithm updates its population on every generation, with a guarantee of better or equivalent fitness strings. For well-behaved functions, 30–40 generations are sufficient. For steep and irregular functions, 50–100 generations are preferred [2]. Considering these factors, a population size of 50, mutation probability of 0.1, and a cross over probability of 0.8 have been selected.

The optimum groove locations for minimum nondimensional friction variable, nondimensional load, nondimensional flow, and mass parameter at different ε are shown in Figures 11, 12, 13, and 14. θ_1 and θ_2 are the starting positions of first and second groove, respectively, in degrees.

From the results shown in Figures 11 through 14, it has been observed that first groove location remains near 0°, whereas the second groove location varies with eccentricity ratios in all the cases. Variations of the second groove location are different for different objective functions.

TABLE 4: The optimum configurations combining all the objective functions at a time.

ε	θ_1	θ_2
0.100	0.346	336.325
0.200	0.141	196.512
0.300	1.238	208.919
0.400	0.469	240.390
0.500	0.281	241.176
0.600	0.382	231.287
0.700	0.785	222.092
0.800	0.476	343.712

Similarly by combining all the objective functions at a time the optimum configurations obtained is tabulated (Table 4).

It has been observed from the tabulated results in Table 4 that the staring position of the first groove at different eccentricity ratios for multiobjective function remains near to 0°, whereas second groove location varies for different eccentricity ratios. This indicates that second groove location is more sensitive compared to the first groove location.

If the three variables, namely, eccentricity ratio (ε), starting angles of the first groove (θ_1), and the second groove (θ_2), are taken as chromosome (Table 5), then the optimum results obtained for friction, flow, load, and mass parameter are shown in Figures 15, 16, 17, and 18. The figures include plots of best fitness as well as mean fitness. Genetic algorithm works on a population of individuals. So, mean is the mean fitness for the entire population at a particular iteration.

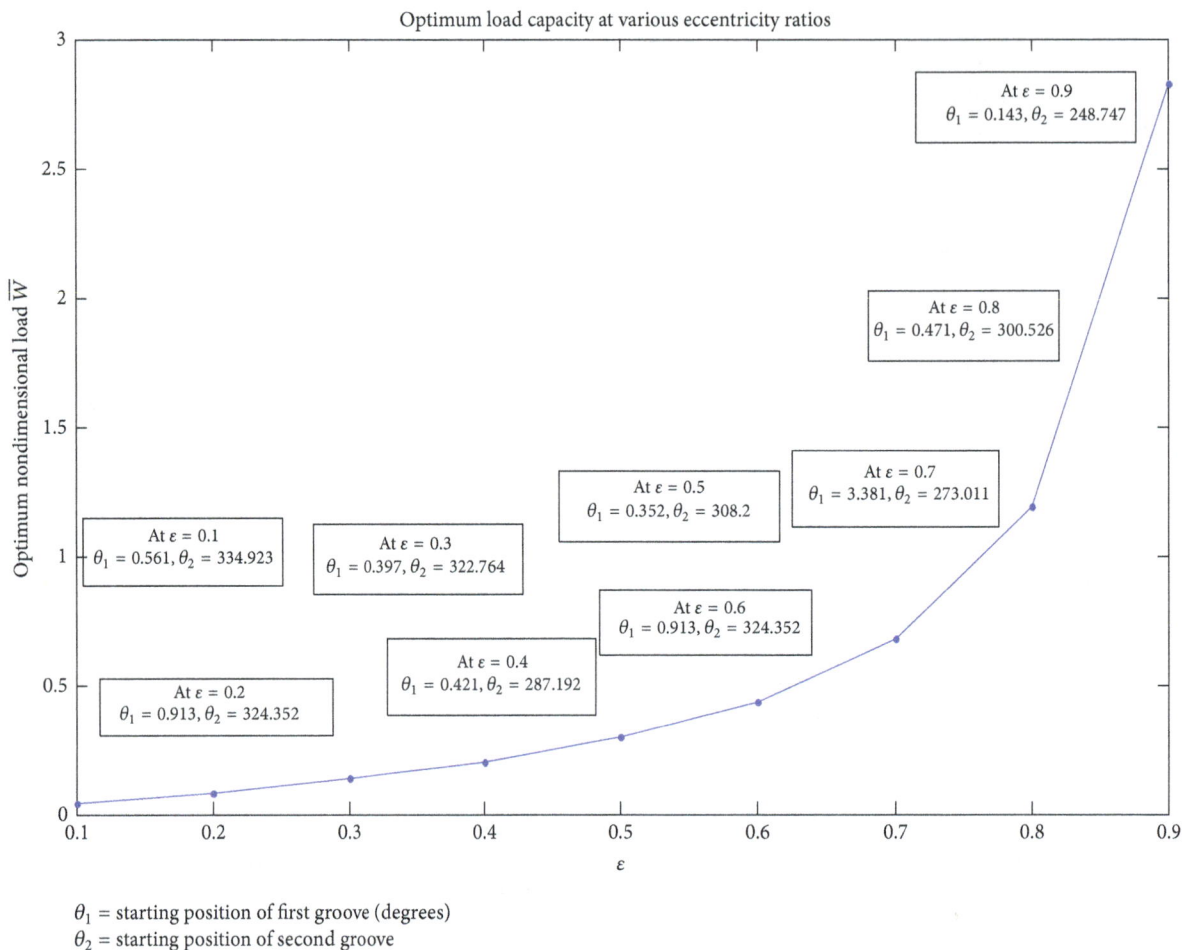

Optimum load capacity at various eccentricity ratios

θ_1 = starting position of first groove (degrees)
θ_2 = starting position of second groove

FIGURE 12: Variation of optimum nondimensional load at different eccentricity ratios.

TABLE 5: Variable bounds for the bearing problem.

Variable	Lower bound	Upper bound
ε	0.100	0.900
Starting angle of first groove (θ_1)	0°	180°
Starting angle of second groove (θ_2)	170°	350°

TABLE 6: Optimum location considering different objectives.

Optimum location for objectives	ε	θ_1	θ_2
Minimum friction variable	0.900	0	232.906
Maximum flow	0.899	5.660	301.960
Maximum load carrying capacity	0.899	0.626	308.230
Maximum mass parameter	0.811	0.890	308.230
Optimization of the combined objectives	0.268	3.670	349.990

Again by combining all the objective functions at a time the fitness value plot has been obtained as shown in Figure 19. Here weighted sum method has been used to combine all the objectives. There are three objectives to be maximized when one has to be minimized. The objectives to

be maximized are made negative, and then the weighted sum of all the four objective functions has been taken making the multiobjective problem of minimization type. Since there are four parameters weights equal to 0.25 is used.

The optimum locations for each objective function including that of multi-objective function have been shown in Table 6. From the above analysis, it has been observed that groove locations for various objective functions are different. The first groove varies between 0° to 5.66°, and the second groove locations for maximum load carrying capacity and maximum mass parameter are the same. Second groove location for minimum friction variable is the least and for multiobjective function is the highest. Another interesting observation is that when the corresponding eccentricity ratios for individual objective functions are high enough, it is much less for multiobjective function.

After carefully looking at the results presented above, it appears that one may get near optimal results by placing a single groove and eliminating the second groove entirely. In view of this, an attempt has been made to find the optimum groove location for a single-grove bearing and compared with two-groove cases for each of the objective functions as presented in Table 7. Since the results are found to be quite interesting, therefore, it would be pertinent to go through

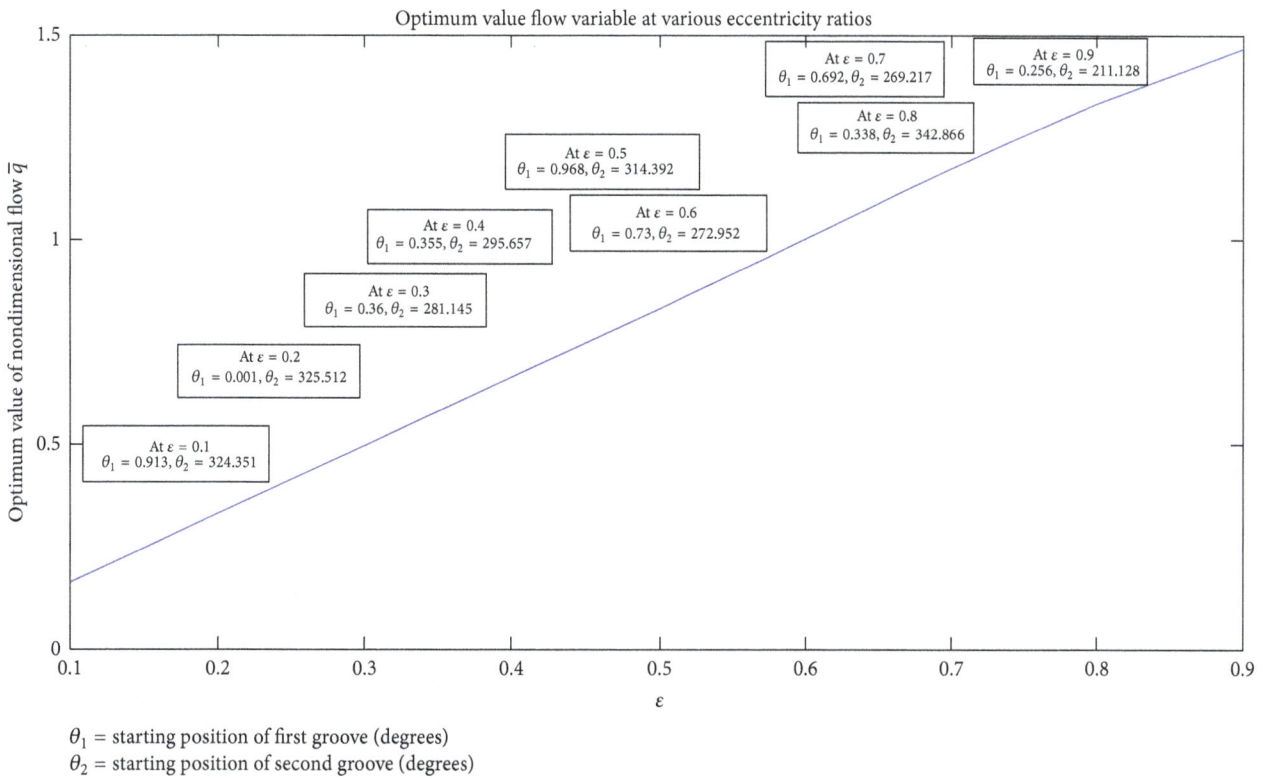

θ_1 = starting position of first groove (degrees)
θ_2 = starting position of second groove (degrees)

FIGURE 13: Variation of flow at optimum grooving location for different eccentricity ratios.

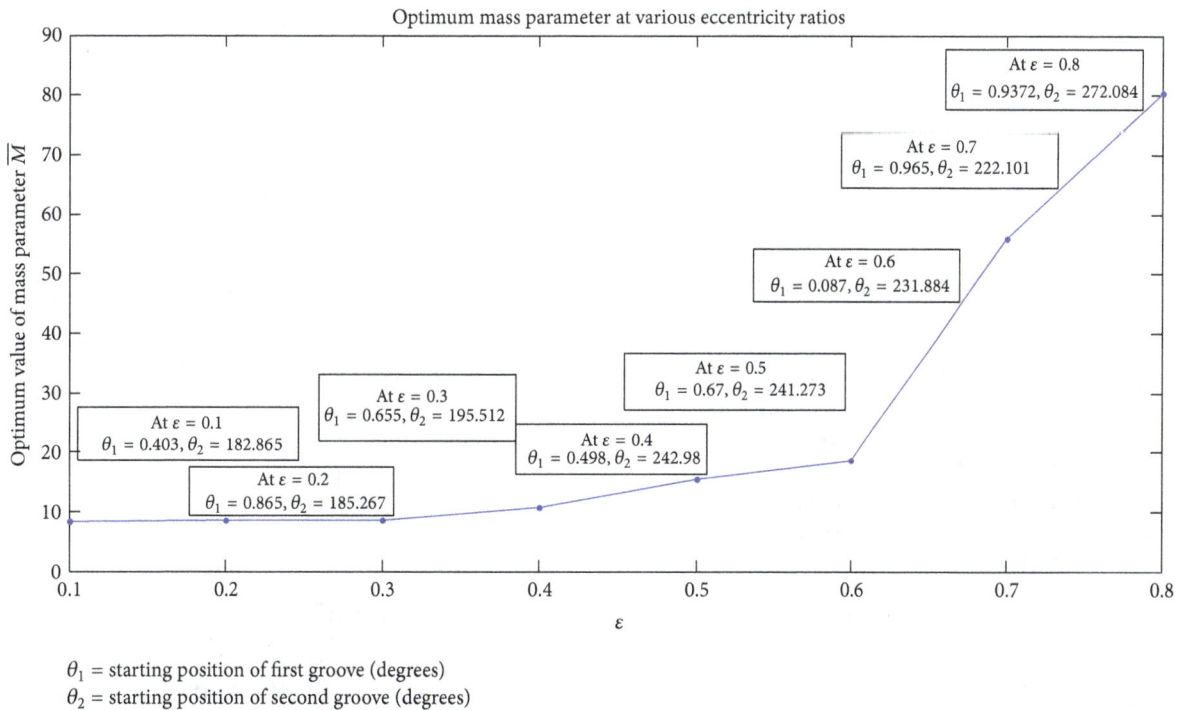

θ_1 = starting position of first groove (degrees)
θ_2 = starting position of second groove (degrees)

FIGURE 14: Variation of mass parameter at optimum grooving location for different eccentricity ratios.

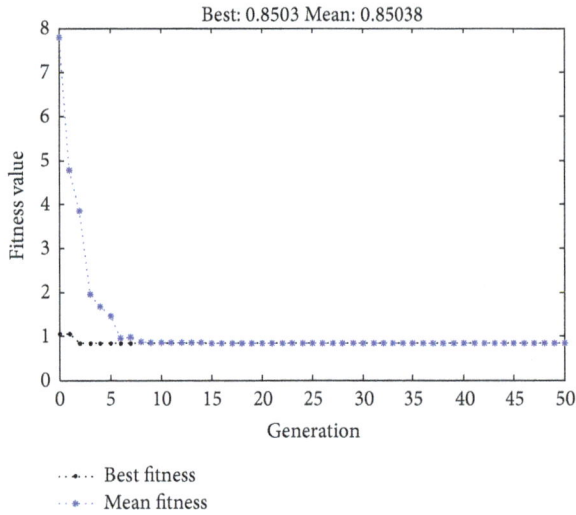

FIGURE 15: Fitness value considering friction variable as objective function.

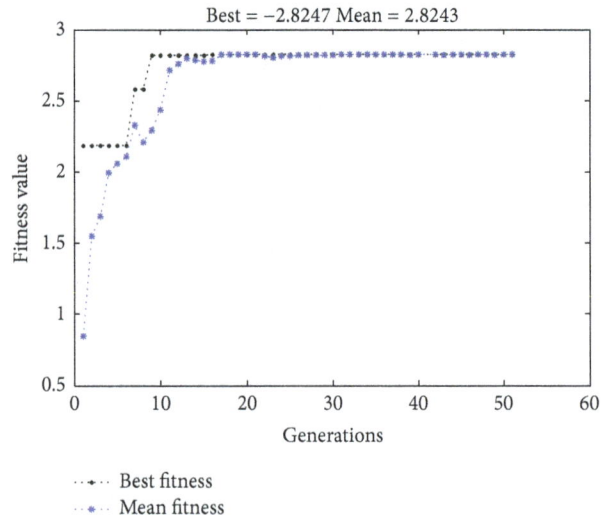

FIGURE 16: Fitness value considering flow as objective function.

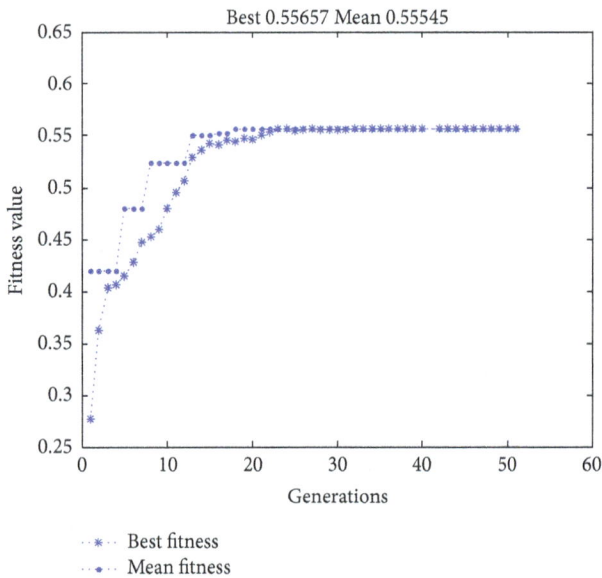

FIGURE 17: Fitness value considering load as objective function.

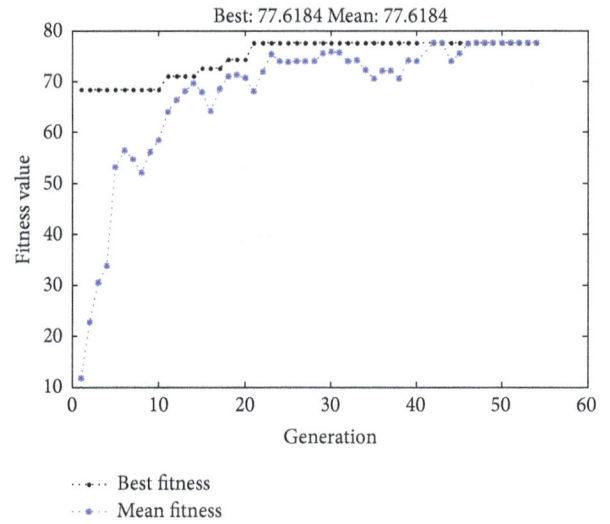

FIGURE 18: Fitness value considering mass parameter as objective function.

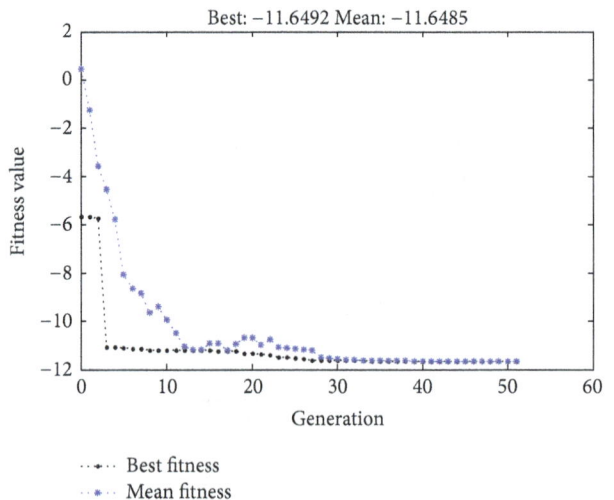

FIGURE 19: Fitness value combining all the objective function.

relevant literature first and then taking up the study to find whether single groove or two grooves would enhance the bearing performance. The authors would like to keep this for future study in detail.

A dimensional example has been shown below to demonstrate how to convert the nondimensional parameters to dimensional parameters.

Let $D = 100$ mm, $N = 3000$ rpm, $W = 15$ kN, $L/D = 1.0$, $C/R = 0.001$.

So, $C = 50 \times 10^{-6}$ m.

Taking minimum film thickness as $h_o = 25 \times 10^{-6}$ m, one gets $h_o/C = 0.5$.

Hence, $\varepsilon = 0.5$.

TABLE 7: Comparison of optimum locations of grooves for two-groove and single-groove bearings.

Comparison	Objective function	ε	θ_1	θ_2
For two groove	Minimum friction variable	0.900	0	232.906
For single groove		0.100	180	—
For two groove	Maximum flow	0.899	5.66	301.96
For single groove		0.100	180	—
For two groove	Maximum load	0.899	0.626	308.23
For single groove		0.657	349.038	—
For two groove	Maximum mass parameter	0.811	0.890	308.23
For single groove		0.704	223.435	—
For two groove	Optimization of all the combined objectives	0.268	3.67	349.99
For single groove		0.657	349.038	—

TABLE 8: Conversion of nondimensional results to dimensional ones.

Eccentricity ratio	Objective function	Optimum groove locations	Present nondimensional result	Dimensional values
0.5	Maximization of flow	$\theta_1 = 0.968$ $\theta_2 = 314.392$	0.8329 (optimum flow variable)	Flow, $Q = 3.271 \times 10^{-5}$ m^3/s
	Minimization of friction	$\theta_1 = 0.492$ $\theta_2 = 278.95$	4.050 (optimum friction variable)	Coefficient of Friction, $\mu = 4.050 \times 10^{-3}$

For $\varepsilon = 0.5$, optimum friction variable and optimum flow variables are shown in Table 8 along with optimum groove locations. These non-dimensional results are converted to dimensional parameters, namely, flow in m^3/s and coefficient of friction by using the above data. These values may further be used to estimate the friction force, temperature rise, and so forth.

5. Conclusion

From the results presented here, it can be inferred that the second groove location is sensitive to the type of objective function whereas the first groove is more or less the same for any objective function. The practice and the notion of convenience of keeping groove positions 180° apart need to be thoroughly looked into as the present results show that optimum groove locations are not 180° apart for any of the objective functions considered in the present work. Experimental verification of the present result may lead to a new approach of production of bearings with optimum groove locations; however, it is beyond the scope of the present work and hopefully experimentalists have a problem in hand.

Appendix

For the purpose of validation of results the steady state characteristics of two-groove oil journal bearing having 20° groove angles placed in horizontal position for $L/D = 1$ are compared with the published results [2] as shown in Table 9. The present results are found to be fairly in good agreement with [2].

TABLE 9: Comparison of present results with [2] for $L/D = 1$ and 20° axial groove for groove in the horizontal position.

ε	S Present [Ref]	ϕ Present [Ref]
0.103	1.453 [1.470]	75.860 [75.990]
0.150	0.980 [0.991]	70.462 [71.580]
0.224	0.629 [0.635]	63.4598 [63.540]
0.352	0.352 [0.358]	56.100 [55.410]
0.460	0.232 [0.235]	49.925 [49.270]
0.559	0.157 [0.159]	45.1075 [44.330]
0.650	0.106 [0.108]	40.120 [39.720]
0.734	0.070 [0.071]	35.432 [35.160]
0.773	0.0562 [0.056]	33.160 [32.860]
0.811	0.043 [0.044]	30.614 [—]
0.883	0.023 [0.024]	25.142 [25.020]

Notations

C: Radial clearance (m)
D: Diameter of the journal (m)
L: Length of the bearing (m)
R: Bearing radius (m)
e: Eccentricity (m)
ε: Eccentricity ratio $= e/C$
η: Coefficient of absolute viscosity of the lubricant (Pa-s)
$\mu, \overline{\mu}$: Coefficient of friction, friction variable $= \mu(R/C)$
N: Speed of the journal in r.p.s
ϕ: Bearing attitude angle
h: Film thickness (m) $= C(1 + \varepsilon \cos \theta)$

$\bar{h}$:   Nondimensional film thickness $= h/C$

θ_1:   Position of starting of the groove

θ_2:   Position of end of the groove

U: Sliding speed

p: Steady state pressure (Pa)

$\bar{p}$:   Nondimensional steady state pressure $= pC^2/6\eta UR$

W: Load carrying capacity (N)

$\overline{W}$:   Nondimensional load carrying capacity $= WC^2/6\eta UR^2 L$

X: Vertical direction

Z: Horizontal direction

W_X: Vertical component (in X direction) of the resultant load

W_Z: Vertical component (in Z direction) of the resultant load

P: Load per unit bearing area $= W/LD$

S: Sommerfeld number $= (\eta N/P)(R/C)^2$

$\bar{q}_z$:   Nondimensional flow coefficient, $(\bar{q}_z = 2Q/ULC)$

$\bar{p}_1, \bar{p}_2$:   Perturbed pressures

ε_1, ϕ_1:   Perturbed eccentricity ratio and attitude angle around the steady state value ε_0, ϕ_0

$K_{XX}, K_{ZZ}, K_{XZ}, K_{ZX}$:   Stiffness coefficients (N/m)

$\overline{K}_{XX}, \overline{K}_{ZZ}, \overline{K}_{XZ}, \overline{K}_{ZX}$:   Nondimensional stiffness coefficients $= K_{ij}C/W$, where $i = X, Z$ and $j = X, Z$

$D_{XX}, D_{ZZ}, D_{XZ}, D_{ZX}$:   Damping coefficient (N·s/m)

$\overline{D}_{XX}, \overline{D}_{ZZ}, \overline{D}_{XZ}, \overline{D}_{ZX}$:   Nondimensional damping coefficient $= C_{ij}C\omega/W$, where $i = X, Z$ and $j = X, Z$

t: Time (s)

ω, ω_p: Journal rotational speed (rad/s), frequency of journal vibration

τ: Nondimensional time, $\tau = \omega_p t$

λ: Whirl ratio $= \omega_p/\omega$

$M, \overline{M}$:   Rotor mass (kg), mass parameter, $\overline{M} = MC\omega^2/W$

$()_0$:   Steady state value.

References

[1] ESDU Items 84031 (and 85028), *Calculation Methods for Steadily Loaded Axial Groove Hydrodynamic Journal Bearings*, with Superlaminar operation, December 1985, 1984.

[2] P. Klit and J. W. Lund, "Calculation of the dynamic coefficients of a journal bearing, using a variational approach," *Journal of Tribology*, vol. 108, no. 3, pp. 421–425, 1986.

[3] D. T. Gethin and M. K. I. El Deihi, "Effect of loading direction on the performance of a twin-axial groove cylindrical-bore bearing," *Tribology International*, vol. 20, no. 4, pp. 179–185, 1987.

[4] W. B. Rowe and D. Koshal, "A new basis for the optimization of hybrid journal bearings," *Wear*, vol. 64, no. 1, pp. 115–131, 1980.

[5] Y. J. Lin and S. T. Noah, "Using genetic algorithms for the optimal design of fluid journal bearing," in *Proceedings of the ASME Design Engineering Technical Conferences*, pp. 12–15, 1999.

[6] H. Hashimoto and K. Matsumoto, "Improvement of operating characteristics of high-speed hydrodynamic journal bearings by optimum design: part I-formulation of methodology and its application to elliptical bearing design," *Journal of Tribology*, vol. 123, no. 2, pp. 305–312, 2001.

[7] S. Boedo and S. L. Eshkabilov, "Optimal shape design of steadily loaded journal bearings using genetic algorithms," *Tribology Transactions*, vol. 46, no. 1, pp. 134–143, 2003.

[8] H. Hirani, "Multiobjective optimization of a journal bearing using the Pareto optimality concept," *Proceedings of the Institution of Mechanical Engineers J*, vol. 218, no. 4, pp. 323–336, 2004.

[9] B. David, D. R. Bull, and M. Ralph, "An overview of genetic algorithms: part 2, research topics," *University Computing*, vol. 25, no. 4, pp. 170–181, 1993.

[10] J. McCall, "Genetic algorithms for modelling and optimisation," *Journal of Computational and Applied Mathematics*, vol. 184, no. 1, pp. 205–222, 2005.

[11] B. J. Hamrock, *Fundamentals of Film Lubrication*, Mc GrawHill, NewYork, NY, USA, 1994.

19

Formation of Composite Surface during Friction Surfacing of Steel with Aluminium

S. Janakiraman and K. Udaya Bhat

Department of Metallurgical & Materials Engineering, NITK Surathkal, Srinivasa Nagar, Surathkal 575025, India

Correspondence should be addressed to K. Udaya Bhat, udayabhatk@gmail.com

Academic Editor: Patrick De Baets

Commercial pure aluminium was deposited on medium carbon steel using friction surfacing route. An aluminium rod was used as the consumable tool. Normal load and tool rotation speed were the variables. Under certain combinations of load and speed the deposition was continuous and uniform. The deposit consisted of Al embedded with fine particles of iron. The interface between substrate material and deposited material was smooth and relatively small. A mechanism is discussed for formation of a composite surface on the steel substrate.

1. Introduction

Steel remains one of the important structural material because of its relatively low cost, high processability, manipulation of the properties using principles of alloy design, heat treatment, and so forth [1]. Unfortunately, it's service properties like corrosion resistance, oxidation resistance, are not very good. This limitation arises because the oxide layer forming on the surface of the steel is a noncompact one [1]. This limitation can be overcome by modifying the surface of the steel appropriately, either by changing the surface chemistry (alloying at the surface) or by deposition of another metal at the surface [2]. The deposited metal may on its own give beneficial properties or after appropriate conversion. On this count, deposition of a thin layer of aluminium on steel is very relevant. Aluminium layer, when it is oxidized, forms a compact oxide layer, protecting the substrate steel from oxidation, corrosion, and abrasion [3, 4]. A thin layer of Al on steel can be obtained by various means, that is, liquid route and solid route. Hot dip aluminising is a predominant method using liquid route, but this route is handicapped by the formation of brittle intermetallics [5]. Chemical routes like pack aluminising do not involve use of liquid state, but they also involve processing at elevated temperatures for long duration, again giving rise to intermetallics and grain growth in the substrate. In this context friction surfacing is

a promising route. It can produce an aluminium layer on the steel substrate [6], and if a compact oxide layer is required top layer can be made to undergo oxidation. In friction surfacing the surface of a component is modified using mechanical energy generated using a friction tool [7]. An alloying element can be added during friction surfacing which will be mixed with the substrate to generate an alloyed surface. In friction surfacing, the tool is a consumable one and depending on the relative strengths of substrate and tool materials, as well as temperature attained, both substrate and tool or only tool material will be undergoing plastic deformation. This will lead to alloying near the surface leading to a change in the surface properties [8]. In friction surfacing of steel with aluminium, steel is used as substrate and aluminium is the consumable tool. If friction surfacing parameters are appropriate it is possible to obtain a uniform aluminium deposition on the steel surface [6]. Aluminium layer will have steel (iron) particles embedded in it producing an iron-aluminum composite layer on the steel surface [6]. This paper discusses formation of such composite layer during friction surfacing (of steel with aluminium).

2. Materials and Experimental Methods

2.1. Materials and Processing. Medium carbon steel plate (C = 0.35, Mn = 0.65, P = 0.03, S = 0.04; all are in wt.%) was

TABLE 1: Processing parameters for various samples and quality of the deposition.

Trial no.	Tool travel speed (mm/min)	Load (kN)	Spindle speed (rpm)	Quality of the deposition
T1	35	3	200	Powdery deposition,
T2	35	3	400	Deposition better than T1, still powdery
T3	35	4	200	Discontinuous and varying width.
T4	35	4	400	Good (uniform, continuous) deposition
T5	35	5	200	Continuous, varying width
T6	35	5	400	Good deposition, Width more than T4

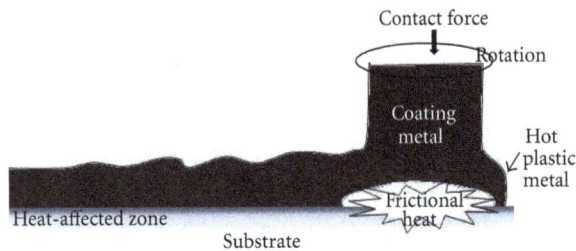

FIGURE 1: Scheme of friction surfacing [8].

taken as the substrate. Substrate dimensions were 150 mm length, 70 mm width, and 8 mm thickness. Controlled roughness on the steel substrate was obtained by milling the substrate using a conventional surface milling machine. Depth of groves produced during milling was measured using cross-sectional microscopy and it was in the range of 25–32 micrometers (um). Roughness of the milled surface was measured using a Veeco optical profile meter. Measured roughness (Ra) was in the range of 5.8 um to 8.3 um. Commercial pure Al (99.6% pure), available in the form of extruded rod, was used as the consumable tool. Extruded rod was machined to a dimension of 100 mm length and 25 mm diameter and it was used for deposition. The friction surfacing was done using the machine made by M/s ETA technologies, Bangalore, India. Figure 1 shows a schematic presentation of friction surfacing [8].

Al was deposited using different processing conditions. Normal load was varied as 3 kN, 4 kN, and 5 kN. This gave a stress level of 6.1 MPa, 8.1 MPa, and 10.2 MPa in the consumable tool. Tool spindle speed was varied as 200 rpm and 400 rpm. Tool plunge depth was fixed at 40 mm. Tool plunge depth is the total depth up to which the tool can be lowered in the machine. For all experiments tool travel speed was fixed as 35 mm/min. For convenience the samples were labelled as T1, T2, ..., T6 and they are listed in Table 1. All the experiments were done in open atmosphere conditions and for 200 s.

2.2. Characterisation of the Deposit. Quality of the deposition was investigated using various parameters, namely, nature of the deposition (powdery or not), continuity, width uniformity. This information is also listed in Table 1. Morphological investigation, composition of the deposit, and cross-sectional microscopy were done using Scanning Electron Microscope (SEM) with an EDS attachment. Phase identification of the deposit was made using X-ray diffractometry (XRD) using $Cu_{K\alpha}$ radiation.

3. Results and Discussion

3.1. Quality of the Deposition. Quality of the deposition was decided based on the macroobservation using normal eye or low-magnification tools (magnification up to 10x). Deposits were either powdery, patchy (discontinuous and varying width), or continuous. Within the continuous group, width could be uniform or nonuniform. For good coverage of the surface, continuous (preferably with uniform width) deposition is essential. Figure 2 shows macroimage of the sample T6. The track length was about 30 cm and width was in the range of 25–30 mm. From the macroimage we could conclude that the deposition was continuous and of almost uniform width. Quality of the deposition is closely related to heat input at the interface and partition of heat between the substrate and the tool. The heat input at the interface (HI) could be written as [7]

$$\text{Heat input (HI)} = \frac{\text{Power input}}{\text{scan speed}}. \tag{1}$$

Extrapolating the concept from friction stir welding [7] power input (PI) could be written as a function of spindle speed (s) and torque (M):

$$\text{PI} = s * M. \tag{2}$$

For the sake of simplicity, the energy losses associated with the drives and transmission systems were neglected. Mechanical energy available at the substrate-tool interface was partitioned in to heat and deformation component required for local plastic deformation [10, 11]. The heat input at the interface gets dissipated predominantly by conduction. The temperature rise at the interface was sufficient to make aluminium near the interface plastic, and the relative sliding of the tool with respect to the substrate leads to transfer of the plastic material as a thin layer. This explains why good deposition was seen only under certain combinations of load and rotation speed (other parameters were kept constant in our experiments). It may be mentioned that only aluminium side became plastic, whereas the temperature rise was not sufficient to make steel plastic.

3.2. Topographical Details. Figure 3(a) shows a low-magnification image observed under SEM in secondary electron mode. It may be noted that topographic information is

(a)

(b)

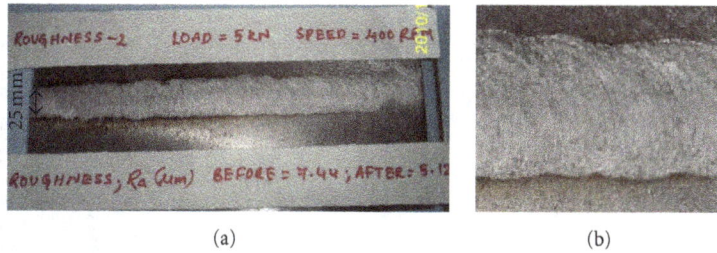

FIGURE 2: Macroimage of the deposition made using a normal load of 5 kN and spindle speed of 400 rpm. On the right side a small region which was magnified from long track is shown.

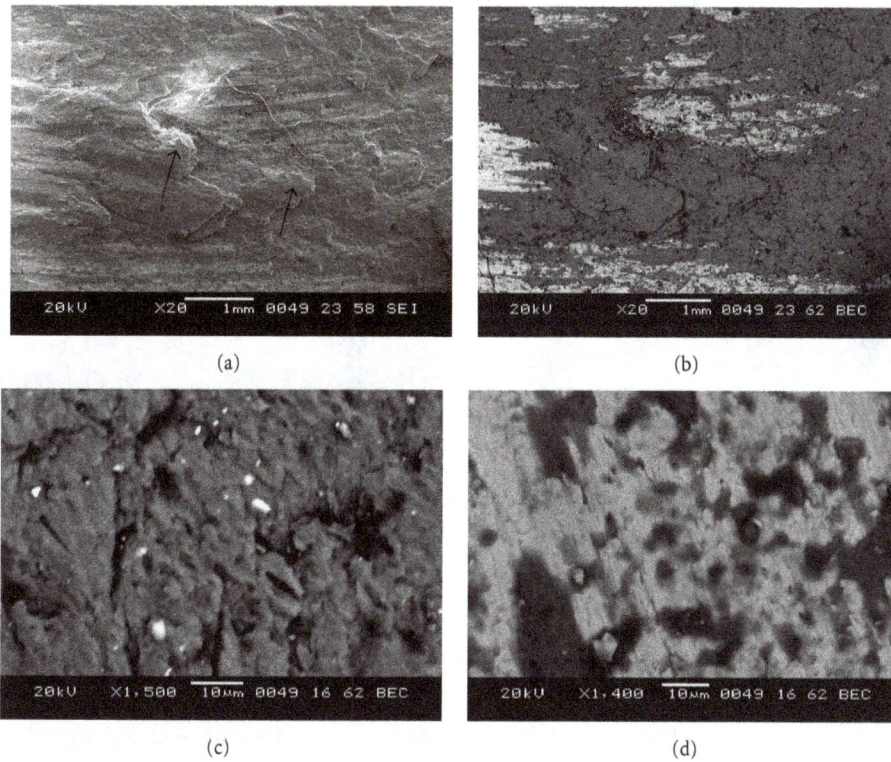

(a)

(b)

(c)

(d)

FIGURE 3: (a) A low-magnification micrograph of friction-surfaced region. (b) SEM: backscattered electron-compositional (BEC) image. (c) and (d) are micrographs taken from Fe-rich and Al-rich regions of Figure (b). In Figure (c) Al-rich region with fine Fe particles is observed. In (d) Fe-rich region has embedded small Al particles.

better revealed in this mode [12]. There is no marking typical of surface milling process. Surface has occasional microhills (indicated by arrow marks) on the flat surface. The backscattered electron-compositional micrograph (BEC image) presented in Figure 3(b) shows two types of contrast, namely, white region which is predominantly iron and grey region which is predominantly aluminium. It must be noted that both micrographs (Figures 3(a) and 3(b)) are taken from same region and under same magnification. By comparing topographic information (Figure 4(a)) and compositional information (Figure 4(b)) we conclude that topographical variations observed in Figure 3(a) is not due to presence of iron-rich and aluminium-rich regions in the deposited layer.

3.3. Cross-Sectional Microscopy. Figure 4(a) shows cross sectional view of the deposited region. Deposition thickness

is fairly uniform, and measurement over 1 mm length gave thickness in the range of 90 to 106 um. Interface is macroscopically smooth, without any profiles created during surface milling. This is more clearly visible in Figure 4(b). Figure 4(b) presents interface between substrate material-deposit material. The interface is relatively smooth and small. The compositional profile across the interface (Figure 4(c)) shows minimum (almost zero) level of mixing of species on either side of the interface. This statement is without considering the material transfer in the form of particles which are visible as white particles (iron) embedded in the deposited aluminium. Figure 4(d) shows a high-magnification micrograph from Figure 4(a). We see iron particles of nanometer scale (arrow pointers in Figures 4(b) and 4(d)) embedded in aluminium matrix. The average spacing between the particles is also very small, indicating that they would contribute

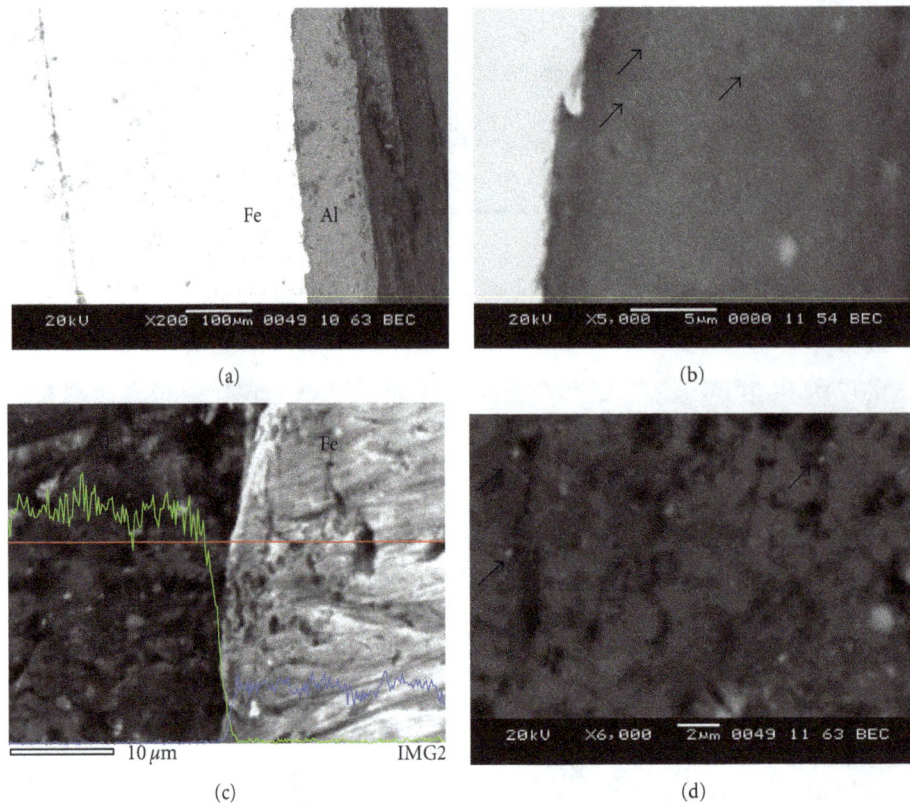

FIGURE 4: (a) Cross-sectional microscopy showing a uniform deposition. (b) Nature of Fe-Al deposit interface. (c) Smooth and thin interface between aluminium and iron. (d) Fine Fe particles embedded in deposited Al.

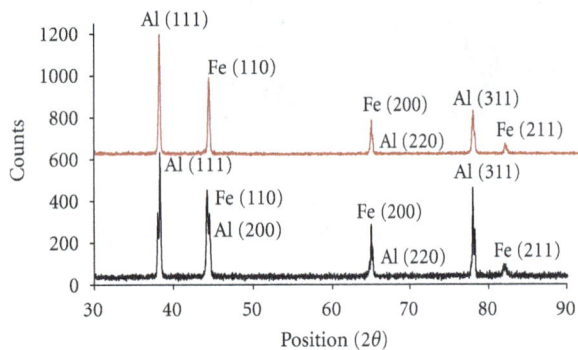

FIGURE 5: XRD plots of two samples, namely, T5 (top) and T6 (bottom).

for particle strengthening [13]. Strong Fe particles are expected to strengthen soft Al matrix.

3.4. Phase Identification. Figure 5 shows the XRD analysis of two samples (viz. T5 and T6). XRD plots for other samples are similar. XRD plot indicates that deposit consists of iron and aluminium. From microstructural observations and XRD results we say that the deposit is a mechanical mixture of aluminium and iron. There are no other phases (Fe-Al intermetallics) formed which could be detected by the XRD.

3.5. Formation of Composite Layer during Friction Surfacing. From XRD and SEM study we can conclude that the deposit is a mechanical mixture of aluminium and iron. Since, the consumable tool was pure Al, during friction surfacing, iron particles must have formed and got mixed with aluminium. This has resulted in the deposition of a composite layer of aluminium and steel. Formation of a composite layer is similar to material transfer during friction conditions [9] and can be explained as follows.

In the beginning of friction surfacing both surfaces have asperities. These asperities have various dimensional scales. This means that only few asperities are in contact with each other forming a contact pair [14]. The effective stress at the contact point may be very high compared to the average stress estimated using normal load and initial section diameter. When there is a relative sliding between two surfaces, the asperities will undergo deformation. Being a weaker material, the plastic deformation will be much more towards Al side than Fe side. Actual strain value will be very high, and it will vary depending on the morphologies of the asperities. Al, though more ductile, may get fractured easily, because of poor strength value. But being fresh surfaces, two Al surfaces have a chance to get rewelded. On the other hand, Fe is a strain hardenable material and at the asperity contact they will become hard, brittle, and get sheared during sliding. Even though fracture surface is clean and fresh, owing to smaller T/T_m (T is the interface temperature,

Shearing direction

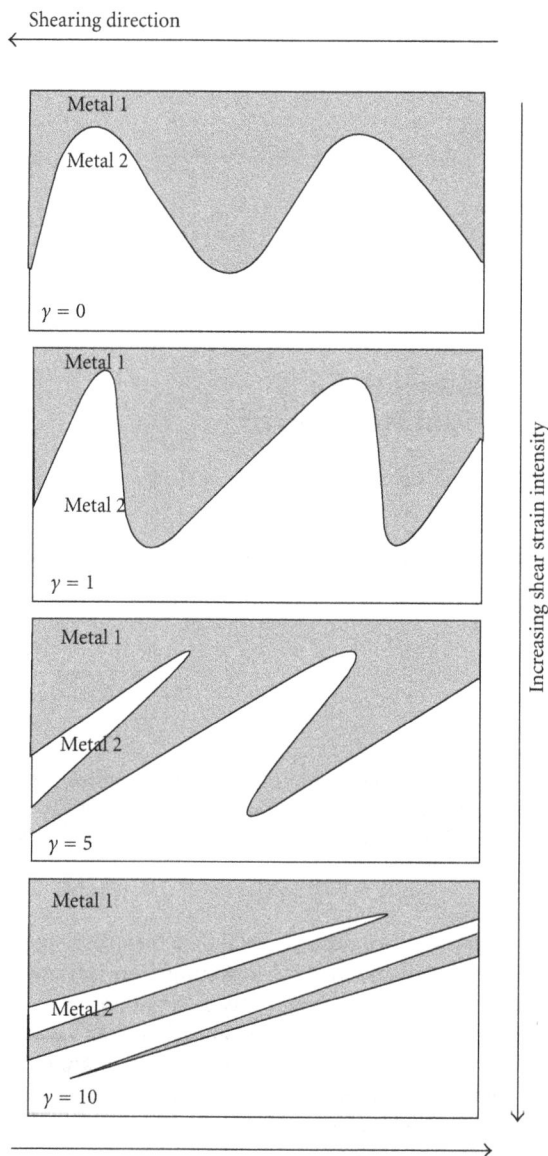

Metal 1

Metal 2

$\gamma = 0$

Metal 1

Metal 2

$\gamma = 1$

Metal 1

Metal 2

$\gamma = 5$

Metal 1

Metal 2

$\gamma = 10$

Increasing shear strain intensity

FIGURE 6: Scheme leading to asperity tip fracturing and incorporation into second material during sliding under friction conditions [9]. γ: shear strain.

T_m is the melting point, both are in kelvins) for steel, the chances of them to get rewelded are small. Figure 6 shows schematic methodology in fracturing of asperity tip at the friction contact. The Fe-Al interface shown in Figure 4(b) shows reduced grove depth (less than 5 um) compared to initial grove depth (25–32 um). This supports the argument that the asperity hills on hard Fe surface get broken during shearing. Broken Fe particles get mixed up in soft Al layer and the mixture gets deposited during friction surfacing.

4. Conclusions

Based on the experimental results, the following conclusions are drawn. A thin layer of Al can be deposited on steel surface using friction surfacing method. Deposited Al consisted of small Fe particles dispersed in it. Deposit is a mechanical

mixture of Al and Fe. The interface between substrate material and deposited material is smooth and relatively sharp. A mechanism for the formation of a composite layer is presented using shearing, mixing, and deposition of plastic material during surfacing.

Acknowledgment

The authors thank the Director of National Institute of Technology Karnataka, Surathkal, India, for the permission, financial assistance, and appreciation, extended to carry out this investigation.

References

[1] W. T. Lankford, N. L. Samways, R. F. Craven, and H. E. McGannor, *The Making, Shaping and Treating of Steel*, USS, 10th edition, 1985.

[2] K. G. Budhinski and M. K. Budinski, *Engineering Materials, Properties and Selection*, PHI Learning Pvt Ltd., New Delhi, India, 9th edition, 2009.

[3] A. Nishimoto and K. Akamatsu, "Microstructure and oxidation resistance of Fe3Al coatings on austenitic stainless steel by spark plasma sintering," *Plasma Processes and Polymers*, vol. 6, no. 1, pp. S941–S943, 2009.

[4] Z. Xiao-Lin, Y. Zheng-Jun, G. Xue-Dong, C. Wui, and Z. Ping-Ze, "Microstructure and corrosion resistance of Fe-Al intermetallic coating on 45 steel synthesises by double glow plasma surface alloying technology," *Transactions of Nonferrous Metals Society of China*, vol. 19, pp. 143–148, 2009.

[5] G. Eggeler, W. Auer, and H. Kaesche, "Reactions between low alloyed steel and initially pure as well as iron-saturated aluminium melts between 670 and 800 degree c," *Zeitschrift fuer Metallkunde*, vol. 77, no. 4, pp. 239–244, 1986.

[6] S. Janakiraman, J. Reddy, S. V. Kailas, and K. Udaya Bhat, "Surface modification of steels using friction stir surfacing," *Materials Science Forum*, vol. 710, pp. 258–263, 2012.

[7] R. S. Mishra, M. W. Mahoney, S. X. McFadden, N. A. Mara, and A. K. Mukherjee, "High strain rate superplasticity in a friction stir processed 7075 Al alloy," *Scripta Materialia*, vol. 42, no. 2, pp. 163–168, 1999.

[8] G. W. Stachowiak and A. W. Batchelor, *Engineering Tribology*, Elsevier, Singapore, 3rd edition, 2005.

[9] J. L. Young Jr., D. Kuhlmann-Wilsdorf, and R. Hull, "The generation of mechanically mixed layers (MMLs) during sliding contact and the effects of lubricant thereon," *Wear*, vol. 246, no. 1-2, pp. 74–90, 2000.

[10] S. Cui, Z. W. Chen, and J. D. Robson, "A model relating tool torque and its associated power and specific energy to rotation and forward speeds during friction stir welding/processing," *International Journal of Machine Tools and Manufacture*, vol. 50, no. 12, pp. 1023–1030, 2010.

[11] Y. J. Chao, X. Qi, and W. Tang, "Heat transfer in friction stir welding: experimental and numerical studies," *ASME Journal of Manufacturing Science and Engineering*, vol. 125, pp. 138–145, 2003.

[12] P. J. Goodhew, J. Humphreys, and R. Beanland, *Electron Microscopy and Analysis*, Taylor and Francis, London, UK, 3rd edition, 2001.

[13] G. E. Dieter, *Mechanical Metallurgy*, McGraw Hill, London, UK, 1988.

[14] H. Zhai and H. Zhang, "Instabilities of sliding friction governed by asperity interference mechanisms," *Wear*, vol. 257, pp. 414–419, 2004.

20

CFD Simulations of Splash Losses of a Gearbox

Carlo Gorla,[1] Franco Concli,[1] Karsten Stahl,[2] Bernd-Robert Höhn,[2] Michaelis Klaus,[2] Hansjörg Schultheiß,[2] and Johann-Paul Stemplinger[2]

[1] *Dipartimento di Meccanica, Politecnico di Milano, Via la Masa 1, 20156 Milano, Italy*
[2] *Lehrstuhl für Maschinenelemente, FZG, 85748 Garching bei München, Germany*

Correspondence should be addressed to Franco Concli, franco.concli@mail.polimi.it

Academic Editor: Philippe Velex

Efficiency is becoming a main concern in the design of power transmissions. It is therefore important, especially during the design phase, to have appropriate models to predict the power losses. For this reason, CFD (computational fluid dynamics) simulations were performed in order to understand the influence of geometrical and operating parameters on the losses in power transmissions. The results of the model were validated with experimental results.

1. Introduction

Efficiency is becoming more and more a main concern in the design of power transmissions, and appropriate models to predict power losses are fundamental in order to reduce them, starting from the earliest stages of the design phase. Power losses of gearboxes are generally classified according to [1], taking into account the machine elements which are responsible for them and their dependency or nondependency from the load. Gear power losses are strongly related to lubrication, with those load dependent coming from the frictional effects in the lubricant film and those load independent mainly deriving from splashing, churning, and windage effects.

Some models, obtained on the basis of experimental tests, can be found in the literature which describe the influence of gear geometric and kinematic parameters on hydraulic losses like for instance those proposed by Mauz [2], who has concentrated on hydraulic losses, or by Dowson [3], who has concentrated on windage losses.

Nevertheless, the authors maintain that a deeper understanding of the physical phenomena responsible of gear losses is still needed in order to improve existing models, and CFD simulation can be an effective approach for such investigation.

Marchesse et al. [4], on the basis of a state-of-the art on the application of CFD to gear power losses, applied CFD models to study windage losses of gears, or have validated their results by means of experimental tests.

A large amount of experimental data on splashing power losses is still available at FZG, resulting from several years of tests [5], with either single discs, single gears and two meshing gears immersed in oil, and this data covers the influence of several parameters like outside diameter, face width, helix angle, temperature, and oil type. In order to improve the understanding of the mechanisms involved in splashing losses, CFD simulations have been run in order to investigate the effect of the same parameters, and the numerical results have been compared with the available test results.

In the first phase of the activities, which is presented in this paper, single discs and gears have been considered. The quite good accordance between CFD simulation and experimental tests has confirmed that CFD represents an effective approach to study splashing power losses, and a simulation program with two meshing gears has therefore been started and is under course.

2. Composition of Gearbox Power Losses

According to [1], the power losses of gears can be subdivided into load-dependent and load-independent losses:

$$P_V = P_{VG} + P_{VG0} + P_{VB} + P_{VB0} + P_{VD} + P_{VX}. \quad (1)$$

The total power losses can be further subdivided according to their origin: the subscripts G, B, D, and X are related to gears, to bearings, to contact seals, and to other factors respectively. The subscript 0 indicates load-independent power losses.

P_{VG} are the load-dependent power losses of gears and arise primarily from the sliding between the flanks of the gear teeth.

P_{VB} are the load-dependent power losses of bearings and are also related to sliding between the rolling elements and the rings.

P_{VB0} are the load-independent power losses of bearings and are, among others, related to viscous effects due to the lubrication.

P_{VD} are the power losses of the seals and are related to sliding.

P_{VX} are other generic power losses.

P_{VG0} are the load-independent power losses of gears due to the interaction between the lubricant and the rotating/moving elements. These losses are the sum of squeezing, power, and splashing power losses. The lubricant squeezing power losses are related to the fact that the gap at the mesh position is changing its volume during the engagement causing an overpressure that squeezes the oil primarily in the axial direction [6–8]. The power and splashing power losses are related to the viscous and pressure effects of the lubricant on the moving/rotating elements [9, 10]. For dip-lubricated power transmissions, this kind of losses cannot be neglected and are an important part of the total losses. For this reason, the authors have studied this kind of losses for a simple geometry and under different operating conditions.

3. Problem Description

In order to understand the influence of the different parameters, the authors have performed some initial simulation with a simplified geometry, consisting of a simple rotating disk in a fully filled case (no free surface is present, both in the experiments and the CFD approach). The presence of one single phase leads to classical mechanical losses generated by moving fluid around a solid. For example, when the fluid is air (or air/oil mixture), these losses are called windage, and the losses approximately evolve with angular velocity power 3. A schematic layout of the analyzed gearbox is shown in Figure 1. The disk is mounted on a cantilevered shaft that is supported by two bearings. A second over-hung mounted shaft is placed in the case parallel to the primary shaft. In a second steps a real gear geometry replaced the disk.

The simulations were conducted under pressure while the gearbox was developed to operate on the bottom of the sea and this pressure is needed in order to compensate the external water pressure at operating depth.

3.1. Oil Properties. In these investigations, different oils were used. The lubricants are mineral based in different viscosity grades. An additional virtual oil FVA3*, with same viscosities as FVA3 and same density as FVA2, was used to evaluate the influence of viscosity on power losses. The detailed oil properties can be seen in Table 1.

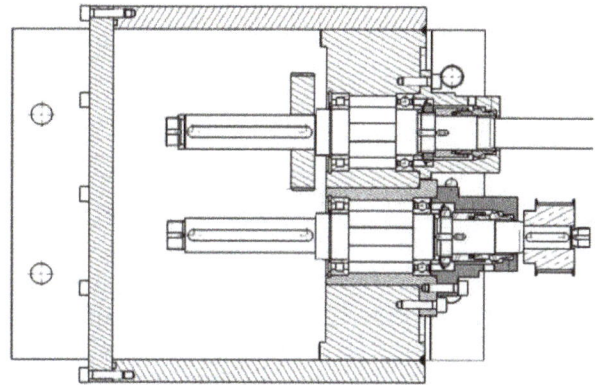

FIGURE 1: Geometry of the gearbox with the gear.

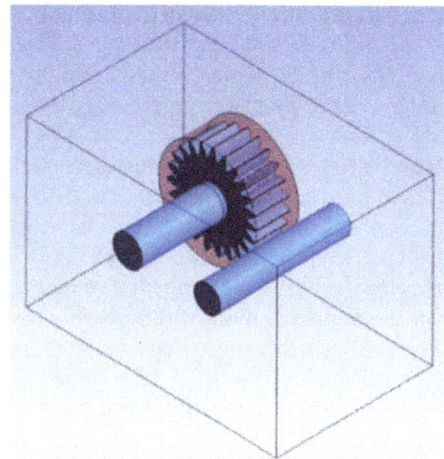

FIGURE 2: Geometry for the numerical model.

TABLE 1: Oil properties of the used lubricants.

Oil	Kinematic viscosity at 40°C ν_{40} (mm²/s)	Kinematic viscosity at 100°C ν_{100} (mm²/s)	Density at 15°C ρ_{15} (kg/m³)
FVA2	29.8	5.2	871
FVA3	95.0	10.7	885
FVA3*	95.0	10.7	871

TABLE 2: Geometrical properties of the disk.

Diameter d_a (mm)	100
Width b (mm)	40

3.2. Geometry. The geometry for the numerical model consists only of the oil volume. As shown in Figure 2, only a part of the case is modelled as also the bearings were neglected.

Both disk and gear are not mounted in the shaft's center; therefore, it is not possible to take advantage of symmetry.

The first analyzed component is a disk; its properties are summarized in Table 2.

Table 3 shows the geometrical parameters for the gear adopted in the different simulations.

TABLE 3: Geometrical properties of the analyzed gears.

	Width b (mm)	Tip diameter d_a (mm)	Helix angle β (°)	Pressure angle α_n (°)	Number of teeth $z_1()$	Normal moduls m_n (mm)
Reference case	40	102.5	0	20	23	4
Width influence	20	102.5	0	20	23	4
Tip diameter influence	40	$96.5 \div 98$	0	20	23	4
Helix angle influence	40	102.5	20	20	23	4

3.3. Mesh. For the disk's case, a simple undeforming mesh was assumed. In order to correctly reproduce the motion of the boundaries, appropriate BC (boundary conditions) were set. Tetrahedral elements have been used. For complex geometries, in fact, the tetrahedral mesh can often be created with far fewer cells than the equivalent mesh consisting of quadrilateral/hexahedral elements. This is because the tetrahedral mesh allows clustering of cells in selected regions of the flow domain. Structured hexahedral meshes will generally force cells to be placed in regions where they are not needed.

Tetrahedral cells are not desirable near walls if the boundary layer needs to be resolved because the first grid point must be very close to the wall, while relatively large grid sizes can be used in the directions parallel to the wall. These requirements lead to long thin tetrahedral elements, creating problems in the approximation of diffusive fluxes. For this reason, some grid generation methods generate first a layer of prisms or hexahedra near solid boundaries (inflation), starting with a triangular or quadrilateral discretization of the surface; on top of this layer, a tetrahedral mesh is generated automatically in the remaining part of the domain (Figure 3).

For the simulations with the gear, an additional feature was used: in order to model the rotation of the gear without mesh deformation, two separate domains were created and meshed separately. The region around the gear is a cylinder, which was defined as dynamic, it rotates around the gear axis. The other region, corresponding to the rest of the domain, was defined as a static zone. This zone has a cylindrical cavity, in which the dynamic mesh is located (as shown in Figure 4).

The two zones have some faces that are intersecting each other but from a numerical point of view they are not connected. In order to link them, an interface zone is defined. The connection ensures that the values of a generic field during the simulations are the same on both sides of the interface.

Due to the rotation of one of the two zones, an additional feature is necessary, sliding mesh. The two cell zones will move relative to each other along the grid interface. The sliding mesh condition does not move the fluid directly but moves the mesh only, allowing to set the appropriate motion of the boundaries inside the dynamic zone, in this case the faces of the gear.

3.4. Solver Settings

3.4.1. Navier Stokes Equations. The CFD is based on some differential equations. For a generic element, it is possible

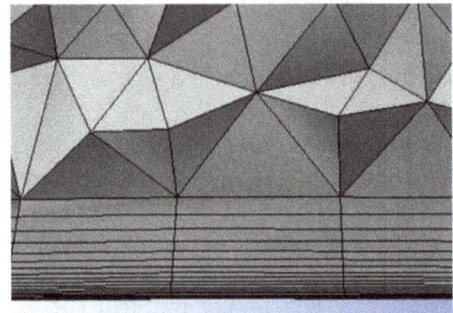

FIGURE 3: Detail of the mesh near the wall.

FIGURE 4: Partition of the mesh: the dynamic zone is marked with light blue, and the static zone is marked with light grey.

to write five equations. The first equation is the averaged mass conservation equation for no-stationary incompressible flows and can be written as follows:

$$\frac{\partial \langle \rho u_i \rangle}{\partial x_i} = 0, \tag{2}$$

where x_i is the Cartesian coordinate and u_i is the velocity component.

The second equation is the averaged momentum conservation equation and can be written as follows:

$$\frac{\partial}{\partial t}\left(\rho \langle u_i \rangle\right) + \frac{\partial}{\partial x_j}\left(\rho \langle u_i \rangle \langle u_i \rangle\right)$$

$$= -\frac{\partial}{\partial x_i}\langle p \rangle + \frac{\partial}{\partial x_j}\left[\mu\left(\frac{\partial \langle u_i \rangle}{\partial x_j} + \frac{\partial \langle u_j \rangle}{\partial x_i}\right) - \frac{\partial}{\partial x_j}\tau_{ij}\right], \tag{3}$$

where p is the pressure, x_i and x_j are the Cartesian coordinates, and ρ is the density. The additional term τ_{ij} that appears in the averaged equation (in comparison to the nonaveraged transport equations) is called unresolved term or Reynolds term. The averaging process produces a set of equations that is not closed. For this reason a turbulence model is needed in order to be able to solve the system of equations.

The unresolved or Reynolds term is expressed by using the eddy-viscosity hypothesis as follows:

$$-\tau_{ij} = -\rho \langle u_i' u_j' \rangle = \mu_t \left(\frac{\partial \langle u_i \rangle}{\partial x_j} + \frac{\partial \langle u_j \rangle}{\partial x_i} \right) - \frac{2}{3} \rho \delta_{ij} k, \quad (4)$$

where k is the turbulent kinetic energy and μ_t the eddy viscosity.

The energy equation was not activated in the given model: the operating temperature was defined as a priori and consequently the properties of the fluid do not change during the calculations.

For the pressure-velocity coupling a simple scheme was adopted as suggested for flows in closed domains [11].

3.4.2. Turbulence Models. Turbulent flows are characterized by fluctuating velocity fields. These fluctuations mix transported quantities such as momentum, energy, and species concentration, cause the transported quantities to fluctuate as well. Since these fluctuations can be of small scale and high frequency, they are too much computationally expensive to simulate directly in practical engineering calculations. Instead, the instantaneous exact governing equations can be averaged to remove the small scales, resulting in a modified set of equations that are computationally less expensive to solve. However, the modified equations contain additional unknown variables, and turbulence models are needed to determine these variables in terms of known quantities.

The simplest "complete models" of turbulence are two-equation models in which the solution of two separate transport equations allows the turbulent velocity and length scales to be independently determined. This model is a semiempirical model based on model transport equations for the turbulence kinetic energy (k) and its dissipation rate (ε) [12].

The RNG k-ε model assumes that the eddy viscosity is related to the turbulence kinetic energy and dissipation via the following relation:

$$\mu_t = C_\mu \rho \frac{k^2}{\varepsilon}. \quad (5)$$

The model transport equation for k is derived from the exact equation, while the model transport equation for ε was obtained using physical reasoning and bears little resemblance to its mathematically exact counterpart [11]. In the derivation of the k-ε model, the assumption is that the

flow is fully turbulent, and the effects of molecular viscosity are negligible:

$$\frac{\partial}{\partial t}(\rho k) + \frac{\partial}{\partial x_i}(\rho k u_i) = \frac{\partial}{\partial x_j} \left[\alpha_k \mu_{\text{eff}} \frac{\partial k}{\partial x_j} \right]$$
$$+ G_k + G_b - \rho \varepsilon - Y_M + S_k$$

$$\frac{\partial}{\partial t}(\rho \varepsilon) + \frac{\partial}{\partial x_i}(\rho \varepsilon u_i) = \frac{\partial}{\partial x_j} \left[\alpha_\varepsilon \mu_{\text{eff}} \frac{\partial k}{\partial x_j} \right] + \frac{C_{1\varepsilon} \varepsilon}{k}(G_k + C_{3\varepsilon} G_b)$$
$$- \frac{C_{2\varepsilon} \rho \varepsilon^2}{k} - R_\varepsilon + S_\varepsilon.$$
$$(6)$$

In these equations, G_k represents the generation of turbulence kinetic energy due to the mean velocity gradients. G_b is the generation of turbulence kinetic energy due to buoyancy effect. Y_M represents the contribution of the fluctuating dilatation in compressible turbulence to the overall dissipation rate. $C_{1\varepsilon}$, $C_{2\varepsilon}$, and $C_{3\varepsilon}$ are constants. σ_k and σ_ε are the turbulent Prandtl numbers for k and ε, respectively. The quantities $\propto_k$ and $\propto_\varepsilon$ are the inverse effective Prandtl numbers for k and ε, respectively. S_k and S_ε are user-defined source terms. R_ε is the additional term that characterizes the RNG-k-ε-model. R_ε is derived from a rigorous statistical technique called renormalization group theory and improves the accuracy of the standard-k-ε-model.

Some other more complex turbulent models exist, but some workers claim that this model offers improved accuracy in rotating flows [13]. For this reason, this turbulence model was selected to perform the simulations.

3.4.3. Spatial Discretization. By default, the solver stores discrete values of the scalar ϕ at the cell centers. However, face values ϕ_f are required for the convection terms and must be interpolated from the cell center values. This is accomplished using an upwind scheme. Upwinding means that the face value ϕ_f is derived from quantities in the cell upstream, or "upwind," relative to the direction of the normal velocity v_n.

In order to solve the differential equations, the authors have chosen to use second-order upwind schemes. When second-order upwinding is selected, the face value ϕ_f is computed on the base of the cell-centered averaged value ϕ and its gradient $\Delta \phi$. For the determination of the gradient, a least squares cell-based evaluation was used. In this method, the solution is assumed to vary linearly between two cell centroids.

3.4.4. Boundary Conditions. All the boundaries were set to a no slip condition. That means that on the wall there is no relative velocity between the walls and the fluid. To describe properly the velocity profile in the normal direction, an enhanced wall treatment was used. This technique calculates the velocity at the elements' center and then reconstructs the velocity profile starting from these quantities. For this reason, it is important to create an appropriate fine mesh near the walls.

In order to evaluate the quality of the results in terms of velocity profile one has to check the y^+ value that is defined as follows:

$$y^+ = \frac{u_* y}{\nu},\qquad(7)$$

and—in case of enhanced wall treatment—should be approximately equal to one. u_* is the friction velocity at the nearest wall, y is the distance from the center of the first cells to the wall and ν is the local kinematic viscosity of the fluid (Figure 5).

3.4.5. Time Steps. The determination of the time step size is based on the estimation of the truncation error associated with the time integration scheme. If the truncation error is smaller than a specified tolerance, the size of the time step is increased; if the truncation error is greater, the time step size is decreased. An estimation of the truncation error can be obtained by using a predictor-corrector type of algorithm according to [14] in association with the time integration scheme. At each time step, a predicted solution can be obtained using a computationally inexpensive explicit method (Adams-Bashford for the second-order unsteady formulation). This predicted solution is used as an initial condition for the time step, and the correction is computed using the nonlinear iterations associated with the implicit formulation. The norm of the difference between the predicted and corrected solutions is used as a measure of the truncation error. By comparing the truncation error with the desired level of accuracy the code, is able to adjust the time step size by increasing it or decreasing it.

4. Operating Conditions

With CFD analysis, the effect of different parameters was investigated. For the simple case with the disk instead of the gear, the only varied parameter was the velocity in a range between 500 and 8000 rpm (Table 4).

For the case with the gear, the investigated parameters were the velocity, the pressure, and the geometry (face width, tip diameter, and helix angle).

Table 4 summarizes the combination of parameters adopted in the different simulations.

The real gearbox used for the experiments was developed to operate on the bottom of the sea and an internal pressure (6 bar) was needed in order to compensate the external water pressure at operating depth.

The experiments have been therefore performed with the operating conditions of 6 bar. The pressurization of the gearbox was made after some experiments with an operating pressure of 0 bar. The choice to operate with 6 bar was taken in order to be sure to have only oil in the gearbox, avoiding the presence of air bubbles. Moreover, the level of the static pressure does not affect the results.

5. Experimental Tests

In order to validate the results of the CFD simulations, some experiments were performed. On the driving shaft, a torque

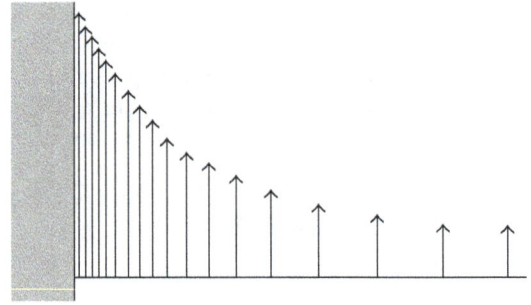

FIGURE 5: Velocity profile near the wall.

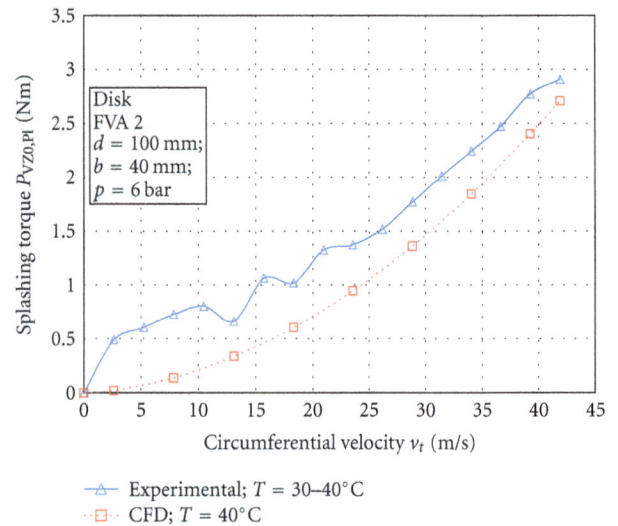

FIGURE 6: Results in terms of resistant torque versus circumferential velocity for the disk.

meter was mounted. This instrument measures the resistant torque and the rotational velocity so it is possible to evaluate the power losses. Due to the configuration of the system, it is not possible to measure directly the load-independent power losses due to power. For this reason, some additional tests were performed without the gear/disk and with air instead of oil in the gearbox so as to be able to measure the power losses generated by the bearings. Making the difference between the total losses and the losses of the bearings, it is possible to evaluate the losses only due to the churning.

6. Results

6.1. Results: Disk. Figure 6 shows the results in terms of no load loss torque $P_{VZ0,Pl}$ versus circumferential v_t velocity for the case with the disk. The continuous line represents the results of the experiments without the losses of the bearings and the dots the results of the simulations. In this simple case, the results of the simulations appear to underestimate the losses. It is however to say that the experimental results that appear in Figure 6 are obtained like the difference between the total measured losses in the working configuration (including therefore the bearing

TABLE 4: combination of parameters adopted in the different simulations.

	Disk or Gear	d_a (mm)	b (mm)	β (°)	n (rpm)	p (bar)	T (°C)	Oil level (%)	Oil
A	D	100	40	/	500	6	40	100	FVA2
B	D	100	40	/	2000	6	40	100	FVA2
C	D	100	40	/	5000	6	40	100	FVA2
D	D	100	40	/	8000	6	40	100	FVA2
1.1	G	102.5	40	0	1500	6	88*	100	FVA2
1.2	G	102.5	40	0	3000	6	84*	100	FVA2
1.3	G	102.5	40	0	4500	6	85*	100	FVA2
1.4	G	102.5	40	0	6000	6	93*	100	FVA2
1.5	G	102.5	40	0	7500	6	107*	100	FVA2
2.1	G	102.5	40	0	500	6	90	100	FVA2
2.2	G	102.5	40	0	2000	6	90	100	FVA2
2.3	G	102.5	40	0	5000	6	90	100	FVA2
2.4	G	102.5	40	0	8000	6	90	100	FVA2
3.1	G	102.5	20	0	500	6	90	100	FVA2
3.2	G	102.5	20	0	2000	6	90	100	FVA2
3.3	G	102.5	20	0	5000	6	90	100	FVA2
3.4	G	102.5	20	0	8000	6	90	100	FVA2
4.1	G	98	40	0	500	6	89+	100	FVA2
4.2	G	98	40	0	2000	6	89+	100	FVA2
4.3	G	98	40	0	5000	6	89+	100	FVA2
4.4	G	98	40	0	8000	6	109+	100	FVA2
5.1	G	98	40	0	500	6	90	100	FVA2
5.2	G	98	40	0	2000	6	90	100	FVA2
5.3	G	98	40	0	5000	6	90	100	FVA2
5.4	G	98	40	0	8000	6	90	100	FVA2
6.1	G	96.5	40	0	500	6	90	100	FVA2
6.2	G	96.5	40	0	2000	6	90	100	FVA2
6.3	G	96.5	40	0	5000	6	90	100	FVA2
6.4	G	96.5	40	0	8000	6	90	100	FVA2
7.1	G	102.5	40	20	500	6	90	100	FVA2
7.2	G	102.5	40	20	2000	6	90	100	FVA2
7.3	G	102.5	40	20	5000	6	90	100	FVA2
7.4	G	102.5	40	20	8000	6	90	100	FVA3
8.1	G	102.5	40	0	500	6	90	100	FVA3
8.2	G	102.5	40	0	2000	6	90	100	FVA3
8.3	G	102.5	40	0	5000	6	90	100	FVA3
8.4	G	102.5	40	0	8000	6	90	100	FVA3
9.1	G	102.5	40	0	500	6	90	100	FVA3*
9.2	G	102.5	40	0	2000	6	90	100	FVA3*
9.3	G	102.5	40	0	5000	6	90	100	FVA3*
9.4	G	102.5	40	0	8000	6	90	100	FVA3*

+ Measured temperature.
* Oil with the viscosity of FVA3 and the density of FVA2.

losses) and the losses measured with only the shaft (no disk) and air instead of oil in the gearbox. In this manner, it is possible to separate the power losses due to the rotation of the disk in the oil from the losses caused by the bearings. It is however to say that in this second measurement, the absence of the lubricant may minimally affect the losses caused by the bearings: the inner race of the bearings, in fact, during the rotation is laterally in contact with air instead of oil, and this leads to a minimal underestimation of the losses of the bearing. This problem becomes less significant in the measurements with the gear where the power losses are significantly higher.

Another reason for the little difference between the numerical and the experimental results can be found in the

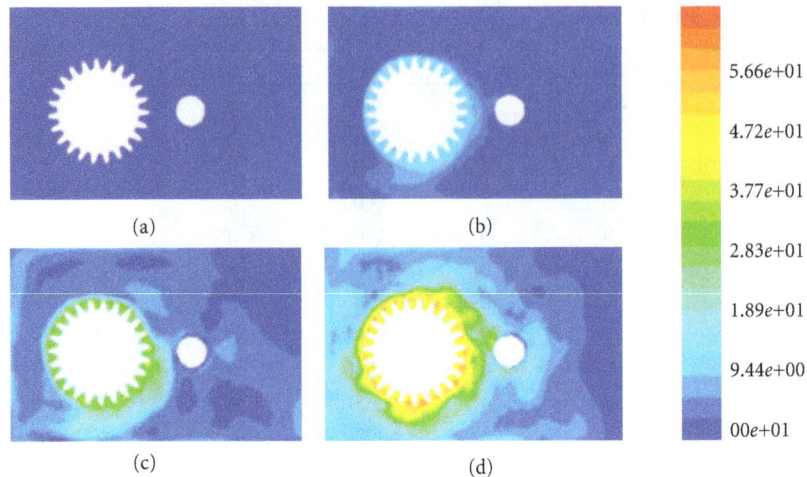

FIGURE 7: Contour plot of the velocity field for four different tangential velocities (m/s) in the symmetry plane of the gear: (a) 500 rpm; (b) 2000 rpm; (c) 5000 rpm; (d) 8000 rpm.

fact that in the simulations, the temperature (and therefore the oil properties) was fixed to 40°C while in the experiments, it was varying due to the fact that the cooling/heating system was not able to compensate immediately a change in the temperature.

6.2. Results: Gear.

Figure 7 shows the velocity field in the gear symmetry plane for four different tangential velocities. It can be seen that the area in which the velocity is appreciable is expanding with the increase of the tangential velocity.

From Figure 8, it is possible to have an idea of the fluxes of lubricant inside the gearbox. An ideal particle of fluid escapes the gear radially and then, after a loop, it comes again in contact with the gear from the axial direction. Due to the rotation of the gear, there are no zones in the gearbox where the lubricant is steady but the complete domain is involved in the lubricant flux.

Figures 9 to 13 show the results in terms of splashing resistant torque $P_{VZ0,Pl}$ versus circumferential velocity v_t for the cases with the gear. The continuous line represents the results of the experiments without the losses of the bearings, the broken line the results of the calculations according to Mauz [2], and the dotted line the results of the simulations.

Mauz proposed some equations derived by some experiments. The model proposed by Mauz has the big advantage that it does not need computational resources but, as shown in the next diagrams, it has also some limits. For example, it is not valid for tangential velocities higher than 60 m/s and it does not take into account some parameters like, for example, the helix angle.

Figure 9 shows the influence of the temperature on the resistant torque. It can be seen that the influence of the temperature on the results cannot be neglected. The experimental tests were performed with a temperature that is not exactly constant, which is due to the cooling/heating system that is not able to compensate immediately a fluctuation of the temperature of the lubrication bath. Some simulations were performed with the same temperature measured in

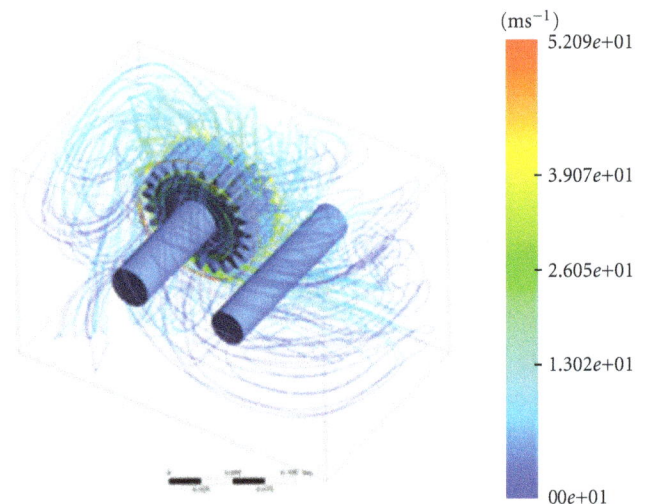

FIGURE 8: Velocitystreamlines; $b = 40$ mm; $d = 102.5$ mm; $T = 90°$C; $n = 8000$ rpm.

the experiments for each rotational speed: the lubricant properties have been changed in each simulation according to the measured temperature at the different rotational speeds. It can be seen that, the CFD results are in very good agreement with the measured ones. Some other calculations, instead, were performed with a constant temperature of 90°C. The CFD results for the constant temperature are below the measured ones only for low rotational speeds where the effective temperature of the oil bath in the test rig was lower than 90°C. For higher rotational speeds the CFD results calculated with a constant temperature of 90°C are over the measured ones where the temperature of the oil bath in the test rig was higher than 90°C.

The diagram shows also the power losses according to Mauz. The results differ significantly from the measured and simulated values.

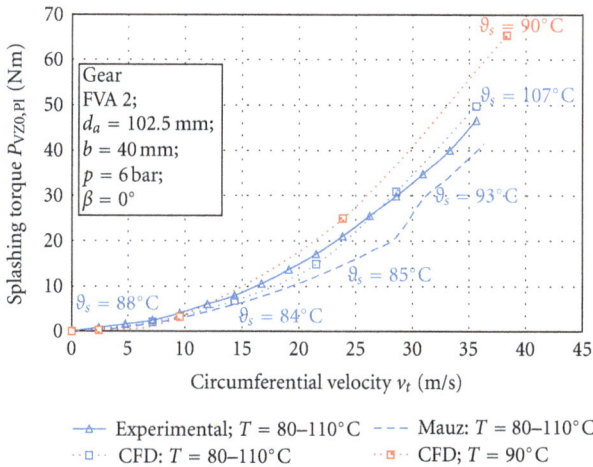

FIGURE 9: Results in terms of resistant torque versus circumferential velocity—effect of the temperature.

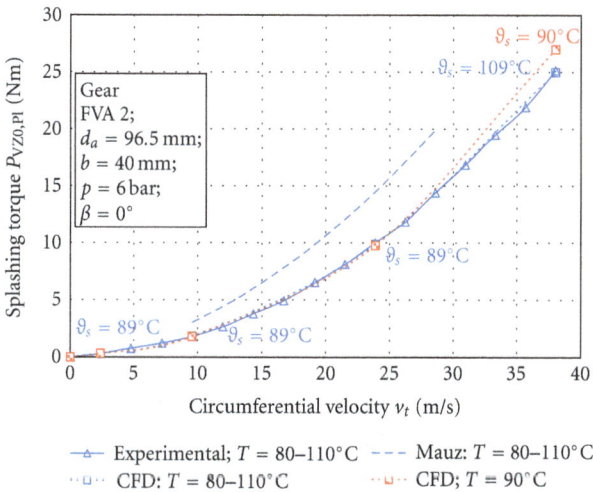

FIGURE 10: Results in terms of resistant torque versus circumferential velocity—effect of the temperature.

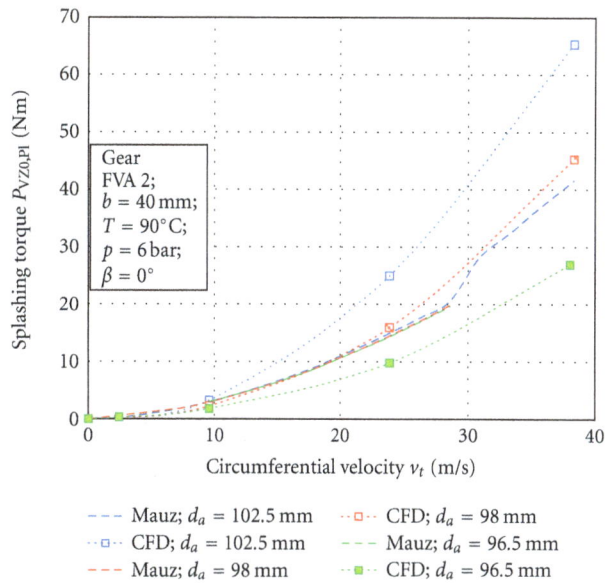

FIGURE 11: Results in terms of resistant torque versus circumferential velocity—influence of the tip diameter.

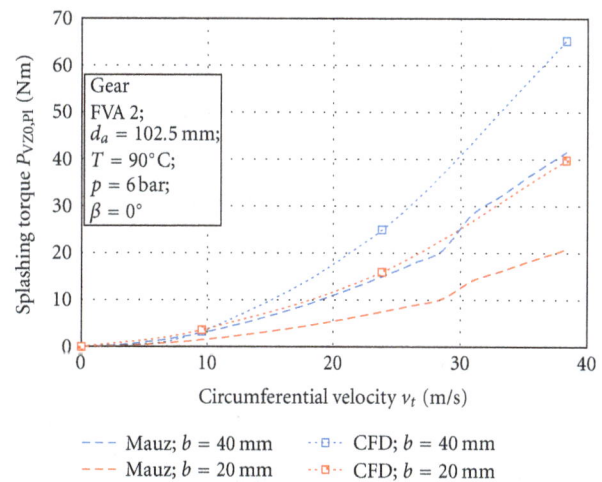

FIGURE 12: Results in terms of resistant torque versus circumferential velocity—Influence of the face width.

Figure 10 shows the similar results as Figure 8 but for another tip diameter. Also in this case, the results of the simulations performed with the same temperature of the experiments overlap the measured results.

Figure 11 shows the effect of the tip diameter on the power losses. Simulations have been performed for a diameter of 102.5 mm, 98 mm, and 96.5 mm. Experiments have been conducted only for the biggest and the smallest diameter (see Figures 9 and 10).

The effect of the tip diameter on the power losses is extremely high. The results according to Mauz did not correctly describe this phenomenon. The results of the simulations, however, are able to well predict the power losses as confirmed by the experiments.

An increase of the tip diameter from 96.5 mm to 98 mm causes an increase of the power losses of about 64%.

Similarly, an increase of the tip diameter from 96.5 mm to 102.5 mm causes an increase of the power losses of about 142%.

Figure 12 shows the effect of the face width on the power losses.

The effect of the face width on the power losses is extremely high. Also in this case the results of the simulations are better than the results obtained by Mauz even if in this case also for Mauz an increase of the face width of 100% gives a significant increase of the power losses.

The doubling of the tip diameter from 20 mm to 40 mm causes an increase of the power losses of about 60%.

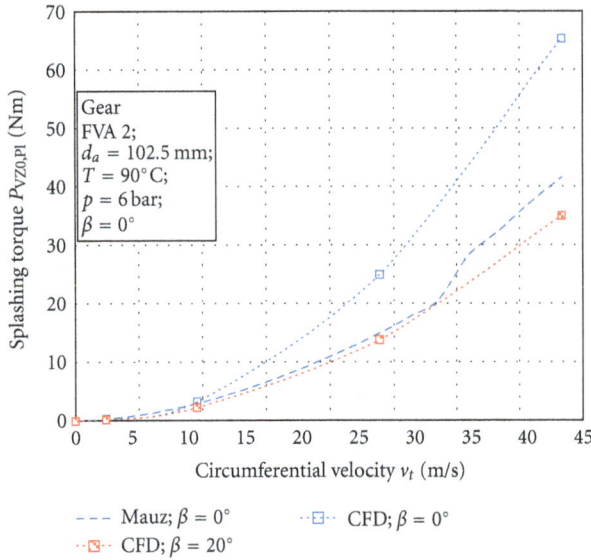

FIGURE 13: Results in terms of resistant torque versus circumferential velocity—Influence of the helix angle.

Figure 13 shows the effect of the helix angle on the power losses. It is possible to appreciate how the losses decrease with the helix angle.

In particular, the losses decrease to about 45% changing the helix angle from 0° to 20°. The model proposed by Mauz does not take into account the helix angle.

Figure 14 shows the effect of the oil viscosity on the power losses, while Figure 15 shows the influence of the oil type (density) on the power losses.

Increasing the density from 819.5 Kg/m^3 to 824.5 Kg/m^3 leads to an increase of the power losses of about 15%.

7. Conclusions

The results of the experiments confirm that the CFD represent a valid method to predict power losses. The error in the predictions for the analyzed cases is lower than 5%.

The simulations were performed for two different components, disk and gear. The power losses of the dip lubricated disk appear significantly lower than the power losses of dip lubricated gear.

Additional simulations were performed in order to understand the influence of different parameters on the power losses. Power losses increase significantly with a larger tip diameter or face depth and decrease with a larger helix angle. Also the temperature has an influence on the results: the losses decrease with a higher temperature and thus reduced lubricant viscosity and density. But while the influence of the density is significant, the viscosity seems to have no significant effects on the power losses.

The trend of the results is in line with those presented by Höhn et al. [5] for the case with two gears. Moreover, as expected, the losses evolve with angular velocity power 3.

Since CFD proves to be a valid tool to predict the load-independent losses, it is planned to investigate other

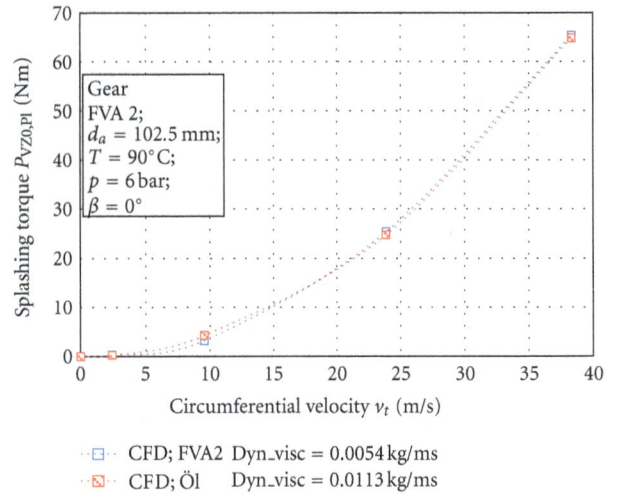

FIGURE 14: Results in terms of resistant torque versus circumferential velocity—Influence of the lubricant viscosity.

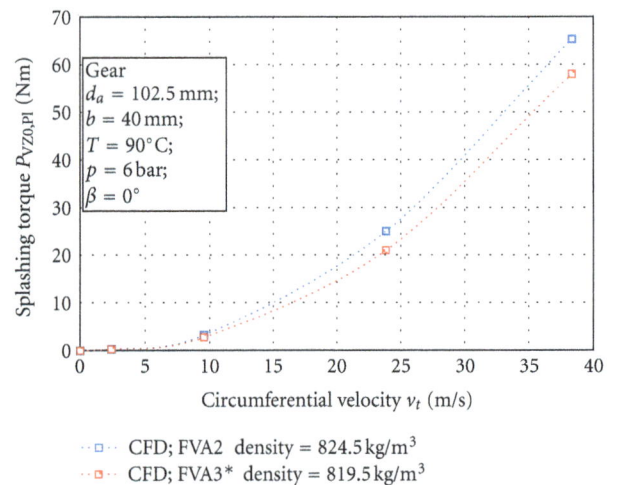

FIGURE 15: Results in terms of resistant torque versus circumferential velocity—Influence of the lubricant density.

components of the load-independent power losses, like, for example, the oil squeezing losses.

References

[1] G. Niemann and H. Winter, *Maschinenelemente Band II—Getriebe Allgemein, Zahnradgetriebe—Grundlagen, Stirnradgetriebe*, Springer, New York, NY, USA, 1983.

[2] W. Mauz, *HydrauliSche VerluSte Von Stirnradgetrieben Bei Umfangsgeschwindigkeiten BiS 60 M/S*, Bericht des Institutes für Maschinenkonstruktion und Getriebebau Nr. 159, Universität Stuttgart, 1987.

[3] P. H. Dawson, "Windage loss in larger high-speed gears," *Proceedings of the Institution of Mechanical Engineers*, vol. 198, no. 1, pp. 51–59, 1984.

[4] Y. Marchesse, C. Changenet, F. Ville, and P. Velex, "Investigations on CFD simulations for predicting windage power losses

in spur gears," *Journal of Mechanical Design*, vol. 133, no. 2, Article ID 024501, 7 pages, 2011.

[5] B. R. Hohn, K. Michaelis, and H. P. Otto, "Influence on no-load gear losses," in *Proceedings of the Ecotrib Conference*, vol. 2, pp. 639–644, 2011.

[6] F. Concli and C. Gorla, "Influence of lubricant temperature, lubricant level and rotational speed on the churning power losses in an industrial planetary speed reducer: computational and experimental study," *International Journal of Computational Methods and Experimental Measurements, Wessex Institute of Technology*. In press.

[7] F. Concli and C. Gorla, *Computational and Experimental Analysis of the Churning Power Losses in an Industrial planetary Speed Reducers*, Advances in Fluid Mechanics IX, WIT Transactions on Engineering Sciences, Wessex Institute of Technology, 2012.

[8] F. Concli and C. Gorla, "Churning power losses in planetary speed reducer: computational-experimental analysis," in *Proceedings of the EngineSOFT International Conference*, 2011.

[9] F. Concli and C. Gorla, "Oil squeezing power losses of a spur gear pair by mean of CFD simulations," in *Proceedings of the Bienall Conference on Engineering Systems Design and Analysis (ESDA '12)*, 2012.

[10] F. Concli and C. Gorla, *Oil Squeezing Power Losses of a Gear Pair: A CFD Analysis*, Advances in Fluid Mechanics IX, WIT Transactions on Engineering Sciences, Wessex Institute of Technology, 2012.

[11] H. K. Versteeg and W. Malalasekera, *An Introduction to Computational Fluid Dynamics—the Finite Volume Method*, Longman Group, London, UK, 1995.

[12] B. E. Launder and D. B. Spalding, "The numerical computation of turbulent flows," *Computer Methods in Applied Mechanics and Engineering*, vol. 3, no. 2, pp. 269–289, 1974.

[13] V. Yakhot, S. A. Orszag, S. Thangam, T. B. Gatski, and C. G. Speziale, "Development of turbulence models for shear flows by a double expansion technique," *Physics of Fluids A*, vol. 4, no. 7, pp. 1510–1520, 1992.

[14] P. M. Gresho, R. L. Lee, and R. L. Sani, *On the Time Dependent Solution of the Incompressible Navier-Stokes Equations in Two and Three Dimensions*, Recent Advances in Numerical Methods in Fluids, Pineridge Press, Swansea, UK, 1980.

21

A Correlative Defect Analyzer Combining Glide Test with Atomic Force Microscope

Jizhong He[1,2]

[1] Institute of Engineering, College of Engineering, Peking University, Nanjing, Jiangsu 210012, China
[2] MicroFocus Technologies, Inc., Wuxi, Jiangsu 214125, China

Correspondence should be addressed to Jizhong He; jz.he@ufocustech.com

Academic Editor: Tom Karis

We have developed a novel instrument combining a glide tester with an Atomic Force Microscope (AFM) for hard disk drive (HDD) media defect test and analysis. The sample stays on the same test spindle during both glide test and AFM imaging without losing the relevant coordinates. This enables an in situ evaluation with the high-resolution AFM of the defects detected by the glide test. The ability for the immediate follow-on AFM analysis solves the problem of relocating the defects quickly and accurately in the current workflow. The tool is furnished with other functions such as scribing, optical imaging, and head burnishing. Typical data generated from the tool are shown at the end of the paper. It is further demonstrated that novel experiments can be carried out on the platform by taking advantage of the correlative capabilities of the tool.

1. Introduction

Media defect control has always been a critical part of the HDD manufacturing process. It has a direct effect on the manufacturing product yield which drives the bottom line of business. In the hard disk drive, high media defect level can also cause reliability problems resulting in unforeseen economic losses. Furthermore, defect-free media is an enabler for implementing new HDD technologies. On the other hand, in order to allow high areal density recording necessary for sustained market growth, the head disk spacing in HDD has been pushed down to an extremely small margin [1–3]. As a result, even defects with very small sizes are now becoming serious performance and reliability challenges.

Defect failure analysis (DFA) which analyzes media defects on rejected disks from the production lines plays a central role in the defect process control as it finds the root causes and provides clues for corrective actions. The DFA is done separately from the line test, for example, the glide test. The normal procedure is to send a small portion of the line rejects to the DFA lab where the technicians try to relocate the defects manually, for example, with an optical microscope, before sending them off for examination with an analytical

tool such as an AFM. As the criteria for the defects of interest become smaller, manual defect relocation becomes a bigger problem. There are often cases of missed defects when doing DFA or not finding the right ones within the contaminations generated during the handling after the line test, leading to long frustrating days with negative impacts on manufacturing progress.

We have developed a tool which combines glide with AFM. We choose the glide test for its unique sensitivity to only asperities with height, as these are more likely to be in the "killer" defect category in the drive [4, 5]. On the other hand, AFM has been widely used in the medial DFA labs [6, 7]. It provides critical topographic data with the nanometer resolution, perfect for analyzing media defects of the high-density HDD media. AFM has a typical maximum scan range of $100\,\mu m$ or less. As a result, the defect has to be located right in the middle of the AFM scan range which is oftentimes proven to be difficult to do it manually. This severely restricts the access of the defect analysis with AFM.

The combination of glide with AFM solves a number of problems. It enables an immediate in situ analysis with the AFM after the glide test because the disk sample stays on the same chuck without losing the coordinates. It enables finding the smaller defects more accurately. The combination also makes the automation possible resulting in orders of magnitude faster throughput for media DFA. We call the tool the Correlative Defect Analyzer (CDA) because multiple tests or analyses are integrated in a single tool to perform on the same correlated defects.

2. Design and Construction

The concept of combining AFM with glide is not new. There was at least one attempt by one of the HDD companies where an AFM module was added to an existing glide tester. No specific technical details are known to the author, but there were reported problems of vibration and weak algorithm, resulting in low AFM image quality and missed defects. (Private communication with relevant engineers familiar with the tool; no commercial product was ever released to the general market.)

The CDA tool described in the paper is designed from the ground up. It consists of four major functional blocks; see Figure 1. A spindle is mounted on a plate secured to a granite base by a pair of high rigid precision linear guides. A high precision ball screw drives the stage in the X direction as the X-stage; see part A of Figure 1. The ball screw provides a stiffness of $15 \, \text{N}/\mu\text{m}$, while the guides secure the lateral movement with a stiffness better than $100 \, \text{N}/\mu\text{m}$. The high stiffness is necessary to control the spindle vibration at high RPM. The granite is chosen for its unique vibration damping characteristics. The AFM is mounted on a miniature ball screw stage bolted on a vertical granite arch; see part B of Figure 1. A Z-stage is also mounted vertically on the arch. The Z-stage is comprised of a pair of linear guides, a ball screw driving mechanism, and a housing for the optics. The housing also provides an attachment plane for an automatic turret; see part C of Figure 1. The glide head mount is installed on the automatic turret along with optical objectives, a burnish head mount, and a scriber; see part D of Figure 1.

The air bearing spindle has a dual mode capable of both high speed spinning for glide test, the G mode, and high precision positioning for AFM imaging, the A mode (Chinese patent pending, 201110142434.1). In the G mode, the spindle operates with the controller set to run the RPM to a very low jitter level. In the A mode, the controller is switched to turn the spindle as a rotary stage capable of a resolution of about 3 arc sec. After the desired location is reached, the lower part of the air support is removed so that the spindle body is pushed down by the upper air support and sits securely to the spindle housing; see Figure 2. The solid contact between the spindle shaft and the spindle housing is necessary for achieving low-noise high-resolution AFM imaging.

Care has been taken during the design phase to select the right components with matching material and functional properties in order to achieve the stringent requirements for long-term stability and high positioning repeatability. To verify the performance of the tool, we have designed experiments to measure the positioning repeatability of the stages. Here in this paper, we only show one example measurement on the X-stage. The measurement method on the other stages is similar.

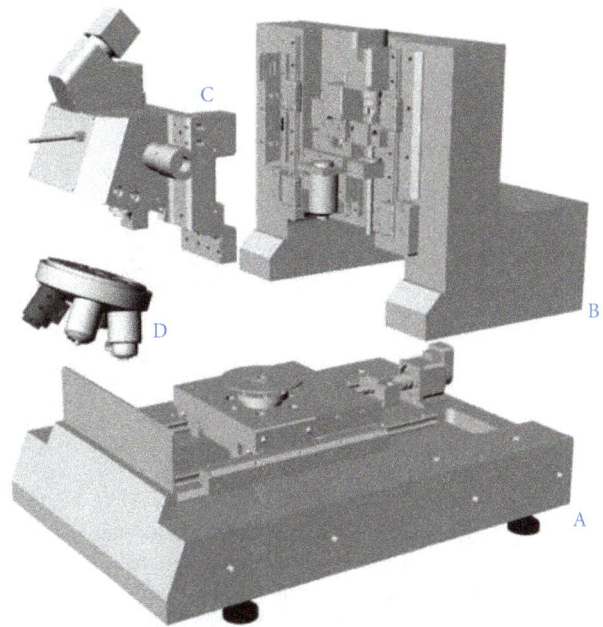

FIGURE 1: The exploded view of the CDA-101A. Part A, the granite base with the spindle mounted on the X-stage. Part B, the vertical granite arch with the AFM stage and the Z-stage. Part C, the optical housing with CCD camera and automatic turret mount. Part D, the automatic turret with glide, burnish, scriber, and optical objectives. Read the text for further detailed description.

A disk with a micrometer sized defect is clamped on a chuck driven by the X-stage. It serves as the target for measuring the repeatability with the stage repeatedly moved to a predetermined command location. A 50x objective is focused on the disk sample surface and takes the images whenever the disk is moved into the predetermined command position. Images are analyzed automatically by designated software to determine the offset of the target defect to the center. From the offset, the equivalent X-stage position is determined, where the image of the defect will be centered under the microscope.

Statistical data is shown in Figure 3 where the frequency occurrence is plotted against the position of the X-stage with a bin size of $1 \, \mu\text{m}$. At these X-stage positions, the target defect is at the exact center in the microscope view. The scatter of the defect location is attributed primarily to the minute shift in the mechanical stage every time it is moved. The two peaks in the plot indicate the two populations of the stage position. For each population, the stage is found to have very high short term repeatability with a standard deviation of about $50 \, \text{nm}$. The reason for the two position population is because that the experiment is carried out at an elevated room temperature first and repeated after overnight cooling at another temperature, resulting in two stable positions at the two temperatures. The overnight temperature swing is about 10 deg Celsius, resulting in a drift of $6 \, \mu\text{m}$. With a characteristic length of 30 cm, this corresponds to 2 ppm/C, a very respectable figure and sufficiently low enough for current applications.

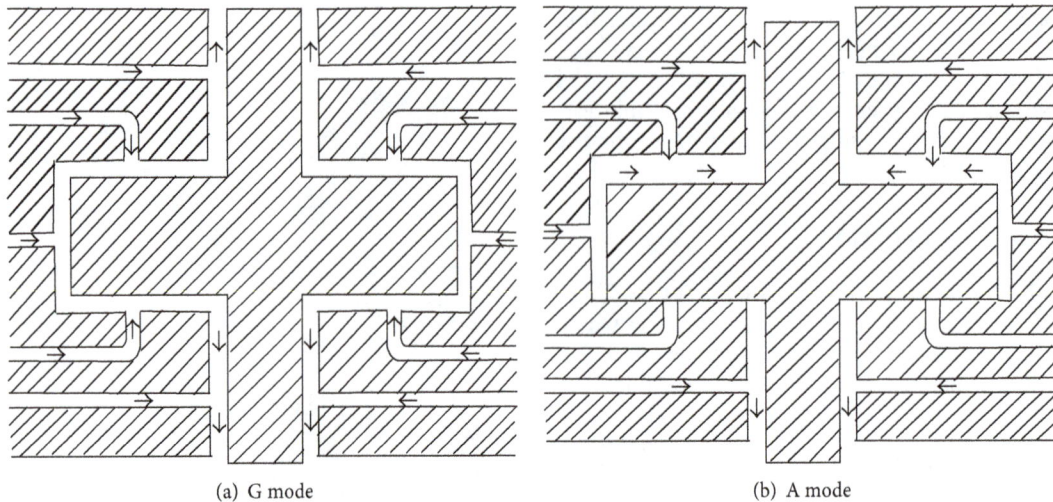

(a) G mode

(b) A mode

FIGURE 2: The dual mode spindle. (a) The G mode. The arrows indicate the flow direction of the air support. In the G mode, the spindle is supported both radially and axially like a conventional air bearing spindle. The spindle is under speed control circuitry to achieve high speed spinning with minimal jitter. (b) The A mode. In the A mode, the spindle is under the positioning control circuitry. It functions as a slow turning rotary table and can be clamped down for imaging when in position. To clamp down the spindle, the bottom axial air support is removed. As the result, the spindle body is pressed against the housing. The working pressure for the air spindle is 5 bars. Note the spindle shown in (b) is already in the clamped position.

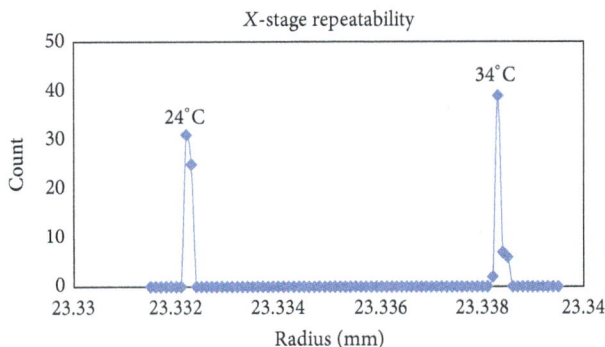

FIGURE 3: The frequency count plot of the X-stage position, with a bin size of 0.001 mm, that the X-stage must be moved to in order for the target defect to be in the middle of the microscope view. The two peaks correspond to the two highly repeatable positions of 23.332 mm and 23.338 mm due to the two stabilized room temperatures. See text for more detailed description.

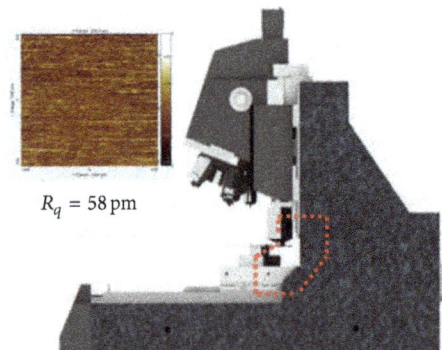

$R_q = 58$ pm

FIGURE 4: The illustration of the mechanical loop in the AFM. The mechanical loop in sequence consists of the sample, the sample chuck, the stage, the linear guides, the granite base, the granite arch, the approach stage, the scanner body, and the cantilever tip. Test image for noise floor measurement is shown in the insert. The z-scale is ±250 pm. The RMS noise is estimated to be 58 pm.

The AFM scanner is an OEM component. It has a very compact body with a size of a typical microscope objective. The optical interferometry detection scheme has a published noise level of 10 pm. To realize the full potential, we mount the scanner with the approach stage on the massive granite arch to maintain high rigidity. The key is to design the system with the shortest mechanical loop possible; see Figure 4. The short mechanical loop reduces the susceptibility of the scanner to external disturbances. We carry out an effectively still scan (scan range is 1 nm) to measure the mechanical noise floor. Since the tip essentially stays at one spot on the sample, the resulting scan is a measure of the vibration the tip experiences. The data taken with the tool on a homebuilt isolation table without an acoustic enclosure and on the fourth floor of a building shows the RMS noise floor to be about 60 pm. The figure of the noise floor is determined with the commercial application SPIP by using the roughness calculator (SPIP SPM Image Processor application by Image Metrology, Horsholm, Demark; the roughness, Sq, calculation at zero scan range by definition is the direct measure of the system mechanical RMS noise floor); see the insert in Figure 4. The image shows overall random signal fluctuation. The streaking in the image is the result of the low frequency noise showing up as an artifact from the raster scan and has no significance towards the calculation of the roughness.

(a) Glide trace panel

1	15	29
2	16	30
3	17	31
4	18	32
5	19	33
6	20	34
7	21	35
8	22	36
9	23	37
10	24	38
11	25	39
12	26	40
13	27	41
14	28	42

(b) Arrangement schematic

R: 28.165 + 0.460

(c) Trace #1

R: 28.110 + 0.460

(d) Trace #29

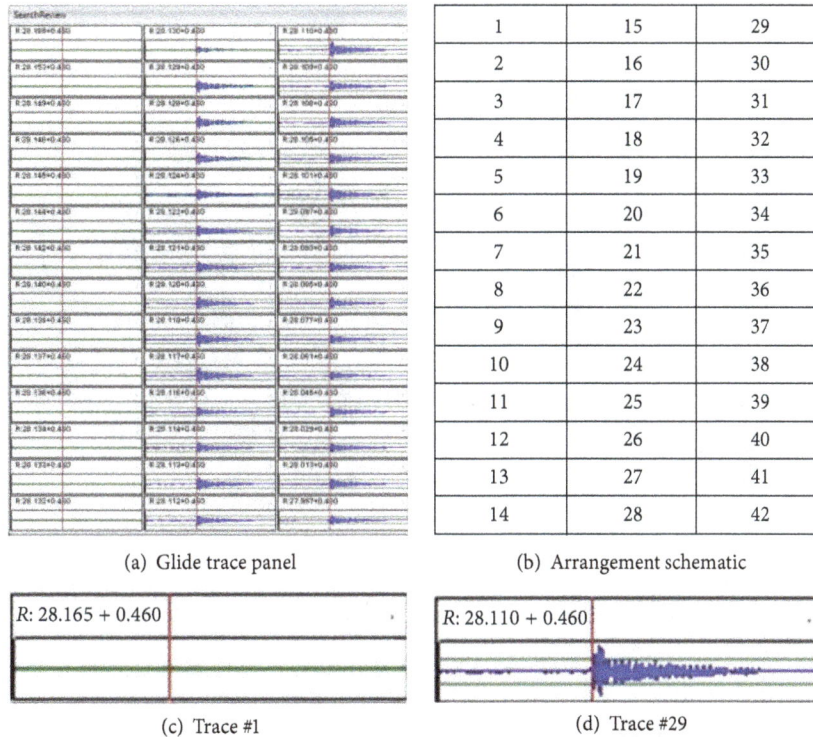

FIGURE 5: Signal analysis panel illustrating the "Glide Edge Finder" algorithm with signal at 42 tracks. The first track with the glide signal is track no. 15 with a radius of 28.130 mm. (a) Glide signal panel showing the signal at each track arranged in descending order of the radius. (b) Schematic of the track arrangement. (c) Signal trace at track no. 1 with the radius equal to 28.165 mm. No signal due to defect is present. (d) Signal trace at track no. 29 with the radius equal to 28.110 mm. Peak signal from defect is present. The starting radius with defect signal defines the edge, 28.130 mm. The value 0.460 mm is the glide head edge offset. The angle is obtained from the averaging of the peak angular position of the data, shown as the red line in (a).

3. Defect Indexing and Unified Coordinate System (UCS)

The width of a typical glide head is about one millimeter. Glide heads with wider sensor area can complete a full surface scan in fewer tracks. This is necessary in order to meet the throughput requirement on the production lines because every disk goes through the glide before shipping. Due to the large sensor width, the uncertainty from the glide test is about a few hundred micrometers in the radial direction and a few millimeters in the track direction. We have developed a "Glide Edge Finder" algorithm that narrows down the location of the defects to within $5 \mu m$. The algorithm moves the glide head track by track in very fine steps in the vicinity of the defect based on results from the prior glide test. It finds the edge of the glide head where the defect first makes contact with the glide head. This method is independent of the shape of the glide air bearing slider. However, the exact shape of a glide head does affect the glide head edge offset. As a result, every new glide head will need to be calibrated for the edge offset when first installed on the tool.

To help explain the process, we show a screen capture of a signal review panel of the CDA software for an actual defect indexing; see Figure 5. The glide signal traces are displayed for each individual tracks. The transition from the tracks with no signal to the ones with signal determines the edge. The angular position is determined by the peak of the signal in relationship to the encoder position of the spindle. For the tracks with the glide signal, the angular positions of the peaks are collected and averaged for statistics. The algorithm has certain logic built-in to handle small defects with marginal signals.

The tool also automatically monitors the glide head position variation, for example, from the time when the glide head is replaced. We have designed an integrated glide head holder which has a set of objective lens group to image the back of the glide head for its exact position. Since it has the optical function of a 5x objective, it is called a glide objective; see Figure 6. The image of the glide head is automatically taken and analyzed by the software for correcting glide head position shift, as much as $100 \mu m$. from head to head. The dimple in the suspension HGA is chosen as the target for position calibration with an estimated uncertainty of about $10 \mu m$. We also integrate the glide channel preamplifier inside the glide objective in order to make the glide head electric connection to the preamp as short as possible in order to achieve maximum SNR.

Since there are multiple tests and sensors correlated in the same tool, there will be multiple position offsets among them to be corrected, for example, between glide and AFM, AFM and optical, AFM and scriber, and so forth. It is one of the goals for the tool to make the offset calibration as transparent

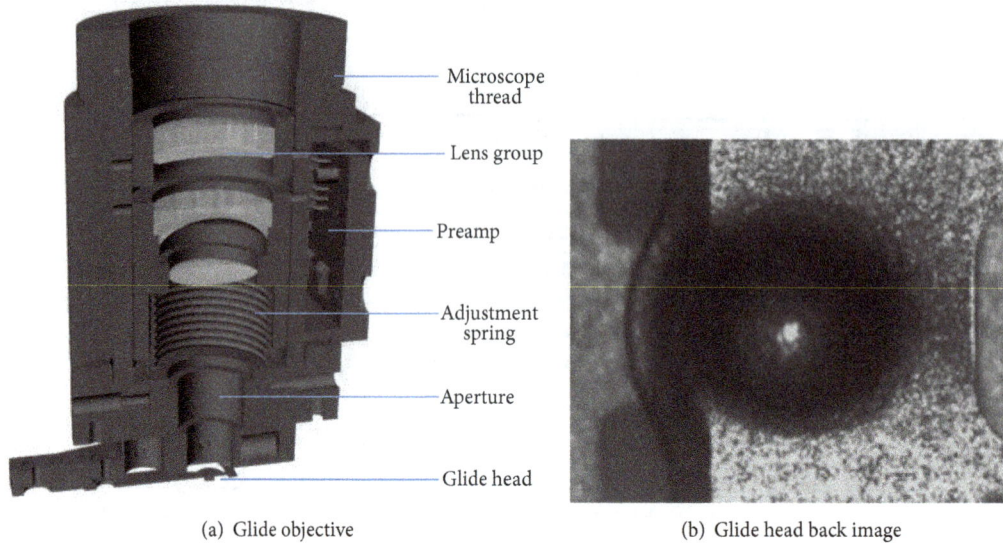

(a) Glide objective

(b) Glide head back image

FIGURE 6: Glide objective illustration. (a) shows the internal construction of the glide objective with a 5x lens group enables the imaging on the back of the glide head. An adjustable spring enables the fine focus on the glide head. The preamplifier is integrated to the glide objective to achieve short wire length to the glide sensor. (b) The image of the back of the glide head. In the field of the view, only the center part of the glide head is visible, showing the dimple of the suspension HGA. Glide head position is calibrated against the dimple position.

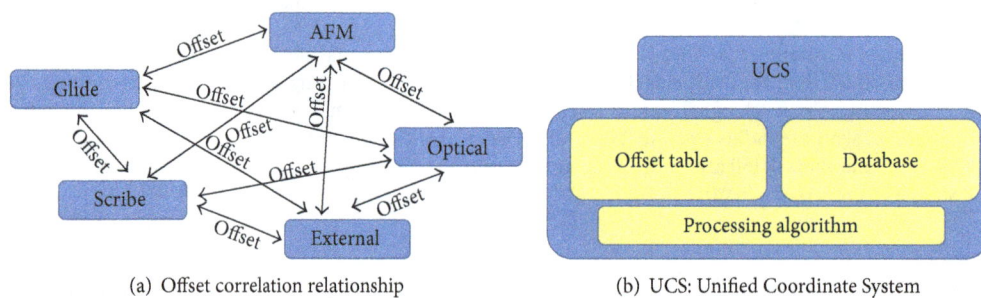

(a) Offset correlation relationship

(b) UCS: Unified Coordinate System

FIGURE 7: Schematic drawing showing the relationship and conversion among various coordinate systems. (a) Offset transformations that need to be handled if done individually. (b) Unified Coordinate System created to handle the transformation among the various coordinates with background algorithm and offset table. The parameters are stored in the database and undergo constant update when necessary.

as possible to the users. We have developed a Unified Coordinate System (UCS) to handle the transformation. The UCS processes the offset parameters from an offset table. The initial and current parameters are stored in the database processed by the UCS algorithm.

Within the CDA tool, there are multiple sets of coordinates: the stage or machine, the AFM, the glide, the burnish, the scribe, the optical, as well as the external and the UCS. The stage or machine coordinate system is the special one. The tool relies on the stage coordinate system to position the sample. Because of the existing offsets, there will be different stage coordinates for the same physical spot under different sensors. This creates problems for the user practically and sometimes conceptually. The UCS coordinate system is a defined one. It has a specific defined offset relationship with the various sensor coordinate systems, determined by measuring the actual offsets. Under the UCS system, there is only one set of coordinates to deal with by the user. The UCS can also interface with external coordinator systems of customized

formats. The function of the UCS is illustrated in the schematics in Figure 7.

4. Example Test

We use a disk from a defunct vintage HDD teardown to test run the CDA tool. Since we have designed a very easy-to-operate user interface, the procedure to run the tool is very straightforward. The sample is loaded and automatically clamped down by a vacuum disk chuck of the conventional design but with increased suction area to meet the specific requirement of the tool. The vacuum chuck is necessary not only for the easiness of mounting samples, but also for the absence of the center obtrusion. This allows a full radius range of the sample for test and analysis and also prevents the possible collision between the chuck and a sensor close to the ID radius.

Following the procedure outlined in Figure 8, the sample goes through the glide test scan first with the CDA in the

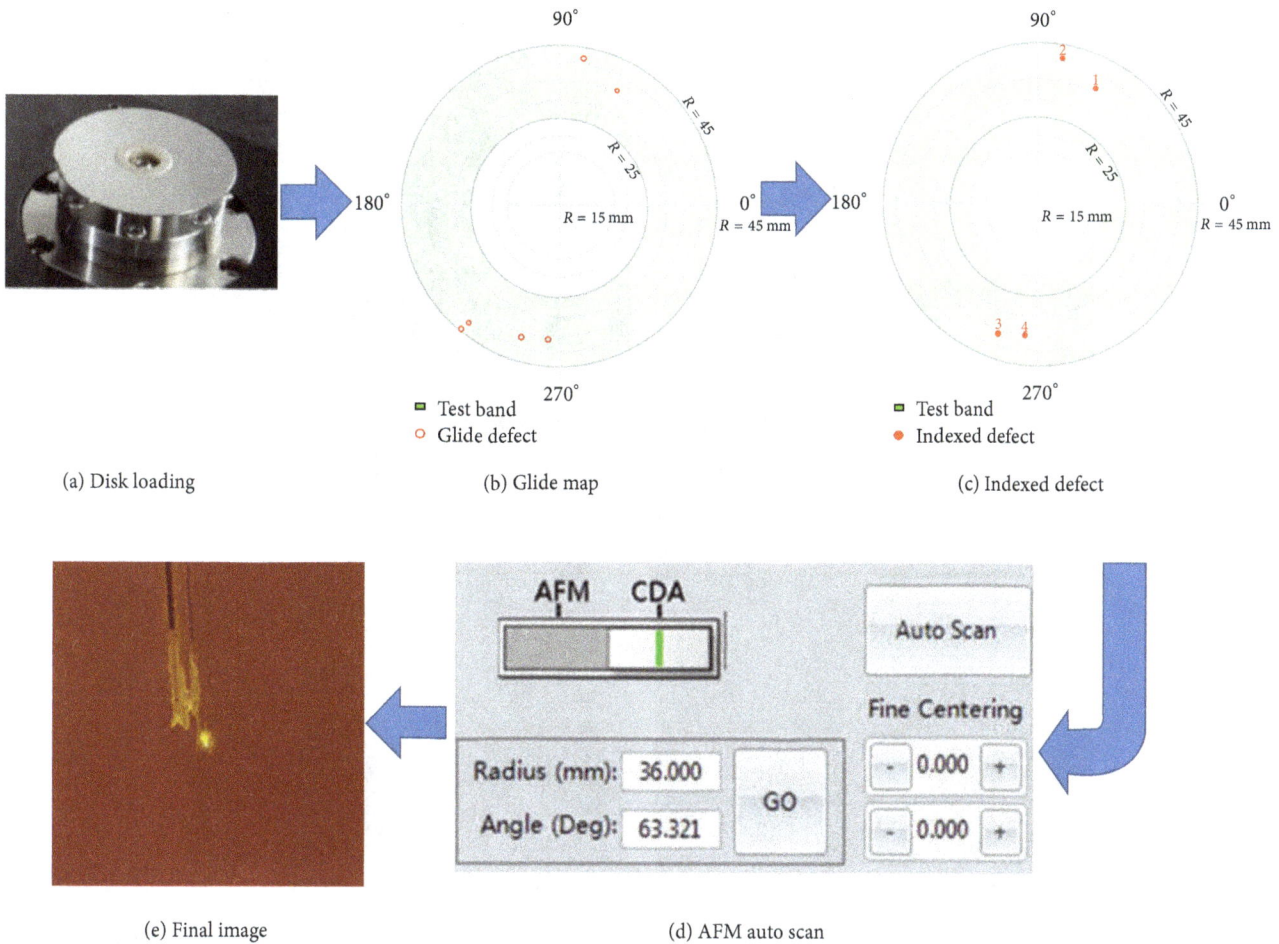

FIGURE 8: Test flow illustration, from sample load to AFM image. (a) Load the disk sample to the chuck. (b) Glide defect mapping from glide test. (c) Map of indexed defect from glide defect search. (d) CDA to AFM switch panel. A simple press of the "Auto Scan" button will start the automatic AFM scan routine. (e) Final AFM image of one of the scanned defects.

G mode. The operator then selects which one, or all, of the defects for AFM imaging. An indexing run follows on the selected defects before the tool is switched to the AFM mode or the A mode. After the indexing run, the selected defects can be relocated with an uncertainty of ±5 μm. The AFM scan can be carried out either manually or fully automatically. The automated AFM scan procedure includes sample positioning, tip approaching, scanning, tip retracting, and data saving.

The time it takes to complete the whole process depends on a number of factors such as disk form factor, glide head velocity, AFM scanning speed, and the number of scan lines. In general, the time for the initial glide test portion is about 100 sec and about 30 sec additional per defect for the indexing. AFM imaging takes about 130 sec with, for example, 256 lines and 2 Hz scan speed, which produces images with reasonable qualities. So the overall time needed from start to end could be under 10 minutes for one defect, a vast improvement over the traditional DFA involving AFM imaging.

With the precisely known positions, defects can be further analyzed with other available means in the tool such as an optical microscope. For the same defect on the sample disk,

the images of the bright field and the dark field optics and the AFM scan are shown together in Figure 9. The defect is visible in the bright field, but the dark field image shows higher contrast and also reveals a long scratch mark originating from the defect. AFM image has much high resolution, showing the defect with a lateral size of about 3 μm and a height of about 30 nm.

Not only the same defect can be analyzed with various sensors, media process specific technology can also be applied to work on the defect correlatively. In our example test, after the initial AFM imaging, a burnish head is used to sweep the defect multiple times. To study the effect of burnishing, the defect is again imaged with the AFM for the second time. The after burnish AFM image is shown in Figure 9(d). The defect retains the same general shape compared with that before the burnish. But the defect height is reduced from 30 nm to 28 nm. We also estimate the apparent volume of the defect before and after with the SPIP software. It turns out the volume number from the software actually increases slightly from $0.080\ \mu m^3$ to $0.082\ \mu m^3$. We conclude within the experiment error, there is no change in defect volume due to the burnish head sweep. Rather, the contamination on the

(a) Bright field

(b) Dark field

(c) AFM image before burnish

(d) AFM image after burnish

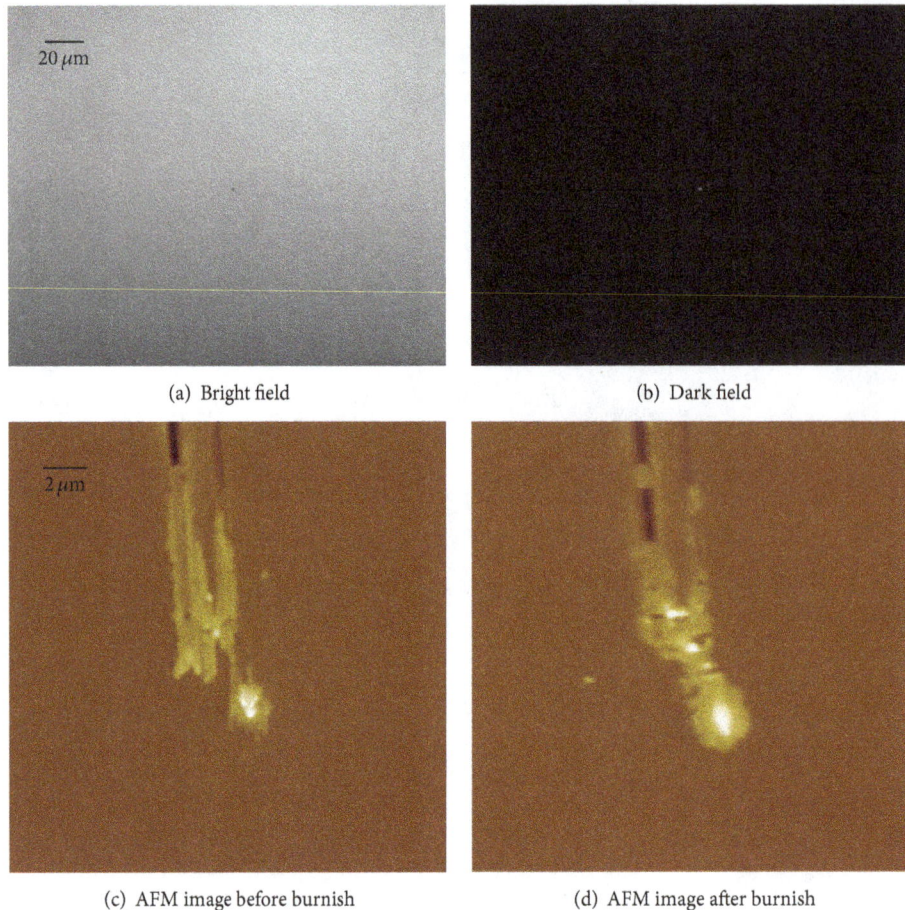

FIGURE 9: Correlated images of a media defect from a teardown disk sample. (a) Bright field image. The defect is very small but can be seen as speck of dark spot in the center of the image. (b) Dark field image. The defect is more visible and a scratch is also revealed against the dark background. (c) AFM images before the burnish sweeps. (d) AFM images after the burnish sweeps.

defect is redistributed. Further more elaborate experiment can be designed and carried out in order to obtain more concrete result on the burnish effect on defects.

5. Conclusion

We have demonstrated a unique tool, combining the glide with the AFM, which can be used to drastically speed up the defect failure analysis process. By performing the AFM analysis in situ, the accuracy of defect relocation and identification is guaranteed. The precision defect indexing algorithm is shown to be robust. High mechanical precision is achieved with the careful selection of components during the design phase. Coupled with the sophisticated software algorithm, a successful correlated defect analysis and a high-resolution low-noise AFM imaging have been demonstrated. With the transparent UCS coordinate system, the users are free from the burden of performing tedious coordinate transformations among the various tests available in the tool. A simple experiment of using the AFM to study the effect of the burnish on the defect shows the advantage of correlative test and analysis. Because of the open frame design and the general layout of the CDA tool, additional processes can be

considered for integration into the tool. This will open up new ways of conducting test and analysis by combining various tools in one single unit correlatively.

References

[1] B. Marchon and T. Olson, "Magnetic spacing trends: from LMR to PMR and beyond," *IEEE Transactions on Magnetics*, vol. 45, no. 10, pp. 3608–3611, 2009.

[2] W. Song, A. Ovcharenko, M. Yang, H. Zheng, and F. E. Talke, "Contact between a thermal flying height control slider and a disk asperity," *Microsystem Technologies*, vol. 18, pp. 1549–1557, 2012.

[3] V. Sharma, S. H. Kim, and S. H. Choa, "Head and media design considerations for reducing thermal asperity," *Tribology International*, vol. 34, no. 5, pp. 307–314, 2001.

[4] J. He, G. Sheng, J. Hopkins, and S. Duan, "Head and media instantaneous contact friction measurement and glide test," *IEEE Transactions on Magnetics*, vol. 46, no. 10, pp. 3767–3771, 2010.

[5] H.-L. Leo and G. B. Sinclair, "So how hard does a head hit a disk?" *IEEE Transactions on Magnetics*, vol. 27, pp. 5154–5156, 1991.

[6] Z. W. Zhong and S. H. Gee, "Failure analysis of ultrasonic pitting and carbon voids on magnetic recording disks," *Ceramics International Journal*, vol. 30, pp. 1619–1622, 2004.

[7] J. Windeln, C. Bram, H. L. Eckes et al., "Applied surface analysis in magnetic storage technology," *Applied Surface Science*, vol. 179, no. 1–4, pp. 167–180, 2001.

Permissions

The contributors of this book come from diverse backgrounds, making this book a truly international effort. This book will bring forth new frontiers with its revolutionizing research information and detailed analysis of the nascent developments around the world.

We would like to thank all the contributing authors for lending their expertise to make the book truly unique. They have played a crucial role in the development of this book. Without their invaluable contributions this book wouldn't have been possible. They have made vital efforts to compile up to date information on the varied aspects of this subject to make this book a valuable addition to the collection of many professionals and students.

This book was conceptualized with the vision of imparting up-to-date information and advanced data in this field. To ensure the same, a matchless editorial board was set up. Every individual on the board went through rigorous rounds of assessment to prove their worth. After which they invested a large part of their time researching and compiling the most relevant data for our readers. Conferences and sessions were held from time to time between the editorial board and the contributing authors to present the data in the most comprehensible form. The editorial team has worked tirelessly to provide valuable and valid information to help people across the globe.

Every chapter published in this book has been scrutinized by our experts. Their significance has been extensively debated. The topics covered herein carry significant findings which will fuel the growth of the discipline. They may even be implemented as practical applications or may be referred to as a beginning point for another development. Chapters in this book were first published by Hindawi Publishing Corporation; hereby published with permission under the Creative Commons Attribution License or equivalent.

The editorial board has been involved in producing this book since its inception. They have spent rigorous hours researching and exploring the diverse topics which have resulted in the successful publishing of this book. They have passed on their knowledge of decades through this book. To expedite this challenging task, the publisher supported the team at every step. A small team of assistant editors was also appointed to further simplify the editing procedure and attain best results for the readers.

Our editorial team has been hand-picked from every corner of the world. Their multi-ethnicity adds dynamic inputs to the discussions which result in innovative outcomes. These outcomes are then further discussed with the researchers and contributors who give their valuable feedback and opinion regarding the same. The feedback is then collaborated with the researches and they are edited in a comprehensive manner to aid the understanding of the subject.

Apart from the editorial board, the designing team has also invested a significant amount of their time in understanding the subject and creating the most relevant covers. They scrutinized every image to scout for the most suitable representation of the subject and create an appropriate cover for the book.

The publishing team has been involved in this book since its early stages. They were actively engaged in every process, be it collecting the data, connecting with the contributors or procuring relevant information. The team has been an ardent support to the editorial, designing and production team. Their endless efforts to recruit the best for this project, has resulted in the accomplishment of this book. They are a veteran in the field of academics and their pool of knowledge is as vast as their experience in printing. Their expertise and guidance has proved useful at every step. Their uncompromising quality standards have made this book an exceptional effort. Their encouragement from time to time has been an inspiration for everyone.

The publisher and the editorial board hope that this book will prove to be a valuable piece of knowledge for researchers, students, practitioners and scholars across the globe.

List of Contributors

Sripathi V. Canchi and David B. Bogy
Computer Mechanics Laboratory, Mechanical Engineering, University of California, Berkeley, CA 94720, USA

Run-Han Wang and Aravind N. Murthy
HGST, a Western Digital Company, San Jose, CA 95135, USA

M. Fallqvist and M. Olsson
Department of Material Science, Dalarna University, 781 88 Borlange, Sweden

R. M'Saoubi and J. M. Andersson
R&D Materials and Processes, Seco Tools AB, 737 82 Fagersta, Sweden

Sukhendu Jana, Sayan Das, Utpal Gangopadhyay and Prajit Ghosh
Meghnad Saha Institute of Technology, Techno India Group, Kolkata 700150, India

Anup Mondal
Department of Chemistry, Bengal Engineering and Science University, Howrah 711103, India

Y. M. Shashidhara and S. R. Jayaram
Department of Mechanical Engineering, Malnad College of Engineering, Hassan, Karnataka 573201, India

G. Leprince
Laboratoire dÉnergetique, Universite de Lyon—ECAM Lyon, 40 Montee Saint-Barthelemy, 69321 Lyon Cedex 05, France
INSA de Lyon, Universite de Lyon, LaMCoS, UMR CNRS 5259, Batiment Jean d'Alembert, 18-20 Rue des Sciences, 69621 Villeurbanne Cedex, France

C. Changenet
Laboratoire dÉnergetique, Universite de Lyon—ECAM Lyon, 40 Montee Saint-Barthelemy, 69321 Lyon Cedex 05, France

F. Ville and P. Velex
INSA de Lyon, Universite de Lyon, LaMCoS, UMR CNRS 5259, Batiment Jean d'Alembert, 18-20 Rue des Sciences, 69621 Villeurbanne Cedex, France

M. Uthayakumar and S. Thirumalai Kumaran
Department of Mechanical Engineering, Kalasalingam University, Krishnankoil, Tamil Nadu 626 126, India

S. Aravindan
Department of Mechanical Engineering, Indian Institute of Technology, New Delhi 110 016, India

Michele Scaraggi
DII, Universita del Salento, 73100 Monteroni di Lecce, Italy

Giuseppe Carbone
DMMM, Politecnico di Bari, 70126 Bari, Italy

M. Sudheer, Ravikantha Prabhu, K. Raju and Thirumaleshwara Bhat
Department of Mechanical Engineering, St. Joseph Engineering College, Mangalore 575 028, Karnataka, India

Biplab Chatterjee and Prasanta Sahoo
Department of Mechanical Engineering, Jadavpur University, Kolkata 700032, India

Keiichi Narita
Lubricants Research Laboratory, Idemitsu Kosan Co., Ltd., 24-4 Anesakikaigan, Chiba, Ichihara-shi 299-0107, Ja

Luke Autry and Harris Marcus
Department of Chemical, Materials and Biomolecular Engineering, Institute of Materials Science, University of CT, Storrs, CT 06269, USA

S. F. Gnyusov
Tomsk Polytechnic University, 634050 Tomsk, Russia

V. G. Durakov and S. Yu. Tarasov
Institute of Strength Physics and Materials Science SB RAS, 634055 Tomsk, Russia

Johannes Kümmel and Katja Poser
Institute of Applied Materials (IAM-WK), Karlsruhe Institute of Technology (KIT), Kaiserstraße 12, 76131 Karlsruhe, Germany

Frederik Zanger and Jürgen Michna
Institute of Production Science (wbk), Karlsruhe Institute of Technology (KIT), Kaiserstraße 12, 76131 Karlsruhe, Germany

Volker Schulze
Institute of Applied Materials (IAM-WK), Karlsruhe Institute of Technology (KIT), Kaiserstraße 12, 76131 Karlsruhe, Germany
Institute of Production Science (wbk), Karlsruhe Institute of Technology (KIT), Kaiserstraße 12, 76131 Karlsruhe, Germany

Bruno Marchon
HGST, San Jose, CA 95135, USA

Thomas Pitchford and Sunita Gangopadhyay
Seagate Technology, Minneapolis, MN 55435, USA

Yiao-Tee Hsia
Western Digital, San Jose, CA 95138, USA

Hirofumi Kondo
R&D Division, Sony Chemical & Information Device Corporation, 1078 Kamiishikawa, Kanuma 3228503, Japan

Y. M. Shashidhara and S. R. Jayaram
Department of Mechanical Engineering, Malnad College of Engineering, Karnataka state, Hassan 573 201, India

Udaya P. Singh and Ram S. Gupta
Department of Applied Science and Humanities, Kamla Nehru Institute of Technology, Sultanpur 228118, India

Lintu Roy and S. K. Kakoty
Indian Institute of Technology Guwahati, Guwahati 781309, India

S. Janakiraman and K. Udaya Bhat
Department of Metallurgical & Materials Engineering, NITK Surathkal, Srinivasa Nagar, Surathkal 575025, India

Carlo Gorla and Franco Concli
Dipartimento di Meccanica, Politecnico di Milano, Via la Masa 1, 20156 Milano, Italy

Karsten Stahl, Bernd-Robert Hohn, Michaelis Klaus, Hansjorg Schultheiß and Johann-Paul Stemplinger
Lehrstuhl fur Maschinenelemente, FZG, 85748 Garching bei Munchen, Germany

Jizhong He
Institute of Engineering, College of Engineering, Peking University, Nanjing, Jiangsu 210012, China
Micro Focus Technologies, Inc., Wuxi, Jiangsu 214125, China

www.ingramcontent.com/pod-product-compliance
Lightning Source LLC
Chambersburg PA
CBHW050446200326
41458CB00014B/5083